BIOGAS SYSTEMS

POLICIES, PROGRESS and PROSPECTS

K.M. MITAL

PUBLISHING FOR ONE WORLD

NEW AGE INTERNATIONAL (P) LIMITED, PUBLISHERS
(formerly Wiley Eastern Limited)
New Delhi • Bangalore • Chennai • Cochin • Guwahati • Hyderabad
Jalandhar • Kolkata • Lucknow • Mumbai • Ranchi
Visit us at **www.newagepublishers.com**

Published by New Age International (P) Ltd., Publishers
Reprint : 2007

ISBN : 81-224-1104-5

Rs. 650.00

C-06-11-1140

2 3 4 5 6 7 8

Printed in India at Ramprintograph (India), New Delhi-110 020

PUBLISHING FOR ONE WORLD

NEW AGE INTERNATIONAL (P) LIMITED, PUBLISHERS
(formerly Wiley Eastern Limited)
4835/24, Ansari Road, Daryaganj, New Delhi - 110002
Visit us at **www.newagepublishers.com**

Dedicated to

the fond and loving memory of my two illustrious uncles, late Mr. Sumat Prasad Jain, a former leading advocate from Muzaffarnagar, a veteran freedom-fighter and a Congress Member of Parliament (1952-67); and late Mr. Dalip Singh Mital, an eminent civil engineer from U.P. Irrigation Department and an alumini (1914) of the erstwhile Thomson College of Civil Engineering, Roorkee; both of whom apart from their respective legal and engineering professions, selflessly worked throughout their lives for the upliftment of poor and down-trodden in true Gandhian and Sarvodaya spirit.

PREFACE

Biologically produced energy sources have been identified as attractive alternative to fast depleting fossil fuel supplies. Within the backdrop of a growing energy shortage started from the first energy shock in 1973 which never vanished completely though the shortages showed several uptrends and downtrends in between, importance of renewable energy and energy conservation never diminished . In fact the need for energy supplies from non-conventional sources was never so much as it is today and is likely to be much more in the years to come. Fuels from biomass also called biofuels which include *biogas*, have enormous potential in rural areas. Biomethanation is an appropriate technology well-suited to the needs of developing countries, saves fuelwood and is eco-friendly. United Nations and its constituent agencies like ESCAP (Economic and Social Commission for Asia and Pacific), FAO (Food and Agriculture Organisation) and UNDP (United Nations Development Programme) have played a pioneering role in promoting biomethanation world over. These agencies have funded several projects including socio-economic surveys for promotion of biomethanation. This technology involves treatment of animal wastes, night-soil, industrial and municipal solid wastes in oxygenless environment to produce valuable by-products, usable energy in the form of methane gas, organic fertiliser and protein. A biogas plant is a decentralised energy system which can lead to self-sufficiency in household heat and power needs and at the same time keeps the environment clean. Matching with the growing need for utilisation of biofuels in general and biogas in particular, growing application of scientific and engineering principles has transformed the latter into a full-fledged discipline.

This book is written with a view to focus on issues relating to planning and implementation of biogas programme in different region of the world. It consolidates the progress of biomethanation that has taken place in several developing and developed countries in past few decades. The book may appeal to both technical and non-technical people alike including general public concerned with the energy needs of the country and available energy resources. It is expected to be of great help to energy planners, policy makers, staff of government departments and turnkey agencies, social scientists, research workers, extension staff, teachers, students, and plant owners concerned with socio-economic benefits of biomethanation. The book includes a total of seven chapters each of which is intended to be reasonably delinked from one another and thus suitable for independent reading. However, while reading a chapter it is advantageous that reader be conscious of the broad content of other chapters. The first chapter provides a detailed overview of biogas technology covering all aspects. The second chapter deals with historical perspectives and progress of biomethanation in select countries where it has made particular mark. The third chapter sums up the history of biomethanation in India, reviews references to biogas in important seminars and conferences, provides estimates of potential, consolidates progress of biomethanation in India and presents data about functional status of biogas plants. The third chapter covers implementation strategies particularly the turnkey approach which is largely responsible for bringing biogas revolution in India judging by the recent unprecedented spurt in the number of biogas plants installed. This chapter also extensively covers economic issues particularly the economic-viability of biogas plants and presents the contributions of several researchers in the field. The chapter five consolidates the findings and results of several socio-economic surveys undertaken in chronological order. In chapter six it has been endeavoured to present case-studies of several community biogas plants which have greatly helped in improving rural economy. The chapter also describes issues and relative merits of community versus family-sized plants. The final concluding chapter seven provides overview of energy needs of developing countries, reviews Integrated Rural Energy Programme (IREP) and the *Urjagram* programmes of the union government as supportive programmes for biomethanation, views biogas programme as a strategy for sustainable development, identifies areas for further studies, and concludes that biomethanation will scale new heights even when the subsidies are withdrawn and market-driven approach governs the biogas programme.

Developments in India have been more extensively covered in this book though national boundaries had no bearing on material covered in different chapters. This was found to be both inevitable and desirable, inevitable following the author's greater familiarity with developments in this country and desirable because biogas technology forms part of select group of disciplines which have been developed and utilised to a far greater degree in countries like India, China, Nepal and Philippines than western world.

Biogas is not an exact fuel, its composition and characteristics vary from source to source. Variation in composition of feedstock from place to place poses problems in having uniform and standard rates of biogas yield. For example, in case of agricultural crop residues, composition depends on type of plant, soil condition, fertiliser used; in case of animal faeces it depends on type of animal and breed, grazing conditions, feed consumed, state of animal health, etc. In view of inconsistencies, yield values and their system of measurement have been reported as in the original work without changing to one common value and system throughout the book. A set of conversion factors are given in the end to help the reader in changing one unit to another.

The author is grateful to Dr. R.S. Venkataraman, Chairman and Managing Director, Engineers India Limited, New Delhi, for providing a progressive and learning environment in the organisation which is conducive for innovation and development. A very sincere gratitude is conveyed to Mr. B.N. Jha, Director (Personnel), EIL, for his moral encouragement and support in writing this book. Hearty gratitude is also conveyed to Mr. R. Srinivasan, ex-Member (Personnel), ONGC; and Dr. S.P. Agarwal, a former ILO expert and *Gita* scholar, both of whom have been path-finders and source of inspiration all along.

Grateful thanks are conveyed to Mr. Anil Kumar, IAS, Secretary, Department of Fertiliser, Ministry of Fertiliser and Chemicals; Mr. Dhanendra Kumar, IAS, Secretary (Culture), Haryana Government; Dr. S.J. Chopra, Director (Technical), EIL, New Delhi; Mr. A.K. Gupta, Executive Director, Railway Board, New Delhi; and Mr. Pankaj Gupta, IPS, DIG Police, Amravati (Maharashtra) for their invaluable suggestions.

Gratitude is conveyed to Dr. Prem Vrat, Dean, Post-Graduate Studies and Research, and Professor of Industrial Engineering, IIT, Delhi; Dr. (Mrs.) Seema Agarwal, Coordinator, UN system Education Programme, UNDP, New Delhi; Dr. Sushil, Professor and Head, Department of Management Studies, IIT, Delhi; Mr. A.K. Agarwal, Asst. GM (BMTC), Bhilai Steel Plant (SAIL); Dr. (Mrs.) Kokila K. Agarwal, Senior Research Scientist, The Futures Group (Washington); and Mrs. Rani Mewara, lawyer and women activist, and her husband Mr. Shailen Mewara, an entrepreneur, from Mhow (Indore), for numerous ideas and suggestions in improving the manuscript. Gratitude is also conveyed to Mr. Rajiv Gupta, and Mr. Piyush Gupta, management experts and entrepreneurs from Secunderabad (A.P.) for their varied assistance and cooperation. Grateful thanks are also conveyed to Mr. Navneet Mital, a leading entrepreneur in paper trading from Muzaffarnagar; and Mr. Ranjan Mital, an advocate and real estate expert from Muzaffarnagar for their varied cooperation.

Gratitude is conveyed to concerned staff of Tata Energy Research Institute (TERI) for allowing to avail their library and documentation facilities. Grateful thanks are conveyed to Mr. Jitendra Bakshi, Information Analyst, TERI, New Delhi; and Ms. Meena Bilgi, Programme Executive, Aga Khan Rural Support Programme, Ahmedabad; for their valuable cooperation. Grateful thanks are conveyed to Mr. Basab R. Shee of UN Asian and Pacific Centre for Transfer of Technology (New Delhi); and Mr. Suresh Chandwani of EIL, New Delhi, for their varied assistance.

Gratitude is also conveyed to Mr. K.K. Verma, Manager (HRD), EIL, New Delhi; Dr. P.R.K. Raju, Dy. Manager, Eastern Region, ONGC Ltd., Nazira; and Mr. Deepak Garg, Senior Human Resource Officer, EIL, New Delhi, for their help in locating valuable sources of information.

Words fail the author to express gratitude to his mother, Mrs. Saraswati Mital, and mother-in-law, Mrs. Sarla Rani Gupta, for their moral and financial support. Hearty gratitude is also conveyed to Mr. Braham Prakash, a well-known educationist from Kanpur and family veteran; and Mr. Hari Mohan Mital, author's brother and leading advocate (Muzaffarnagar) for their all-round support and encouragement.

Very hearty gratitude is conveyed to my brother-in-law and family stalwart, Lt. Col. I.C. Gupta (Retd.), who inspired and kept my spirits high all throughout the period of book-writing and was looking forward for its release but suddenly died on March 26, 1996.

In a book-writing project the family of an author has to invariably bear the burden. A book often – indeed always – interferes with family-life. Throughout my career, my wife, Renu, has amazed me with her love and cooperation. My son, Vivek, had to subsist on less than the rightful share of my attention during the long and arduous period the book-writing was in progress. I wonder if I would ever be able to make up their lost weekends and evenings.

K.M. Mital

New Delhi
August 15, 1997
50th anniversary of
India's Independence

CONTENTS

Chapter 1

AN OVERVIEW OF BIOGAS TECHNOLOGY

1.1 Introduction

Biogas can be produced by fermenting organic materials in absence of air (or oxygen) with the help of bacteria (microorganisms) to break down materials to intermediates such as alcohols and fatty acids and finally to methane, carbon dioxide and water. This process is called *anaerobic fermentation* and was known to exist from quite long time back. Biogas has also been known as the swamp gas, sewer gas, fuel gas, marsh gas, wet gas, and in India more commonly as *gobar* gas. Natural gas is also produced by the action of anaerobic bacteria on plants that grew thousands of years ago. Biogas and natural gas are therefore very similar to one another. The main fuel component in both is methane gas. However, over the years, pressure and temperature of underground rocks have converted part of methane in natural gas to other gases such as ethane, propane, butane and condensate. In contrast biogas is produced in a digester by anaerobic fermentation. A period of about 15 days or so enables anaerobic bacteria to convert organic matter to biogas which however is too short for conversion of methane to other gases like ethane, propane, butane, etc. [35]. The enormous potential of the smallest living organisms such as bacteria, yeasts and fungi to transform organic wastes into valuable source of fuel and enriched fertiliser through a simple process of anaerobic fermentation has been widely recognised. Anaerobic fermentation is a simple and low-cost process which can be carried out economically in rural areas where organic wastes are generated aplenty which otherwise pollute environment and pose health hazards [15]. Animal and human wastes are excellent feedstock for biomethanation which are available in plenty/abundantly all over. Wastes in large quantities on renewable basis are also available from agricultural crops and residues, fruit and vegetable plants and municipal refuse. The potential for generating gaseous fuel and enriched fertiliser through biomethanation is enormous which can bring economic and environmental gain to a vast population [12].

Anaerobic fermentation as described earlier is a biological process which takes place in absence of air. A number of stages are involved in this process; initially organic material is hydrolysed by enzymes into simple sugars, alcohol peptides and amino acids. These are then converted to volatile fatty acids, hydrogen, carbon dioxide, water and small amounts of methane. Methane forming bacteria then convert fatty acids into methane, carbon dioxide, and water. The process is temperature–dependent and slows down considerably below 30°C, the optimum being 35-38°C which is known as the *mesophyllic* range. Above this temperature the process slows between 40°C - 45°C and then increases to a peak between 55°C - 60°C known as the *thermophyllic* range. The pH of digester contents also affects fermentation as the methane bacteria is less active below pH of 6 while the ideal pH is 7. Fermentation breaks down waste products at 35°C under controlled conditions that produces a mixture of methane and carbon dioxide in proportion of about 70 to 30 per cent. The resulting gas can be used directly for heating and lighting purposes or for use in an engine-driven generator to electricity [2]. Anaerobic fermentation produces biogas on renewable basis, eliminates smell and reduces harmful bacteria of organic wastes, improves nitrogen and phosphate content of resulting sludge yielding fertiliser, and reducing biological oxygen demand (BOD) and chemical oxygen demand (COD).

1.2 Historical Background of Biomethanation

One of the early persons to observe the phenomenon of biomethanation was Alessandro Volta of Italy. Way back in 1776 he wrote to a friend that 'combustible air' was being produced continuously in lakes and ponds in the vicinity of Como in northern Italy. Volta observed that when he distributed the bottom sediment of the lake, bubbles of gas would rise to the surface. He also noticed that more bubbles came up when the sediment contained more plant material. In 1806, William Henry showed that Volta's gas was identical with methane gas. Humphrey Davy in the early 1800s observed that methane was present in farmyard manure piles [11]. In 1808, Davy conducted the first laboratory experiment to produce methane by anaerobic fermentation of wastes. In earlier periods, anaerobic fermentation was carried out mainly as a municipal waste treatment process and resulting energy recovery was not primary concern. In 1895, biogas from a waste treatment plant in Exter in England was collected and used to light nearby streets [21]. Interest in biogas as a fuel received fillip during Second World War. French scientists took particular interest in promoting biogas technology in forties and installed large number of plants in French colonies in Africa. During this period fuel-starved French and Germans used biogas as a fuel for vehicles and farm tractors. Following war, several nations including England, USA, Canada, Russia, Japan, China, Kenya, Uganda, South Africa, New Zealand and India showed interest in biomethanation but later this interest waned following cheap availability of fossil-fuels for over next three decades. However, series of energy shocks which rocked the world economy from 1973 onwards coupled with concern for environment protection revived interest in biomethanation. In view of its vast potential, large number of community and family-sized plants have been set up in recent years in countries like China, India, Philippines and Nepal [29].

1.3 An Overview of Biomethanation in India

A number of estimates have been made from time to time about potential of biogas generation in India. With livestock population of over 300 million, India can set up over 30 million individual biogas plants and 560,000 community plants at a simple norm of minimum one in each village of the country. With a modest norm of 10 biogas plants per village, it is feasible to set up some 5.5 million plants in country's 5.6 lakh villages. If this becomes reality, it would provide at 1981-82 price level fuel equivalent to 5341 million liters of kerosene valued at Rs. 5484 million and conserve organic manure to the extent of 89 million tonnes valued at Rs. 4464 million otherwise burnt as dung cakes [32, 33]. Effective utilisation of organic wastes can yield gas equivalent of 285.5 million tonnes of coal equivalent of energy consumed in the country during 1978-79 [20]. A small biogas plant of 4 cubic meter gas per day capacity set up at a cost of Rs. 6000 with Rs. 2200 availed as central subsidy, operating for 300 days an year can save 4.16 tonnes of fuelwood equivalent to 744 liters of kerosene, and produces manure worth Rs. 1656 per year at the subsidised market price of fertilisers [4, 7].

As per some broad estimates, over 30 per cent of 980 million tonnes of cattle dung generated annually on renewable basis is directly burnt as fuel without harnessing its full potential as gas and rich fertiliser following fermentation [9]. Assessed in quantitative terms, it is equivalent to a third of India's chemical fertilisers used. If the entire quantity of 980 million tonnes of cattle dung is first used for biomethanation, it can meet domestic fuel needs of 437 million people, raise manure production by 350 million tonnes and susbstantially improve hygienic conditions in rural areas [10]. If the entire quantity of dung produced is available for biomethanation, it can produce about 6375×10^7 cubic meter of gas having 3825×10^7 cubic meter of methane with net heat value as 312×10^{12} kcal [31]. In addition, the resulting plant effluent can generate one million tonne of nitrogen, one million tonne of phosphate, and half million tonne of potash that can be used as enriched fertiliser.

Based on several designs initially developed by scientists like N. V. Joshi, S. V. Desai, S. C. Biswas, Y. N. Kotwal, C. N. Acharya, and J. J Patel of the Agricultural College in Pune developed a biogas plant which later on came to be known as the '*Gramlakshmi*' Gas Plant. Distinctive features of this plant covered integration of gasholder and digester in one unit, underground positioning of the gasholder, daily feeding of slurry, facility of automatic overflow of the digested slurry, and gasholder rotation to break scum formation. Based on pioneering studies carried out by S. C. Dasgupta, Swami Vishwakarmanand and C. N. Acharya in early fifties, original design of '*Gramlakshmi* Gas Plant' underwent further changes and finally in the year 1954 a much modified design emerged. This improved design had two chambers, one to act as primary and the other as secondary digester with facility for gasholder to have rotary as well as vertical movements. Its digester made of brickwork was kept underground and fitted with two pipes one for feeding and the other for removing digested slurry. The gasholder was designed in such a way as to maintain uniformity of pressure at 3 to 4 inch of water column but without using couterweights as was used in earlier design. This modified design of the '*Gramlaxmi* Gas Plant' became the major landmark for biogas programme in the country. In later years it was adopted by the Khadi and Village Industries

Commission (KVIC) for its countrywide popularisation during sixties and seventies [27]. Apart from the KVIC, a Lucknow-based agency called the Planning Research and Action Division (PRAD) also adopted it for promotion. The main pioneer at the PRAD in promoting biomethanation was Mr. Rambux Singh who has virtually become a legend in this field. The KVIC and PRAD also tried to promote plants of the IARI design. However, plants as per the IARI design with pulley and counterweights created alignment problems often causing tilting of gasholder sideways which resulted in uneven and interrupted gas supply. The KVIC therefore preferred to popularise plants which had no attached pulley and counterweights. Gas pressure was instead sought to be built-up with the weight of gasholder. KVIC also initiated R & D activities during sixties and seventies. Based on a Chinese design and field trials carried out at the Gobar Gas Research Station, Ajitmal, Etawah in U. P., PRAD developed a fixed dome type plant which has come to be known as the *Janata Biogas Plant (JBP)*.

From 1962 onwards, the KVIC launched a well-coordinated biogas programme in the country under which over a lakh biogas plants were set up till the beginning of eighties when the programme was finally taken over by the DNES (Department of Non-Conventional Energy Sources) [25,26]. During the Sixth Plan (1980-85), the government set up the National Biogas Development Board (NBDB) with a view to advance grants, subsidies and loans to organisations associated with implementation of the biogas programme. It introduced the National Project for Biogas Development (NPBD) to give a renewed thrust to biomethanation in the country. The DNES under the Ministry of Energy was assigned the task to pursue the NPBD in September, 1982 [8]. The DNES also introduced several measures for the success of the NPBD which included upward revision in the central subsidy for plant construction, availability of more effective post- installation follow-up services, creation of large cadre of trained manpower in the field, and supply or raw materials [22, 23]. During the Seventh Plan (1985-90), it was planned to set up 15 lakh biogas plants with their annual break-up for successive years as 2.5, 2.75, 3, 3.25 and 3.5 lakh respectively [10, 13]. Under the NPBD, over 23.60 lakh plants have been installed by 1995-96 capable to generate gas equivalent of over 75 lakh tonnes of fuelwood valued at Rs. 375 crores per annum and 345 lakh tonnes of enriched organic manure. Based on phenomenal success achieved the eighth plan targets were revised from 10 lakh to 12.5 lakh plants, and ambitious targets have been drawn for the Ninth Plan (1997-2002).

1.4 Basics of Anaerobic Fermentation

Anaerobic fermentation process can be broadly divided into two phases, *acid forming* and *methanogenesis*. In the first phase, acid-forming bacteria break down and liquefy volatile solids of feed into fatty acids. In the second phase, methane forming bacteria convert these volatile acids into methane and carbon dioxide which are the main constituents of biogas [5]. These bacteria are sensitive to changes in their environment. Rapid digestion and efficient biogas generation occur within narrow ranges of temperature and influenced by raw material composition [24]. Consistency in temperature, pH and use of fresh organic matter as feed helps maximise gas yield. Temperature is maintained at approximately 95°F (35°C) but different temperature can also be used provided temperature variations are controlled. For every 20°F (11°C) decrease, gas production is nearly halved. Thus a constant temperature is essential for higher yield. Temperature variations of as little as 5°F (2.7°C) can inhibit methane forming bacteria to cause acid accumulation and possible digester failure [20].

Biogas plant requires no special skill or scientific knowledge to operate. However, acidity of digester contents needs to be occasionally checked. This can be accomplished by dipping a strip of litmus paper in effluent and noting the colour. Increasing acidity indicates an imbalance in various bacteria which can be corrected by adding burnt lime until the effluent is neutral. Generally maximum biogas production requires daily loading of the digester. When the demand for the biogas is less, loading once a week or even less would be adequate if warm water is used for slurry preparation. When demand increases, output can be quickly increased by returning to daily loading [19].

A simple apparatus or plant is enough to produce biogas. Desired capacity and life of a plant affects cost and complexity of a biogas system. A simple batch loaded digester requires an oxygen free container, relatively constant temperature, a means for collecting gas and some mixing device. A comparison of different types of commonly used digester systems is given in Table 1.1. As biogas is explosive gas, adequate safety precautions need be observed while handling it [20].

Size of biogas plant digester is controlled by a number of factors such as number, size and type of animals providing feed, slurry dilution and retention time. The most easily controllable factor dependent on digester size is retention time. Ten days is minimum but a longer period can be used. The longer the retention time, the larger is the size of a digester required. Longer retention time allows more complete decomposition of wastes. Generally fifteen days is a frequently used retention time [20].

Loading rate gives an indication of weight of volatile solids fed to a digester per day. Volatile solids are that portion of total solids which are organic in content. Biological organisms utilise a portion of this material as a substrate to produce gas. The degree of utilisation of this material is an important parameter in determining gas production rate. Not all volatile solids present in feed are biodegradable and normally around 50 per cent of volatile solids are utilised depending upon feed composition. The concentration of volatile solids in digester determines the rate of gas production. For instance, a digester loaded with four units by weight of volatile solids produces twice the gas as the same digester loaded with only two units. Gas production rate is quite often expressed in terms of volume of gas produced per unit weight of volatile solids added or destroyed [24, 28].

Proper mixing of digester contents provides intimate contact between micro-organism and the substrate, maintains uniform temperature, uniform distribution of bacteria and volatile solids throughout slurry. It also minimises sludge formation and prevents a crust or scum formation on top of slurry which interferes with release of biogas. Mixing also prevents accumulation of

Table 1.1 Comparison of Different Types of Digester Systems

S. No.	*Digester Type*	*Suitable Wastes*	*Volatile Solids Content*	*Typical Retention Periods (days)*	*Degree of mixing*	*Operating Temperature*	*Gas Yield*	*Degree of Control Required*	*Remarks*
1.	Batch	Agricultural Iregular or seasonal, fibrous or difficult to digest	Low volumes up to 25% solid	60 or more	Little needed	Usually 30-35	Irregular and discontinuous	Little once started	Messy and time consumming to start
2.	Plug-flow horizontal/vertical	Agricultural continuous or regular flows, less fibre content	Larger volumes 5-15 per cent solids	30-60	Occasional	30-35	Continuous	Simple	Loading and scum removal can be messy
3.	Conventional sewage works	Continuous sewage sludge	Less than 5 per cent solids	30-60	Occasional	30-35 or unheated	Continuous	Simple	Not very effective
4.	High Rate Dister Sewage								
4.1	Primary	Sewage sludge	4-10 per cent solids	10-30	Regular	30-35	Continuous	More sophisticated	Can be automated
4.2	Secondary	From primary	4-10 per cent solids	20-60	None	Unheated	Non collected	Simple	
5.	High rate	Agricultural industrial	4-15 per cent solids	5-20	Continuous	30-35	Continuous	More sophisticated	Can be automated
6.	Anaerobic contact	Industrial Agricultural	Low solids	0.3-5[1]	Continuous	30-35	-do-	Sophisticated	Auromatic
7.	Anaerobic filter	Industrial	Low solids (low organic content)	0.5-5[1]	None needed	30-35	-do--	-do-	-do-

Note : 1. *Liquid retention time.*

undigested feed in digester. Scum is a collection of light weight, inert material that collects on liquid surface in digester. This accumulation tends to markedly reduce effective volume of digester and obstruct release of gas from digester contents unless it is broken up and removed periodically. Methods of mixing or agitation include slurry recirculation pumps and mechanical paddles. However, mechanical components exposed to slurry are susceptible to corrosion which make them difficult to repair without disrupting digestion by opening the digester. Mixing can also be accomplished by recycling gas through diffusers at the bottom of digester tank. An electrically driven vacuum pump draws biogas from storage under the roof and injects it at the bottom of both stages of digester. Recirculation of biogas helps agitation [24,28].

Amount of water required in slurry preparation is an important requirement of digestion process. Considering that optimum total solids (TS) concentration in an anaerobic digester is normally 7 to 9 per cent, considerable dilution is often required since animal faeces may contain 10 to 25 per cent solids. Furthermore, as far as possible, total solids in digester content should be maintained fairly uniformly and allowing only gradual changes whenever it becomes necessary [28].

Very little reduction of feedstock occurs in a plant by volume. In fact, there occurs an increase in volume of effluent on account of slurry preparation with water. Digested slurry after anaerobic fermentation has far less odour. As it still contains most of the original nitrogen, phosphorus and potassium and is still highly polluted, effluent cannot enter a sewer stream after it leaves the digester. Lagoons are commonly used to hold the wastes in most European countries until it can be disposed of by either hauling or pumping onto agricultural land. As a result of fermentation, there is hardly any or marginal increase in the amount of nitrogen, phosphorus or potassium although it ensures their larger availability.

Biogas plants can be fed with slurry continuously or in batches. A batch loaded digester is filled to capacity, sealed until it has produced all it can, emptied, and filled again. Gas production is uneven because bacterial digestion starts slowly, peaks and then tapers off with growing consumption of volatile solids. This difficulty can be overcome by connecting the batch loaded digesters in series loaded at different times so that adequate gas is available daily. This method uses feed efficiently but is less efficient in terms of digester space. In a continuous type biogas plant fresh slurry is added daily. Gas production in this method is consistent because the bacteria always have a fresh supply of volatile solids to digest. The continuously loaded digester uses expensive digester space efficiently although it may not produce as much gas per kg of feed as a batch type may do [24].

Incoming slurry displaces an equal amount of processed slurry from the digester each time it is fed since maximum allowable volume of slurry is fixed for a plant. In view of digester volume being constant, fraction of digester's liquid volume replaced each day determines retention period. For example, if slurry equal to one-tenth of digester's liquid volume is added daily, digester slurry has an average retention time of 10 days. Considering that retention period depends on digester's volume, its selection to suit a particular feedstock and location determines size or capacity of a digester. Other factors that have bearing on plant capacity include number, size and type of animals providing feed and optimum amount of dilution water required for slurry preparation. A very brief retention time does not allow bacteria enough time to digest and hence low gas production results. Long retention time allows more complete decomposition of wastes and hence increased gas production rate follows. However, if retention time is made very long, it does not furnish enough fresh slurry to promote bacterial growth, and hence again a low gas production rate results [20, 24, 28].

Considering that gas supply and demand do not always match, some gas storage either within gasholder or outside is necessary. Low pressure storage systems usually suffice but if the gas is to be used for such applications as vehicle fuel, high pressure storage is required. At low pressure of 100 to 150 mm of water column, polystyrene or butyl rubber bags can be used for storage. However, for storing several days gas output, volume needs become excessive. For example, to store biogas equivalent of 200 litre drum of petrol, a low pressure storage system requires a volume of about 300 cubic meters which roughly corresponds to the volume of an average-sized house. High pressure systems use a three or four stage compressor to compress biogas to pressures of around 3500 psi which can be stored only in high pressure cylinder [35]. Even at such high pressures which are expensive to attain, stored biogas has only one sixth of energy density of liquid fuels.

Methane which is the principal constituent of biogas cannot be liquefied by any pressure at ambient temperature. The critical pressure of methane is 4710 kPA at (–)82°C. In contrast, propane has a critical pressure of 1290.3 kPA at 38°C and can thus be liquefied at ordinary temperature (Critical temperature of a gas is that temperature above which it is not possible to liquefy it by any pressure. Critical pressure is the pressure needed to liquefy gas at critical temperature). Liquefaction poses further problems following presence of large quantities of carbon dioxide and other impurities. In case of natural gas it is possible to store it by absorption on propane or absorption on activated charcoal. If this method is tried on biogas, it may pose several difficulties such as high initial cost of absorbent, large storage needs, and also release of gas from absorbent. Propane tanks, have been successfully used for biogas storage at pressure up to 1400 kPA (200 psi) on farm digester installations in Canada. The Biomass Energy Institute at Manitoba (Canada) demonstrated that this pressure is easily achievable by using a conventional cold storage compressor [28, 30].

For obtaining needed gas yield, it is advisable to have 8-10 per cent of total solids concentration in feed. Cow dung which has a solid concentration of about 20 per cent needs to be diluted. For achieving 10 per cent solid concentration, it is advisable to mix dung and water in 1 : 1 ratio. One kilogram of dung produces about 40 litres of biogas, and thus in order to produce two cubic metres (2000 litres) of gas daily for meeting cooking needs of an average family of four to five members, a mix of 50 kg of dung and 50 litres of water needs to be fed daily [45, p. 4].

Biomethanation provides nearly three times more useful energy than direct burning of dung and also produces nutrient-rich manure (Fig. 1.1). Biogas is

25 kg Fresh Wastes

Direct Burning		Biogas
5 kg dry wastes	Product	1 cub. m biogas
10460 kcal	Gross energy	4713 kcal
10 per cent	Device Efficiency	55 per cent
1046 kcal	Useful energy	2592 kcal
None	Manure	10 kg air-dried manure

Fig. 1.1: Comparative advantages of biogas vis-a-vis direct burning of agro-wastes
Source: TERI, Biogas: A Source of Rural Employment, New Delhi, 1994, p. 5.

convenient to use and saves time in the kitchen. Biogas stove has an efficiency of about 55 per cent which is comparable to that of an LPG stove. Apart from being eco-friendly, use of biogas saves women time in fuelwood collection. Based on equivalent effective heat produced, a 2 m^3 biogas plant can replace in a month fuel equivalent of 26 kg of LPG contained in standard gas cylinders, or 37 litres of kerosene or 88 kg of charcoal or 210 kg of fuelwood or 740 kg. of animal dung. On life-cycle basis according to TERI studies [45, p. 6], biogas is cheaper than traditional non-commercial fuels such as animal dung, fuelwood and crop-wastes; and LPG (Fig. 1.2) and is only marginally expensive than kerosene.

High conversion efficiency is an important requirement for achieving economic viability of a plant. Gas yield is the most important economic parameter for assessing the performance of a biogas plant. Two basic approaches for increasing gas yield are controlling feedstock composition and improving digestion systems. Biological sensitive steps which are easy to control in waste-treatment plants are difficult to regulate during biomethanation. Recourse to multi-phase digestion and use of appropriate biological materials are helpful in regulating operations. In multiphase digestion of biomass, the number of phases are generally large to control temperature, pH, microbial life-cycles, oxidation-reduction potentials, introduction of unique microorganisms and production and removal of inhibitory levels of organic acids. Rate control may be used, to keep various components of fermentation in balance, to maximise overall rate of the fermentation and to alter the end products mainly comprising biogas with maximum methane content [42].

Amount of gas available from a given quantity of substrate in a continuous-fed digester can be estimated using the following relationship [46]:

$$V = \frac{G_e . L_r . V_{gen} \, [1 - e^{-k.HRT}]}{\text{HRT}}$$

where

V=Volume of gas produced per day.

G_e=Maximum gas production expected per unit of VS or TS of substrate used.

L_r = Loading ratio expressed as either VS or TS as long as both G_e and L_r are expressed in the same way.

V_{gen}= Volume of the digester.

e = 2.718

k = Reaction velocity constant.

HRT= Hydraulic retention time of the digester.

1.5 Movable Drum Versus Fixed Dome Type Plant

The conventional movable drum type model comprises a masonary digester with an inlet on one side for feeding slurry and an outlet on the other side for removing digested slurry. The gas collects in a steel gasholder which is inverted over the slurry and moves up and down depending upon accumulation and discharge of gas guided by a central guide pipe (Fig. 1.3). The main drawback of this design is that metal construction costs are very high. The movable gasholder made of mild steel alone accounts for about 40 per cent of the total costs of plant. The maintenance costs of this type of plant are also high considering the fact that vital components like gasholder are corrosion-prone which need to be given anti–corrosive painting at least once a year. Moreover, several construction materials are not readily and locally available at plant sites [6]. JBP which is fixed dome plant dispenses with movable steel drum type gasholder, and its gasholder is built with bricks and cement as immovable and fixed cover of the well like digester itself (Fig. 1.4). When biogas is formed inside the digester, it rises towards the top of the dome and pushes slurry down. Displaced level of slurry provides necessary pressure for release of the gas up to the point of its use. The first pilot drumless JBP was developed by the Gobar Research Station, Ajitmal in early 1977. The depth and diameter were 2 meters and 3 meters respectively with the digester so designed that it could hold about 50 days feed.

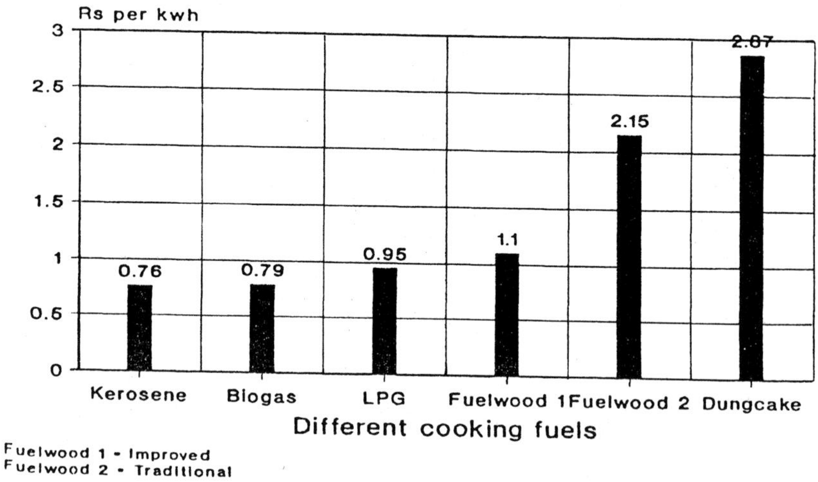

Fig. 1.2.: Comparative costs of different cooking fuels (Rs. per kwh)
Source : TERI, Biogas: A Source of Rural Employment, New Delhi, 1994, p. 6.

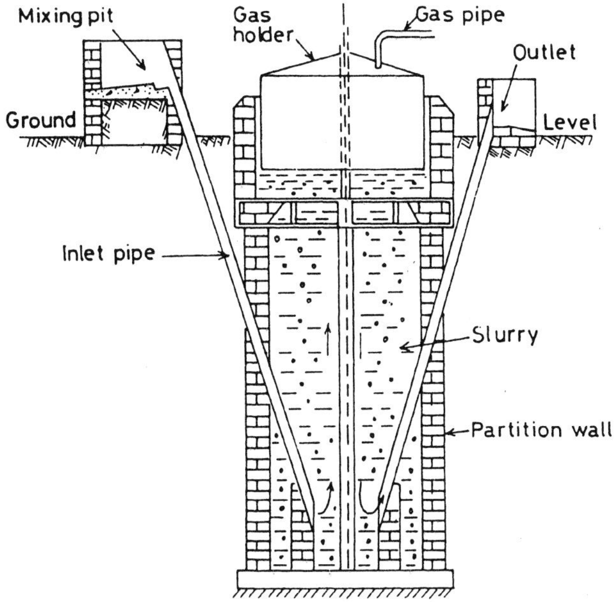

Fig. 1.3 : Cross-sectional view of floating drum type biogas plant promoted and popularised by the KVIC.
Source: UN, Guide-book on Biogas Development, New York, 1980, p. 3.

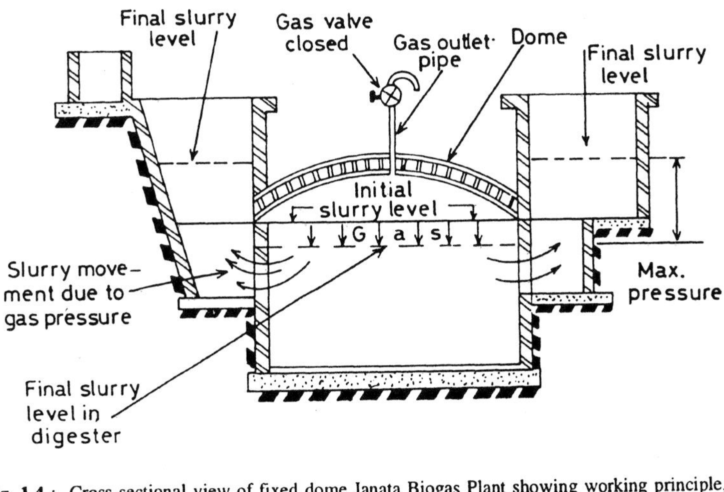

Fig. 1.4 : Cross-sectional view of fixed dome Janata Biogas Plant showing working principle, initially developed by the PRAD, Lucknow and later widely promoted in India.
Source : Rajput, R.K., A Treatise on Janata Model Biogas Plant, Changing Villages, 6(4), 1984, p. 265.

The JBPs have no moving parts and hence the wear and tear and consequent maintenance problems are relatively few. Accordingly these plants have longer working life. Service life of a biogas plant can be expected to range anywhere between 15 to 25 years. In a movable drum type plant, digester is generally expected to last for 40 years, gasholder for 10 years and pipes and accessories for about 20 years. Gas yield of JBPs is considered to be higher (0.04 m^3 per kg of cattle dung at 24-27°C) than those of other plants. However, gas yield per unit volume of the digester is more for a movable drum type than fixed dome type plant. Operating costs per cubic meter of biogas is generally more for a movable drum type than a fixed dome type biogas plant. Effluent obtained from JBPs has the same carbon and nitrogen content and it is richer in phosphorus and potassium contents by 2.5 to 6.5 per cent and 7.2 to 18.5 per cent, respectively as compared to those from other plants. On account of JBPs being underground, rate of gas yield from them shows relatively uniform pattern during days and nights due to their being less susceptible to temperature variations. One side advantage of JBPs is that the space available above the plant can be gainfully used for a variety of applications such as providing a cattleshed. A comparison [5, 18] between fixed dome type (JBP) and movable drum type biogas plant is summarised in Table 1.2.

There occurred spurt in number of biogas plants in India from mid-seventies onwards which received further fillip with the launching of the NPBD in 1981 under the Union Ministry of Agriculture. The NPBD was transferred to the erstwhile DNES in September, 1982. In July 1982, DNES was upgraded as the Ministry of Non-Conventional Energy Sources (MNES). Under the NPBD, eight designs have been approved by the MNES for dissemmation/promotion. All these models barring FLXI model are derived from one of the two basic designs, KVIC floating drum type or Janata fixed dome type.

After the initial high success of the KVIC and Janata models, most valuable addition is low-cost spherical-shaped fixed dome *Deenbandhu* model. Eight

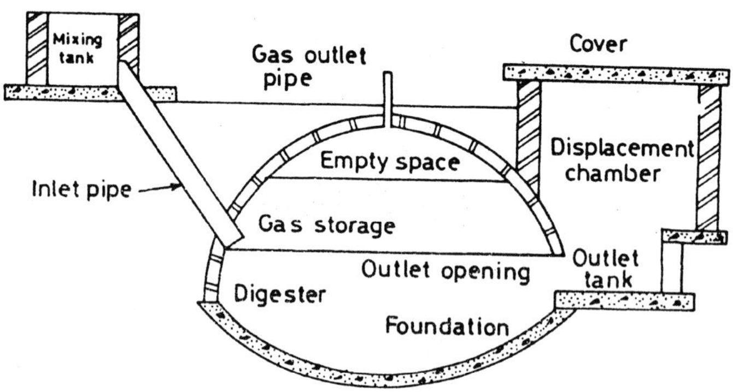

Fig. 1.5 : Cross-sectional view of Deenbandhu Model.

Table 1.2 A Comparison between Fixed Dome Type and Movable Drum Type Biogas Plants

S. No.	*Comparison Factor*	*Fixed Dome Type Plant*	*Movable Drum Type Plant*
0	System	Regular filling, and irregular discharge	Regular filling and regular discharge through over-flow.
1.	Material	Concrete blocks, bricks, sand, quarry-stones, lime, cement, and steel.	Concrete blocks, bricks, sand, quarry-stones, lime, cement, and steel.
2.	Main cost factors	Cement and bricks.	Steel, cement and bricks.
3.	Construction aspects	Suitable to construct locally but considerable masonary skills required.	Possible to construct locally but gasholder is normally required to be constructed in a workshop.
4.	Thermal insulation aspects	Since constructed underground, it ensures uniformity of temperature and heat insulation.	Considerable digester heat is lost through steel gasholder which is difficult to insulate; it is, therefore, less suitable for use in colder regions.
5.	Gas-tightness	Gas storage dome is required to be given special treatment for gas tightness and painted periodically.	No such problems.
6.	Feedstock	Agricultural waste, other organic matter (including fibrous material), animal and human faeces.	Animal and human faeces, chopped agricultural wastes as additive.
7.	Biogas yield	Digestion period normally between 30-60 days, gas yield 0.15-0.35 m^3 per m^3 of digester volume per day which can go up to 0.4-0.6 m^3 if plant is absolutely air-tight.	Digestion period normally between 40-60 days, gas yield 0.4-0.6 m3 per m^3 of digester volume per day.

Table 1.2 (Contd.)

S. No.	*Comparison Factor*	*Fixed Dome Type Plant*	*Movable Drum Type Plant*
8.	Maintenance aspects	If fibrous material is used as a feedstock, plant is required to be cleared once or twice a year for maintenance purposes.	Regardless of feed, gasholder is required to be painted every year and replaced after every 10 year.
9.	Cost considerations	Since construction is more labour-intensive, hence it is relatively inexpensive.	Due to greater use of steel, it is more expensive
10.	Scope for possible improvements	Scope exists for providing facilities for dome-tightness, slurry-mixing and digester-heating.	Scope exists for improving thermal insulation facilities, and improving rust-proofness of gasholder.

Source: 1. BORDA, Biogas: Manual for the Realisation of Biogas Programmes, Bremen Overseas Research and Development Association, German Association for Technical Cooperation (GTZ), GmbH, Eschoborn, Germany (erstwhile FRG). 1980, p. 61-62.

2. GATE, Biogas Plants: Building Instructions, German Appropriate Technology Exchange, German Agency for Technical Cooperation, Eschborn 1, Germany 1980, p. 15.

MNES approved models are : KVIC floating drum, *Janata* fixed dome, *Janata* model II (modified), *Deenbandhu* fixed dome, *Pragati* floating dome, Ferrocement digester, Fibre Reinforced Plastic (FRP) gasholder type and FLXI portable model [43, p. 9]. FLXI model made of rubberised nylon fabric is particularly appropriate in hilly areas and rough terrain where the cost of transporting building material is usually high. Some broad details of five MNES approved models are given below.

Deenbandhu Model : This model was designed and introduced by the Action For Food Production (AFPRO), a national Non-Governmental Voluntary Organisation (NGVO) under the dedicated supervision and control of Raymond M. Myles, J. B. Singh and other AFPRO professionals. Being a low-cost plant with cost nearly half that of the KVIC floating drum type plant, *Deenbandhu*

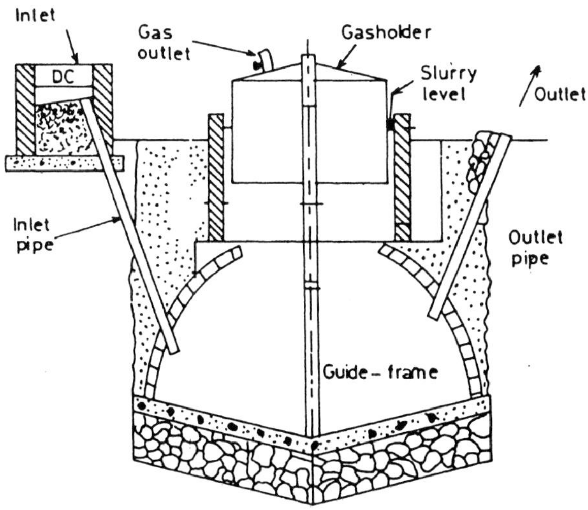

Fig. 1.6 : Cross-sectional view of Pragati Model
Source: TERI, Managing the Biogas Programme, New Delhi, 1994, p. 11.

model is most popular and is ahead of other MNES approved models. Advent of *Deenban*dhu model is most significant development in India's biogas programme. It is a fixed dome plant similar to the *Janata* model and Chinese design (Fig. 1.5). The cost reduction is achieved by minimising the surface area by joining the segments of two spheres of different diameters at their bases. The fixed masonry dome requires skilled workmanship and quality material to eliminate chance of leakage [45, p. 11]. More details about the *Deenbandhu* model can be obtained from the AFPRO (25/1A, Institutional Area, D-Block, Pankha Road, Janakpuri, New Delhi 110 058).

Pragati Model : This model was developed by the United Socio-Economic Development and Research Programme (UNDARP), a Pune-based NGO and was approved for inclusion under the NPBD in 1987. This model combines features of the KVIC and the *Deenbandhu* models (Fig. 1.6). Similar to *Deenbandhu* model, the lower part of the digester is semi-spherical in shape with a conical bottom. The upper part instead of being a fixed dome, has a floating gasholder similar to the KVIC model [45, p. 11]. The plant has gained sufficient popularity in Maharashtra, the state of its origin.

Ferrocement Digester : This model was developed in seventies by the Structural Engineering Research Centre (SERC), Roorkee, which since then has been shifted to Ghaziabad. This model achieves 10-15 per cent reduction in cost over a KVIC digester which results from the use of locally available materials and ferrocement in digester construction, and reduced labour input and maintenance [45, p. 12]. Ferrocement is a versatile form of reinforced mortar in which closely spaced layers of wire mesh reinforcements are impregnated with rich cement mortar mix. Ferrocement can be cast into sections as thin as 1 cm and is thus well-suited for precast products (Fig. 1.7). Apart from the SERC which developed the ferrocement digester under the 'All India Coordinated Project on Biogas', later the Cement Research Institute of India (CRII),

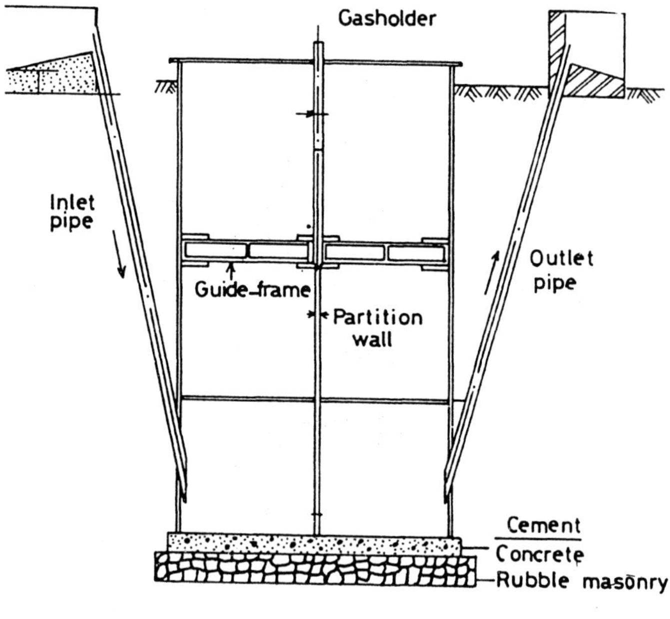

Fig. 1.7 : Cross-sectional view of Ferrocement Digester
Source : TERI, Biogas Technology: A Promise for the Future, New Delhi, 1994, p. 19.

Ballabhgarh (Haryana), also developed ferrocement gasholder for medium-capacity biogas plants in early eighties. Ferrocement possesses high resistance to cracking and permeability which is a requisite characteristic for a biogas plant. Ferrocement has low thermal conductivity and accordingly rate of gas production is fairly uniform in all seasons. A semi-mehcanised process for casting thin cylindrical units of ferrocement has been developed and patented by the SERC. This process is labour-intensive and does not require power or expensive machinery and is thus appropriate for use in rural areas.

Fibre Reinforced Plastic (FRP) Gasholder : It is movable drum type plant in which instead of metal, FRP is used as a gasholder material. FRP is manufactured by using 'Contact Moulding Process', a moulding technique which does not require any external pressure and hence it is a low-cost process which can be carried out locally. Being non-metallic, FRP gasholder does not require any painting and thus eliminates recurring expenditure on maintenance [45, p. 12].

Flexible Portable Model (FLXI) : Conventional biogas plants are not appropriate in hilly areas where high transportation costs of construction materials substantially adds to their costs [45]. Conventional underground digesters are also not appropriate for thousands of miles along the coastlines in India, Pakistan and Bangladesh and several other countries in similar situations. Recurement floods in the vicinity of *Ganges*, *Indus* and *Brahmaputra* river banks adversely affect effectiveness of biogas programmes in these densely populated regions. In hilly and coastal regions balloon type flexible digesters made of neoprene-rubber can provide an effective solution. FLXI plants can be easily installed above ground, conveninently transported to isolated locations, and quickly installed at any time at low cost. In India, *Swastik* Company, Pune produces flexible digesters called FLXI digesters. These digesters are also sometimes referred to as *Swastik* plants. *Swastik* plants need not be installed at locations where conventional plants are appropriate. They should be seen more as an alternative option where conventional plants cannot be built underground.

Matching with the average cattle holding size of rural families, emphasis in recent years has been shifted towards installing 2 m^3 family sized plants. Very recently MNES has also encouraged dissemination of 1 m^3 capacity plants to benefit weaker sections in rural households who may not be having sufficient number of cattle to own a 2 m^3 plant. This measure is expected to go a long way for socio-economic upliftment of the rural poor [44, p. 26].

Cost of a plant depends on type of plant design, plant capacity, hydraulic retention time (HRT), materials used, labour input, workmanship skills involved, and plant location. A plant on hills may cost more because of added cost of material transportation. HRT refers to the number of days the feedstock is needed to remain within the digester to enable biogas production. Choice of HRT depends upon the overall climate conditions of the area where the plant is to set up. In India, different HRTs are recommended for three different temperature zones (Table 1.3). For all approved models under the NPBD, plant dimensions and material requirements are specified for different HRTs. Plant cost increases with increasing HRT [45]. Cost estimates of KVIC, *Janata*,

Deenbandhu, Pragati and FLXI models are given in Table 1.4. It is to be seen that cost-wise fixed dome plants are cheaper than floating-drum plants [4].

NPBD began with an outlay of Rs. 50 crores earmarked during the sixth plan. During eighth plan Rs. 320 crores were allocated for the NPBD which represented 37 per cent of the MNES budget. The NPBD aims to achieve following 5 points objectives. [44, p. 10].

Table 1.3 Specification of HRT for Different Temperature Zones in India.

Zone	*Mean Ambient Temperature*	*HRT (Days)*
Zone I Andhra Pradesh, Goa, Karnataka, Kerala, Maharashtra, Tamil Nadu, Pondicherry and A & N Islands.	> 20°C	30
Zone II Bihar, Gujarat, Haryana, Jammu region of J & K, Madhya Pradesh, Orissa, Punjab, Rajasthan, Uttar Pradesh and West Bengal.	>15-20°C	40
Zone III Himachal Pradesh, North-Eastern states, Sikkim, Kashmir region of J & K and Hill districts of Uttar Pradesh.	< 15°C	55

Source : TERI, Biogas: A Source of Rural Employment, New Delhi, 1994, p. 5.

Table 1.4 Comparative Cost Estimates of MNES Approved Biogas Plants for Different HRTs (In Rupees)

Plant	*KVIC*			*Janata*	*Deenbandhu*		*Pragati*	*FLXI*
Capacity (m^3)	30 *days*	40 *days*	55 *days*	55 *days*	40 *days*	55 *days*	55 *days*	55 *days*
1	7600	8500	11900		5800	8100		9450
2	10500	11700	16500	11500	6800	9400	13500	14000
3	14200	15700	22000	14000	8200	11500	17000	17300
4	16300	18100	25400	17000	10500	14800	20000	25000

Source : TERI, Biogas: A Source of Rural Employment, New Delhi, 1994, p. 10-12.

— To provide energy in a clean and unpolluting form.

— To make available enriched fertiliser as a by-product for supplementing and optimising use of chemical fertilisers.

— To reduce pressure on dwindling fuelwood supplies and prevent indiscriminate deforestation.

— To eliminate the smoke-filled cooking environment, reduce drudgery and prevent eye ailments.

— To achieve improvement in rural sanitation.

For implementing the NPBD, MNES is involved with the following type of activities.

— To advance grants, subsidies and loans to state agro-industries corporations, state renewable energy agencies, KVIC, dairy-development boards, NGOs, etc. [38].

— To promote and popularise biogas programme by providing requisite training and extension services [38].

— To liaise with national and international organisations and to act as focal point for disseminating needed information among all concerned in India.

— To sponsor research projects at agricultural universities, energy institutes and research laborotories for advancement of biogas technology.

— To institute socio-economic surveys for evaluation of biogas programme from time to time and apply course-corrections.

— To arrange R & D activities on different aspects of biomethanation covering plant design, use of locally available materials in plant construction and operation, and distribution of biogas enriched fertiliser.

— To develop, test and popularise low-cost model of biogas plants developed from time to time by various agencies.

— To liaise with banking and other financial institutions for arranging loans for setting up biogas plants.

— To liaise with the government departments for release of special quotas of cement, steel, etc. earmarked for the programme.

NPBD is characterised by a number of activities like short term and long term goal–setting, promotion and awareness activities including dissemination of technical information, training of extension staff, mobilising financial resources, alleviating difficulties in plan implementation through in-depth analysis and interactions along all concerned. For proper monitoring, biogas programme needs to be evaluated from time to time. Periodic evaluation of the programme can be carried out with following objectives [39, 40].

— Socio-economic evaluation of the impact of biogas programme on living standard and quality-of-work life of the affected population [4].

— Evaluation of programme impact on deforestation following reduced dependence on fuelwood as household fuel.

— Programme evaluation as a means for meeting rural energy needs.

— Programme evaluation as a source for meeting fertiliser needs in rural areas.

— Programme evaluation as a public health and eco-friendly measure for treating night-soil and animal wastes.

1.6 Factors Affecting Yield of Biogas Plants

Regardless of specific feed used, biogas generation proceeds most efficiently when raw materials fed to the digester have a desired pH and carbon-to-nitrogen ratio (C:N). Bacteria thrive in a slurry with a pH value of around 7 and consequently if the input slurry has a pH close to 7, fermentation proceeds more smoothly. Under normal conditions, fermentation process balances excess acidity or alkalinity on its own. Carbon is the major chemical element in organic wastes which bacteria digest releasing methane and carbon dioxide in the process. However, microorganisms require certain amount of nitrogen in feed in performing their function. Ratio of carbon-to-nitrogen is thus crucial for anaerobic fermentation.. A high C:N ratio means that the nitrogen will be exhausted before carbon is digested. Coversely, a low C:N ratio i.e., too much nitrogen in relation to carbon results in high ammonium concentrations which tend to become toxic to anaerobic bacteria. It is possible to adjust the C:N ratio in a biogas plant by adding another material to counter the material already in the plant. For instance, saw dust which has a high C:N ratio could be added to poultry droppings which has a low C:N ratio. Dairy cow wastes has a C:N ratio just slightly below that required by bacteria [4].

Total biogas production rate depends on organic material being digested, digester loading rate and environmental conditions in the digester. Under ideal conditions (95°F or 35°C temperature and proper pH), it is possible to produce about 45 ft^3 (1.4 m^3) of biogas at atmospheric pressure from one day's faeces of a 100 pound (54 kg) cow. Not all of the biogas produced is available for utilisation as part of it goes as waste [20]. In very aggregate and general terms, 1 m^3 of slurry fed to biogas plant produces an average of 0.15 to 0.2 m^3 of biogas daily; one cubic meter of biogas produces about 5500 kcal of heat, and each member of a family requires 1.5 to 2 m^3 of digester space to meet his mean requirement of lighting and cooking [18, p. 12]; 32 kg of cattle dung per day, 20 kg of pig faeces per day, and 12 kg of poultry droppings per day produces 1 m^3 (34 ft^3) of biogas every day under normal conditions [36, p. 8].

1.7 Commonly Used Feedstocks for Biomethanation

Common materials used as feedstock include animal wastes, crop-residues and urban-wastes including night-soil. These constitute broad categories under which a variety of materials can be covered. Some of these waste materials are directly used as fuel or fertiliser. Utilisation of these materials for biogas generation harnesses additional benefit while retaining their existing benefits which are anyway tapped. The list of materials which fall under the above three broad categories can be further expanded to include dairy cow wastes, swine manure, poultry litter, sheep and goat droppings, crop-residues of various plants, sugarcane trash, bagasse, rice-husk, rice-bran, tobacco wastes and seeds, cotton dust from textile mills, tea wastes, fruit and vegetable wastes, marine algae, seaweeds, water-hyacinth, pressmud from sugar plants, twigs, bark, tree branches plant leaves, municipal refuse and night-soil, etc. [30, p. 10].

Another development in the area of anaerobic digestion is application of industrial wastes which constitute an important category of feedstocks. Wastes available from food and drink industry are particularly useful for this purpose. Vegetables, fruit-processing and canning units produce large quantities of carbohydrates-rich wastes which are easily amenable to anaerobic fermentation. Drink manufacturers use enormous quantities of biomass in the form of barley, apples, grapes, etc. and residues from these plants can be anaerobically fermented. Methane gas production from vegetable wastes has in past been particularly popular in countries like Australia and South Africa. Wastes from sugar distilleries which have high carbohydrate contents are also a valuable feedstock. Several sugar factories world over have established fermentation units. One manufacturer has successfully completed trials for a digestion plant utilising wastes from an olive-oil factory in Spain which besides generating usable gas also solved problems of pollution-control.

1.8 Properties and Characteristics of Biogas

Biogas which consists of 60-65 per cent methane, 35-40 per cent carbon dioxide and traces of hydrogen sulphide, ammonia and other impurities is an explosive and toxic gas. Methane which is the principal constituent of biogas is colourless, odourless and tasteless. Methane is also combustible in a concentration of 5 to 15 per cent in air and, therefore, potentially hazardous. Considering that biogas is warm when it leaves digester, it contains water vapour which condenses when exposed to colder temperatures. Condensation traps in gas pipelines are necessary to prevent water from blocking low points in pipelines. Biogas following condensation of water vapour or even without condensation, can be used as a fuel for cooking, lighting and several other applications. However, in stoves gas-burner nozzle needs to be enlarged to compensate for low heat value of biogas. Tests using biogas as the only fuel in a petrol engine showed that energy present in 200 ft^3 (5.7 m^3) of biogas is equal to energy present in one gallon of petrol. In diesel engines biogas replaces most of diesel fuel but about 20 per cent of diesel is still needed as a secondary fuel. For this purpose engine adjustment and biogas conversion kit are necessary which, however, do not cost much. If biogas is used as an engine fuel, the engine should be stationary, close to the digester and used frequently. In this way engine's cooling water can be gainfully used to heat digester and thereby increase system's efficiency [24].

Calorific value of biogas ranges between 500 to 700 BTUs per ft^3 (0.03 m^3). In comparison, natural gas has a heat value of 850 BTU per ft^3 (0.03 m^3), petrol contains nearly 120,000 BTUs per gallon (3.8 litres), diesel oil 133.000 BTUs per gallon and propane 92,000 BTUs per gallon. Biogas is difficult to store, compress or liquefy. Biogas unlike liquefied petroleum gas (LPG) cannot be converted to liquid under normal temperatures. A temperature of (–) 260°F (-162°C) is needed to liquefy methane at atmospheric pressure. Liquefaction of methane requires pressures of nearly 5000 psi (LPG liquefies at 160 psi) at a temperature of (–)117 (-83°C) and is thus not practical. If biogas is compressed to a pressure of 1000 psi, it requires about 1320 BTU of energy to put 6350 BTUs into a storage container. As biogas cannot be liquefied economically, it is best-suited for such applications as cooking, lighting, grain-drying or operating stationary internal combustion (IC) engines. Biogas is normally produced on continuous basis. As most applications are intermittent in nature hence some kind of storage facilities are needed. Amount of gas storage is dependent on storage time and pressure. High demand applications, such as grain-drying are normally impractical following excessive storage capacity requirement. Although biogas can be used for variety of applications including crop-drying, refrigeration and running of diesel engines, so far it is mainly used as cooking and lighting fuel in most developing countries [20, 24].

Methane content in biogas is affected by digester conditions. Biogas contains traces of hydrogen sulphide which is highly corrosive. This can be removed by passing the gas through steel wool or iron filings. Following condensation of water vapour, it can be used as fuel for any standard application. Biogas contains nearly forty per cent carbon dioxide which can be removed by bubbling it through lime water. Lime water is a mixture of 1.8 kg burnt lime in 1000 litres of water. One 1000 litres of lime water removes 560 litres of carbon dioxide. For removing 35 per cent carbon dioxide in biogas, 625 litres of lime water is needed to purify 1 m^3 of biogas. Lime water becomes milky following absorption of carbon dioxide [36, p. 83].

1.9 Potential of Biogas Plant Sludge as Enriched Fertiliser

Effluent or sludge resulting from a biogas plant is a resource and contains almost all of nitrogen present in raw manure prior to digestion. The effluent is an excellent fertiliser because nitrogen in the effluent is more readily absorbed by plants than nitrogen in raw manure [1]. The sludge that collects at the bottom of a digester should be regularly removed because accumulated sludge tends to reduce active digester space. The end result of applying anaerobically digested sludge to soils has the same beneficial effect as that obtained from applying any other kind of organic matter. The humus materials formed in sludge tend to improve soil physical properties such as aeration, moisture-holding and water-infiltration capacity. Furthermore, sludge serves as a source of energy and nutrients for development of microbial population which directly or indirectly improve solubility and availability of essential chemical nutrients contained in soil minerals [24, 30].

The organic fraction of sludges from an anaerobic digester operating on plant and animal waste may contain 30 to 40 per cent lignin, undigested cellulose and lipid materials on dry weight basis depending upon lignin type of raw material used. The remainder consists of substances originally present in the raw material but protected from bacterial decomposition by lignin, newly synthesised bacterial cellular substances and relatively small amount of volatile fatty acids [24, 30]. The amount of bacterial cell mass is small (less than 10-20 per cent of substrate is converted to cells). Thus there is a less risk of creating odour and insect-breeding problems when anaerobically digested sludges are stored or spread on land than when untreated or partially treated organic waste materials are similarly handled, indiscriminately disposed off or stored. The sludge is an almost odourless, homogenised liquid which does not attract rodents or flies.

1.10 Biomethanation as an Aid for Environment Improvement.

Biogas plants help to bring improvements in ecology and environment. Biomethanation provides a non-traditional way for disposal of sewage, animal and human faeces in urban and rural areas. In biogas plants as these wastes are no longer exposed to open air, menace of mosquitoes, flies and hookworms is controlled to a large extent. Following check on hookworms, biomethanation also helps to control disease like *Schistosomiasis, hepatitis* and intestinal infections. It improves household hygiene as it prevents occurrence of polluting smoke that emanates from traditional *fuelwood* stoves (*Chulhas*). Faeces contain countless parasitic ova and pathogenic oraganisms which often cause ailments like typhoid, dysentry, *hepatitis* and *poliomyelitis* [30]. A summary of diseases caused by inept handling of night-soil (human faeces) and their causative micro-organisms is given in Table 1.5. Chances of health hazards posed, basically depend upon three factors, namely, proportion of viable pathogenic organisms in faeces, survival rate of these organisms in sludge, and length of time sludge is stored prior to use.

Survival rate of pathogenic micro-organisms varies from situation to situation and there are no fixed rates for this purpose. Survival or conversely kill rate of some common enteric micro-organisms is given in Table 1.6. Most of these bacteria do not survive beyond a period of 14 days. Anaerobic fermentation process is one of the most effective ways of killing enteric micro-organisms which is highly beneficial to public health. The survival of enteric micro-organisms has been the subject of many investigations the outcome of which depends on numerous factors such as the type of organism, ambient temperature, type of soil, moisture, and pH of digester contents. Free ammonia which is released during anaerobic fermentation is another factor that influences survival and growth of micro-organisms. The usual proportion of ammonia in digester contents is of the order of 0.07 per cent. If free ammonia level is around 0.2 per cent, the ova of *Schistosoma* can hardly survive beyond six days and hookworm and *Ascaris* only marginally higher. Free ammonia reduces the survival period of micro-organisms as it tends to permeate egg shell and cell membrane and in the process kills ova and bacteria. Addition of certain plant materials like algae to sludge checks the growth of some bacteria. For instance, algae when added to effluent, retards the activity of *Anopheles* larvae following change in pH level and in the process makes the bacteria ineffective [30]. In spite of the fact that certain pathogens and parasites do survive even after anaerobic fermentation, and some remain active much longer, no cases of outbreak of diseases are as such reported from any quarter. This can be explained by the fact that most of the rural folk are anyway vulnerable to higher health risks, and biomethanation reduces these health hazards significantly, if not completely.

Table 1.5 A Set of Disease Caused by Improper Handling of Night-soil and their Causative Microorganisms

Disease Category	*Disease Description*	*Causative Microorganisms*
Viral	Infectious *hepatitis, gastroenteritis,* respiratry illness, *poliomyelitis*	*Adenovirus reovirus Entenovirus (poliovirus)*
Bacterial	Typhoid fever, *salmonellosis,* Bacillary dysentery (*Shigellosis*), Cholera, *Salmonella typhosa Salmonella spp. Shigella spp.*	
	Tuberculosis	tuberculosis
Protozoan	*Amebrasis* (Amebic dysentery)	*Entemoeba histolytica*
Helminthic	Roundworm Pinworm Whipworm Tapeworm Hookworm	*Ascaris lumbricoides* *Oxyaris vermicularis* *Trichurus trichiura* *Taenia saginate* *Ankylostoma duodenale* *Necator americanus.*

Source : NAS, Methane Generation from Human, Animal and Agricultural Wastes, Report of an Ad. Hoc Panel of the Advisory Committee on Technology Innovation, National Academy of Sciences, Washington, D. C., 1977.

Table 1.6 : Kill-Rates of Enteric Microorganisms During Anaerobic Fermentation

Microorganisms	*Temperature (^0C)*	*Retention Time*	*Kill Rate of Pathogens (Per cent)*	*References*
Polivirus	35	2	98.5	1, 2
Salmonella ssp.	22-37	6-20	82-96	3
Salmonella typhosa	22-37	6	99	3
Mycobacterium tuberculosis	30	Not reported	100	4
Ascaris	29	15	90	3
Parasite cysts	30	10	100	5

Source: 1. NAS, Methane Generation from Human, Animal and Agricultural Wastes, Report of an Ad. Hoc Panel of the Advisory Committee on Technology Innovation, National Academy of Science, Washington, DC, 1977.
2. Bertiucci, J., et al., Inactivation of Viruses During Anaerobic Sludge Digestion, Journal Water Pollution Control Federation, 49(7), 1642-1651 (1977).
3. Foster, D. H. and Engelbrecht, R. S., Microbial Hazards in Dispersing of Wastewater on Soil, in 'W. F. Sopper and L. T. Kardos (eds), Recyling Treated Municipal Wastewater and Sludge through Forest and Cropland, Pennyslvania State University Press, P. A, 1973'.
4. Briscoe, J., Public Health in Rural India: The Case of Excreta Disposal, Ph. D. thesis, Harvard University, Boston, MA, 1976.
5. Muller, G. Tubercle Bacteria Present in Sludge Produced in Mehcanical and Biological Sewage Treatment Plants. Abstracted by 'W. J. Muller, Sewage and Industrial Wastes, 32(9), 1030 (1960).'

Fig 1.8 : Biogas burners
Source : Gate, Biogas: Manual for the Realisation of Biogas Programmes, BORDA, Bremen, Germany, 1980

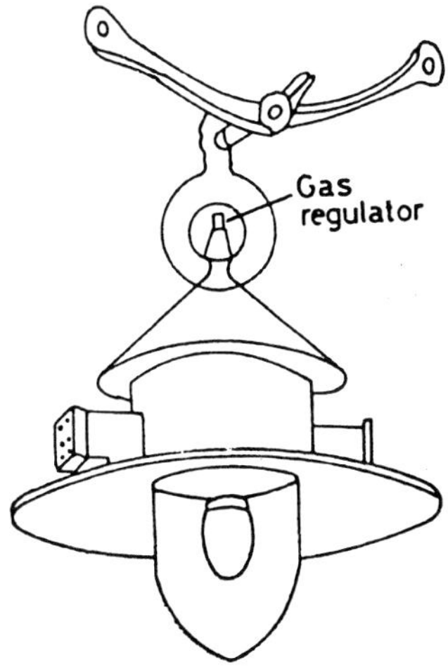

Fig. 1.9 : Single mantle biogas lamp
Source : UN, Guidebook on Biogas Development, New York, 1980.

1.11 Applications of Biogas

Biogas provides a clean fuel for cooking and lighting purposes. A biogas burner (Fig. 1.8) and lamp (Fig. 1.9) are structurally similar with a nozzle, an air inlet and a mixing chamber as common in both. The uncommon items are mantle in case of lamp and fire-sieve plate in case of stove. A schematic plan for using biogas as a cooking and lighting fuel is given in Fig. 1.10. Biogas lamp works on simple principle that when gas is burnt, mantle of the lamp glows and emits light. Mantle in a biogas lamp is similar to the one used in a 'coleman' or propane lamp. Brightness of biogas lamp mainly depends on factors like gas

Fig. 1.10 : A Schematic plan for using biogas as a cooking, lighting and engine fuel
Source : Buren, A. V., Chinese Biogas Manual, Intermediate Technology Pub. Ltd., London, July, 1979, p. 109.

pressure, gas-air ratio and through mixing. Proper nozzle adjustment is necessary to achieve desired light intensity. Low gas pressure causes poor light intensity but higher pressure though it improves brightness, lowers mantle life. Biogas lamps can be either suspended (hanging type) or table (standing type). As biogas is almost half as light as air, and as hot air rises upwards, brightness of a standing biogas lamp is somewhat greater than that of a hanging type. Biogas lamps are generally designed to produce 100 candle power and consume 0.11 to 0.15 m^3 (4 to 5.5 ft^3) biogas per hour. Several designs of biogas lamps are popular but most common ones are Indian, Chinese and Pakistani types.

Biogas stoves normally operate at gas pressure of 75 to 90 mm (3-3.5 inch) water column. As biogas passes through the nozzle, air is allowed to be drawn in mixing chamber. Nozzle adjustment is generally necessary to achieve desired flame temperature. Brightness and combustibility of gas is regulated by controlling gas pressure and air-fuel ratio which is generally maintained as 10:1. Combustibility of gas is maximum when the flame appears to be slightly yellow, a bit bright and burnished. When the flame turns blue and smokeless after air jet adjustment, it burns at a temperature of about 800°C. For meeting cooking needs of an individual 0.28 to 0.42 m^3 (10 to 15 ft^3) of biogas is needed daily. Cooking appliances in wide range are available both in single and double burner category with each burner consuming anywhere between 0.25 to 1.25 m^3 of gas per hour. Biogas cannot be burnt on stoves designed for the LPG/natural gas at it tends to lift off the stoves on account of its flame speed factor being slower than these gases. This problem can be somewhat overcome by reducing gas pressure but this causes higher gas consumption and reduces burning efficiency.

Biogas provides reasonably efficient fuel for both spark-ignition (petrol) and compression ignition (diesel) engines. Whereas diesel engines run on biogas as a dual-fuel comprising both oil and gas, petrol engines can be made to run on cent per cent biogas. However, in the latter case, it is initially necessary to use little petrol for starting-up. While running IC engine on biogas, some modification in the fuel injection system in required. In a partially converted system, a gas tap is fitted onto the inlet manifold to the engine. The engine is initially started on liquid fuel and little gas supply, and later the proportion is reversed when engine starts running at desired speed. Opening of gas tap and adjustment of air supply is carried out simultaneously. In steady engine running, dual-fuel helps to achieve about 80 per cent saving in diesel consumption (Fig. 1.11).

Biogas can also be utilised for electricity generation. As per some studies carried out in the past, 1 kwh of electricity can be generated from 0.7 m^3 of gas which can light about 15 electric bulbs of 60 watt rating for one hour. In contrast if 0.7 m^3 of gas is directly burnt to light gas lamps, only 5 gas lamps of 100 candle power (CP) rating equivalent to 60 watt electric bulb can be lighted for one hour duration. This shows that power route is more efficient for lighting than by direct gas burning. However, electricity generation from biogas is economical only from large–sized plants and it requires high initial capital investment. Asynchronous generators of 3, 5.5 and 7.5 kw rating coupled with IC engines running on biogas have been successfully tried and used.

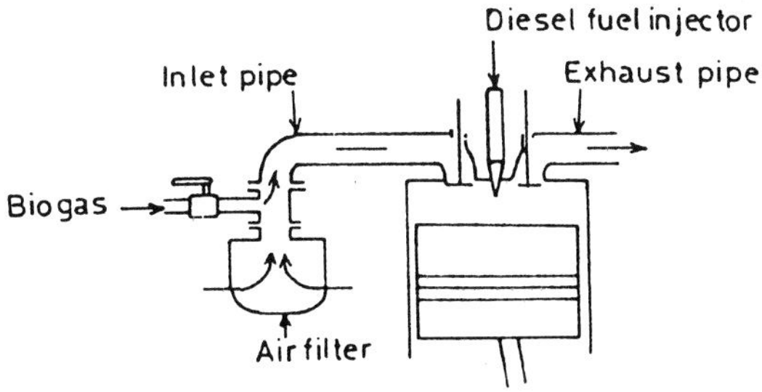

Fig. 1.11: Diesel engine running on dual-fuel
Source : UN, Guidebook on Biogas Development, New York, 1990, p. 84.

Biogas can also be used as a vehicle-fuel. It has been successfully used on a commercial scale in countries like Italy and New Zealand. Conversion kit suitable for biogas is virtually the same as for compressed natural gas (CNG) which is marketed as an automotive-fuel in many parts of the world [34]. For use as a vehicle-fuel, biogas is first compressed in a three or four stage compressor up to a pressure of 2500-3500 psi and stored in a high pressure gas storage cascade. Cascade helps to facilitate quick refuelling of storage cylinders. The number of cylinders in cascade depends on size of digester and the number of cylinders fitted to each vehicle. Being gaseous fuel, biogas provides excellent cold starting for vehicles because unlike petrol it does not first have to be vapourised [35]. Once started, the engine performs well immediately with no sluggish warm-up. This is because unlike petrol, methane does not condense on cold intake manifolds, cylinder heads and pistons. Biogas is carried in compressed state in cylinder heads and pistons. Biogas is carried in compressed state in cylinders which are fitted on vehicles. Each cylinder is about 1 meter long, 0.3 meter in diameter and weighs close to 50 kg. Two cylinders can be easily mounted in a biogas driven car and four in a truck.

1.12 Community Versus Family Size Plants

Biogas plants ca be constructed either as individual family units or as community plants. Whereas individual plants are appropriate for families, community plants are set up to meet fuel and fertiliser needs of groups, institutions and village as a whole. Community plants are particularly appropriate where individuals do not have adequate manure and fiance to set up family units. Community plants can be operated by a group of families, cooperatives or members of a locality. Utility of a community plant is further enhanced if some small scale industry is made to run on biogas as a fuel which may also create some employment opportunities for rural youth. In case of community plants, common utilisation of gas in the direct neighourhood of the plant such as for running a community kitchen is generally desirable. Many energy planners argue that individual plants should be constructed only when idea of setting up a community plant is infeasible. It is often seen that a farmer who is reluctant to invest a large sum of money for his own private plant may not find it difficult to pay a small deposit and monthly charges as his share in running of a community plant. Community plants provide economies–of–scale following savings in initial capital cost, land utilised and labour charges needed in running plants. Community plants make large-scale composting of organic wastes feasible by mixing them with plant sludge.

Community biogas plants are socially more egalitarian. Out of some 100 odd million households in India, only about 20 million own three to four heads of cattle needed to feed family-size biogas plants. According to the report of the Planning Commission's Working Group on Energy Policy (1979), even if all those households which can set up family-size plants were to do so, enough cattle would still be available to feed 560,000 community plants at a simple norm of one plant per village. In view of their vast potential, community plants received major thrust from the sixth plan (1980-85) onwards when ambitious target of 100 plants was made [8]. Plant economics can be further improved if night-soil from community latrines is also used as valuable feedstock. Animal wastes strewn on village roads and grazing lands which otherwise go as waste can also be gainfully utilised in community plants. Community plants ensure higher gas yield per kg of feed on account of economic viability of providing facilities such as warming up slurry during winter season which is not cost-effective in family units. Higher biogas supply from community plants besides meeting individual household fuel requirement can be used for generating power, fuelling rural industrial units, operating irrigation pump sets, and lighting streets.

Community plants are considered to be more cost-effective. The capital cost to set up a 500 ft^3/day (CFD) capacity plant is nearly half that required to install 50 plants of 100 CFD each. However, running of these plans is not without problems. For instance, due to countrywide social and economic disparities, task of dung-collection and gas distribution poses practical difficulties. Similarly, evolving a uniform pricing policy for gas and fertiliser produced which meets requirement of both rich and poor farmers alike is a difficult exercise.

1.13 Integrated Farming System

Integrated farming system may comprise a biogas plant, energy farming, wind-mills; solar heating and photovoltaic systems; food, fodder and algal cultivation and pisciculture (Fig. 1.12). Effluent from the biogas plant provides a nutritious protein-rich feed for animals particularly fish. Solids are recovered from sludge by allowing it to settle in tanks or by draining and drying. These solids along with other additives after being mashed and detoxified can be used as animal feed. After separation of solids from sludge, remaining liquid can be utilised as a nutrient for algal growth. *Chlorella*, a single-celled high protein can be harvested with the liquid separated from sludge in a shallow pond. The pond is usually lined with concrete metal or plastic material which makes harvesting easier and minimises contamination. *Chlorella* in proportions up to 10 per cent of animal feed can be used to replace soyabean as a meal for protein supplementation. With the availability of fish meal from sludge, it is advisable to have a fish-pond as part of the integrated farming system [5]. Algae produced can be utilised in two ways. Part of it can be used as feedstock for producing biogas and part as fish meal along with other additives. A large portion of biogas plant sludge is used as fertiliser for growing food, fodder and energy crops. Biogas plant sludge contains variety of products including nitrogen which is mostly lost during aerobic composting. Nitrogen is one of the most important nutrients for plant growth. Nitrogen content can be increased by adding animal urine to influent slurry fed to a biogas plant. For maximising fertilising effect, slurry should be ploughed into the soil about one week before sowing is planned. In integrated farming system, sludge is stored till the time of its spread. In a farm it can be stored in two ways, as a liquid fertiliser in a covered pit and as solid manure in a compost pit [30]. As per rough norm one cubic meter of slurry can fertilise about 100 sq. m of land annually.

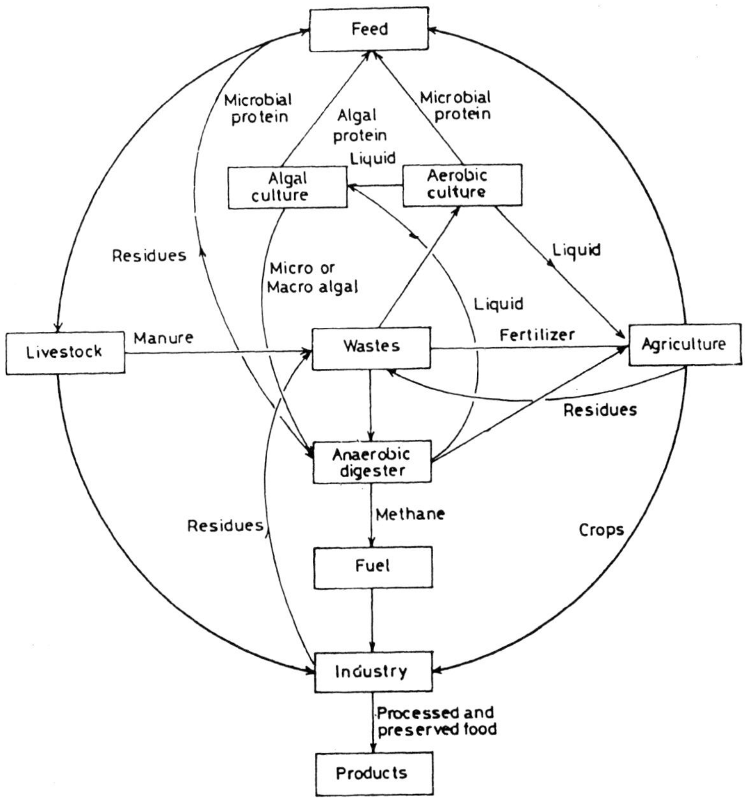

Fig. 1.12 : Elements of Integrated Farming System
Source : Huang, C. J, *et al.*, Design and Operational Model for the Integrated Swine Waste Managements System in Hawali, Proc. ASAE Pacific Region Annual Meeting, Hawaii, March 18-20, 1980.

1.14 Concluding Remarks

Biogas systems are appropriate for promotion as small scale decentralised energy systems which are very relevant for the country's pattern of rural and agriculture based economy. Considering that organic wastes are anyway generated on renewable basis in any nation, it is always advantageous to view biogas as a supplementary fuel which can go a long way in meeting part of the energy needs of a nation [15]. It should be realised that production of biogas from various types of wastes will not completely solve the energy problems of a nation even if all wastes are available for biomethanation. With a high rate of deforestation which must be prevented for ecological reasons, potential for utilising renewable energies in rural areas is sufficiently high. Success in biomethanation on a large scale is linked with success in afforestation. Biogas programme cannot make significant headway unless biomass availability is also increased simultaneously so that it can substitute for cattle dung which is preferred to be directly used as cooking fuel [2]. An integrated energy system, recycling of wastes and biomass agro-forestry for meeting rural energy needs can also help to prevent soil-erosion; and improve rural health, environment and ecology [3, 25]. Reliance on biogas as a fuel requires neither any import nor dependence on diminishing supply of indigenous energy resources. Whereas fuelwood, charcoal, kerosene, petroleum and coal resources may not last forever, biogas shall be available as long as cattle plays a major role in agriculture and milk production [14]. In view of these factors, it is necessary that all all-out efforts be made to popularise biogas programme and enrich related technology.

REFERENCES

1. Balasubramanian, P. R. and R. Kasturibai, Biogas Plant Effluent in Aquaculture, *The Hindu,* Visakhapatnam, Sept. 4, 1991, p. 19.
2. Bembalkar, N. L., Energy Crisis: Indian Response, *Indian Express (Science Express)*, Visakhapatnam, Feb. 26, 1991, III.
3. Bhushan, B., Biogas Plan Under Attack; Devaluing Decentralised Energy, *Times of India,* New Delhi, July 31, 1986, p. 8.
4. Biswas, T. D., *Biogas Technology and Rural Development,* IARI, New Delhi, 1979.
5. BORDA, *Biogas: Manual for the Realisation of Biogas Programmes*, Bremen Overseas Research and Development Association, Bremen, Federal Republic of Germany (erstwhile), 1980
6. Brown, N., Biogas Systems in Development, *Appropriate Technology,* **14(3)**, 57 (1987).
7. Chawla, O. P., *Advances in Biogas Technology*, Information Division, ICAR, New Delhi, 1986, p. 17-18.
8. Chengappa, R., National Biogas Project: Raising a Stink, *India Today*, July 31, 1986, p. 138.
9. Directorate of Economics and Statistics, *Indian Agriculture in Brief,* Ministry of Agriculture, New Delhi, 1974.
10. DNES, *A Personnel Comminication*, New Delhi, Feb. 16, 1989.
11. Dodson, C. E. and P. J. Meynall, *Anaerobic Treatment of Organic Waste*, Genera

Information and Biomass Advancement Company Ltd., Marlborough, Wiltz, SM8 2HZ, 1981.

12. *Economic Times*, Untapped Potential (Editorial), New Delhi, Saturday, May 4, 1985, p. 5.
13. *Economic Times*, Biogas Technology: India's Achievement Lauded, New Delhi, Nov. 10, 1985, p. 5.
14. *Economic Times*, Time-bound Programme for Non-Conventional Energy Sources Urged, New Delhi, Jan. 29, 1986, p. 4.
15. *Economic Times*, Energy Minister Reiterates Commitment to Alternative Energy Sources, New Delhi, Jan. 13, 1990, p. 4.
16. *Economic Times*, Economic Survey: 1990-91, New Delhi, July 21, 1991, p. 14,
17. *Finanacial Express*, Wider Use of Solar Heating Systems Favoured, Madras, Aug. 11, 1991, p. 13.
18. GATE, *Survey of Biogas Plants*, German Appropriate Technology Exchange, Postfach 5180, D-6236 Eschborn 1, Federal Republic of Germany (erstwhile), 1978.
19. Golueke, C. G., and P. H. Mc Gauhey, Waste Materials, in '*Annual Review of Energy*', Vol 1, Palo Alto, California, 1979, p. 257-277.
20. Hansen, R., *Methane Generation from Livestock Wastes*, Circular No. 5002, Colorado State University Extension Service, Fort Collins, Colorado, April 15, 1977.
21. HEL, *Why Anaerobic Digestion of Farm Wastes*, Hamworthy Engineering Ltd., Fleets Corner, Poole, Dorset BH 177 LA, UK.
22. *The Hindu*, Role of Renewable Sources of Energy Stressed, Visakhapatnam, Dec. 27, 1990.
23. *The Hindu*, Biomass: A Potential Energy Source, Visakhapatnam. May 14, 1991, p. 17.
24. Homan, E., *et al.*, Biogas from Manure, Special Circular 260, College of Agriculture Extension Service, Pennsylvania State University, University Park, Pennsylvania, 1980.
25. Khandelwal, K. C., Biogas for Better Life, *Changing Villages*, **6(4)**, 274-283 (1984).
26. Khandelwal, K. C. and S. S. Mahdi, *Biogas Technology: A Practical Handbook*. Tata Mc Graw Hill Publishing Co. Ltd., New Delhi, 1986.
27. KVIC, *Gobar Gas on the March*, A Report, Mumbai, March, 1979.
28. Lapp. H. M., *et al.*, Methane Production from Animal Wastes-1: Fundamental Considerations, *Canadian Agricultural Engineering*, **17(2)**, Dec., 97-102 (1975).
29. Leach, G., Household Energy in South Asia, *Biomass*, **12(3)**, 155-184 (1987).
30. NAS, *Methane Generation from Human, Animal and Agricultural Waste*, Report of an Ad. Hoc. Panel of the Advisory Committee on Technology Innovation, National Academy of Sciences, Washington, DC, 1977.
31. Neelakantan, S., Assessment of the Availability of Cowdung for Potential Gobar Gas Production in India, *Agricultural Situation in India* **30(9)**, 633-636 (1975).
32. NSS, *National Sample Survey*, 17th Round, 1961-62, New Delhi.
33. Parikh, K. S., *Energy*, Second India Statistics, Macmillan Co of India Ltd., Delhi, 1976, p. 66-98.
34. Ramarao, B. V., Energy-related Options in Transport Sector, *The Hindu*, Visakhapatnam, April 11, 1991.
35. Stewart, D. J., *Energy: Biogas Production*, Media Services, FPP 375, MAF, Box 2298, Wellington, New Zealand.

36. United Nations, Guidebook on Biogas Development, Energy Resources Development Series No. 21, New York, 1980.
37. Huang, C. J., *et al.*, *Design and Operational Model for the Integrated Swine Waste Management System in Hawaii,* paper presented at the ASAE Pacific Region Annual Meeting, Hilo Lagoon Hotel, Hilo, Hawaii, March 18-20, 1980.
38. Myles, R. M., *A Practical Guide to Janata Biogas Plant Technology,* A Report, AFPRO, New Delhi, 1985.
39. Panditrao, Y. A., Energy Crisis and Biogas Special Supplement on Energy, *Economic Times,* New Delhi, Jan. 17, 1981, p. 11.
40. Pani, N., Energy Planning: Fresh Approach Needed, *Economic Times,* New Delhi, Mumbai, June 13, 1988, p. 11.
41. Shah, A. C., Impact of Drought on Biogas programme, *Economic Times*, Mumbai, June 13, 1988, p. 11.
42. Smith, W. H., *et al.,* Methane from Biomass and Waste: A Programme Review, *TIDE,* **2(1)**, 1-18 (1992).
43. TERI, *Managing the Biogas Programme* (Sponsored by the MNES), New Delhi, 1994.
44. TERI, *Biogas Technology* (Sponsored by the MNES), New Delhi, 1994.
45. TERI, Biogas: *A Source of Rural Employment* (Sponsored by the MNES), New Delhi, 1994.
46. House, D., *Biogas Handbook,* Peace Press, California, 1978.

Chapter 2

ADVANCING FRONTIERS OF BIOMETHANATION IN WORLD

2.1 Introduction

With growing uncertainties in conventional energy supplies and dwindling reserves of fossil fuels, renewable energy sources including biofuels have assumed overriding significance. Renewable energy sources occur in nature in the form of energy flows of indefinite duration unlike mineral deposits of finite magnitude. Whereas fossil fuel deposits yield energy relatively in concentrated and portable forms suitable for large scale industrial and urban use, biomass sources are most economically exploited on decentralised basis and thus well-matched on the needs of dispersed rural populations particularly that exist in developing countries [2]. Technology and equipment needed for exploiting renewable energy sources are appropriate for even less developed of the developing nations. The need for harnessing renewable energy supplies is overriding in view of fossil fuels particularly oil becoming increasingly expensive and scarce with each passing year and population doubling every 30 to 40 years. In this depressing scenario, nonrenewable energy sources which remain bound in depleting reservoirs, may not provide long term solution for the problem of dwindling energy supplies. Biofules comprising solid, liquid and gaseous fuels are an important category of renewable energy sources. Solar energy represents by far the most abundant source of renewable energy and in fact it is the source for others including energy from biomass. Organic wastes generated adundantly on recurring basis can be harnessed to generate biogas by a simple process of anaerobic fermentation [9]. Biomethanation has made an impressive start in world community though achievements like several other undertakings are far short of the ultimate potential [17, 25]. In this chapter, it has been endeavoured to summarise progress of biomethanation in select regions and countries of the world where it made most impact. [On account of developments in India where progress has been phenomenal and known to the author in greater depth on account of his being from this country, these are separately covered in Chapter 3].

2.2 Biogas Potential in World

Biogas potential from organic waste materials and especially cultivated biomass for harnessing energy is unlimited. Agricultural and forestry residues provide abundant biomass resources though some of them on account of their low bulk densities are difficult to collect, store, transport and feed. Biomethanation is a naturally occurring process that is being used since the eighteenth century. The technology has traditionally been used as a waste treatment process and energy recovery is only a recent motivation. Initially, interest in this process peaked during forties but later waned following cheap availability of fossil fuels. However, successive energy shocks during eighties once again revived interest in the process when it started to be viewed more as a source of gas recovery rather than simply a waste treatment process [18]. In view of abundant availability of wastes, need for their treatment, and shortage of fuels, biomethanation has made significant strides in several parts of the world particularly developing countries. A profile giving distinct activities of biogas programme in different countries are summarised in Table 2.1 which was originally compiled by UNESCO [6]. World livestock population is given in Table 2.2 and 2.3 which serves as an indicator of biogas potential [23, 34].

Table 2.1 Distinct Activities of Biogas Programme in Different Countries

Country	*Activities*
Algeria	R & D projects in experimental stage.
Australia	Fuel production using animal and crop residues.
Bangaladesh	Energy and fertiliser production by family and community plants.
Barbados	Biogas and fertiliser production.
Canada	R & D projects and biogas generation using farm and industrial wastes.
P. R China	Fertiliser and biogas generation by family and community plants.
Cook Islands	Integrated biogas farming systems.
Costa Rica	R & D projects
Equador	Biogas utilisation in cooperative farms.
EL Salvador	Biogas production from coffee wastes.
Fizi	Protein and fertiliser production by integrated farming system with fish-pond using algae and farm wastes.
France	Methane and fertiliser production using animal and municipal solid wastes.
Guatemala	Biogas from straw and piggery wastes.
Guyana	Biogas and fertiliser production.
Honduras	Biogas generation as a substitute to fuelwood.
India	Biogas and fertiliser production by family and community plants for meeting rural household needs. Development of new plant designs and implementation strategies.
Indonesia	R & D projects, demonstration-cum-research activities, and solid wastes treatment for improving environment and gas production.
Iran	Public health and environmental control by promotion of community plants.
Jamaica	Biogas production.
Japan	R & D projects, treatment of industrial and city wastes for environmental improvement, and gas and fertiliser production.
South Korea	R & D projects, gas and fertiliser production by family and community plants.
Malaysia	Biogas production tapoica residues, rubber-effluent and palm-oil sludge.
Mauritius	Biogas production using bagasse as feed material.
Nepal	Privatisation in plant construction, biogas and fertiliser production by individual and community plants.
The Netherlands	Methane production using piggery wastes

(*Contd.*)

Pakistan	Biogas and fertiliser production by family-sized plants using animal wastes, biogas appliances design.
Papua New Guiena	Biogas and fertiliser production using animal and city wastes.
Philippines	Gas and fertiliser production using algal oxidation ponds, piggery wastes, distillery-wastes and filter-cake; and development of new plant design and construction technology.
Rwanda	Biogas generation and utilisation for cooking.
Senegal	Biogas utilisation for household cooking and lighting.
Singapore	R & D and training activities.
Sri Lanka	R & D projects; biogas production from animal wastes, water-hyacinth and seaweeds; integrated energy systems in rural areas comprising solar, wind and biogas plants.
Taiwan	Biogas and fertiliser production using hog manure, integrated energy systems combining biogas plans with algal and fish cultivation, development of a flexible biogas plant made of red-mud-plastic (RMP).
Tanzania	Biogas and fertiliser production from individual and community plants.
Thailand	Biogas and fertiliser production by individual and community plants.
Turkey	R & D projects, biogas and fertiliser production by family and community plants.
UK	R & D projects, biogas production using industrial and municipal solid wastes, gas utilisation for meeting industrial fuel needs.
USA	R & D projects: biogas production from cattle, swine and poultry waste; gas utilisation for meeting dairy-farm and industrial fuel needs; mathematical modelling of anaerobic fermentation process.
Zambia	Biogas generation from dairy and family farms.
Zaire	Experimentation in laboratories and biogas production by family units.

Source : Buren, A. V., Transferring Biogas Technology to the Rural Areas of the Third World, International Institute for Environment and Development, London, 1980. Based on de Silva/UNESCO (1979).

Table 2.2 Livestock Population in World

S. No.	*Livestock*	*Approximate Population in the World*
1.	Cattle	100×10^6
2	Sheep	1000×10^6
3.	Goats	350×10^6
4.	Buffaloes	100×10^6
5.	Camels	11×10^6
6.	Pigs	550×10^6
7.	Horses	64×10^6
8.	Mules	18×10^6
9.	Asses	40×10^6

Table 2.3 Cattle Population in Select Countries

S. No.	*Countries/Continents*	*Cattle Population*
	Countries	
1.	India	203,800,000
2.	United States	108,813,000
3.	CIS	97,100,000
4.	Brazil	91,093,000
5.	Argentina	42,983,000
6.	Mexico	23,627,000

(Contd.)

Table 2.3 *(Contd.)*

7.	France	21,417,000
8.	Columbia	18,830,000
9.	Australia	18,500,000
10.	Turkey	15,175,000
11.	West Germany(erstwhile)	13,981,000
12.	South Africa	12,145,000
13.	United Kingdom	12,018,000
14.	Canada	11,755,000
15.	Italy	9,700,000
16.	New Zealand	8,200,000
17.	Philippines	5,805,000
18	Ireland	5,020,000
19.	Japan	3,050,000
	Continents	
20.	Asia	422,400,000
21.	South Africa	190,600,000
22.	North America	158,700,000
23.	Africa	134,000,000
24.	Europe	123,400,000
25.	Oceania	26,700,000
26.	Australia	18,500,000

Source : The World Agricultural Production and Trade, New York, 1968.

J. A. Eusebio and B.I. Rabino [10] while working at the University of Philippines at Laguna (Los Banos) made estimates of biogas potential in South-East Asian countries (Table 2.4). Based on the livestock and poultry population census undertaken by FAO in 1975 for countries of South-East Asian region, if 60 per cent of organic wastes of farm animals excluding goats are collected and utilised for biogas generation, countries of this region can save an equivalent of 8.9 billion litres of petroleum products and reduce their import-bill by over 15 per cent.

2.3 Biogas Programme in Latin American and Caribbean Countries

Latin American and Caribbean countries have achieved considerable success in promoting biomethanation. Promotion of alternative sources of energy including biogas is being largely done by Organisation Latinamericana de Energia (OLADE) which was set up by the Organisation of American States (OAS) in 1978. OLADE is funded by Inter-American Development Bank, and OAS, including oil exporting countries of inside and outside the region. OLADE in past organised several seminars and workshops for advancement and promotion of biomethanation. An important seminar was held in Santo Domingo, the Dominican Republic from Aug. 18 to 22, 1980 which adopted to popularise three types of biogas plants in the region. The first type was based on fixed dome Chinese model which is semi-continuously fed . The second type called OLADE–Guatemala type was batch-fed and provided with a separate gasholder.

Table 2.4 Estimates of Energy and Nitrogen Yields from Livestock and Poultry Wastes in South-East-Asian Region

Country	*Livestock and Poultry Population × 10^3*				*Fresh Faeces Output (Tons/days)*	*Petroleum Equivalent per Year (litres)*	*Nitrogen Content (Tonnes/day)*
	Carabao	*Cattle*	*Pigs*	*Poultry*			
1. Burma	1,909	7,875	1,961	22,551	49,115.00	890,408,602	38800.85
2. Indonesia	2,842	6,687	4,378	131,142	55,907.91	1,247,884,589	44167.25
3. Khmer Republic	929	2,459	1,208	6,615	18,221.06	328,418,846	14394.65
4. Republic of Korea		1,260	1,452	19,305	7,517.02	175.276,817	5983.44
5. Lao PDR	1,260	464	1,368	14,228	11,085.44	224,498,852	8757.50
6. Malaysia	302	359	1,095	39,000	5,844.46	187,697,980	4617.12
7. Philippines	5,240	2,250	1,700	54,274	57,974.05	1,141,888,892	45799.50
8. Singapore			1,332	15,765	1,992.50	73,477,925	15574.85
9. Thailand	5,378	4,939	4,962	73,500	64,998.54	1,281,658,636	51348.85
10 . Others		43,454	4,267	54,048	188,978.40	3,338,852,507	149292.94
Total	17,650	69,747	31,747	430,428	461,634.38	8,890,063,655	364691.17

Note: Other at S. No. 10 include: American Samoa, Cook Islands, Fiji, French Polynesia, Gilbert and Ellice Islands, Guam, Naurio New Caledonia, New Hebrides, Nive Island, Pacific Islands, Papua New Guinea, Solomon Islands, Tokelau Island, Tonga, Western Samoa. Data was derived from the FAO Monthly Bulletin of Agricultural Economics & Statistics, Vol. 24-25.

Its design was jointly developed by the OLADE and a private research organisation in Guatemala. The third one called Xochicall-Mexicao type was basically a horizontal plant. During 1981, OLADE set up over 60 biogas demonstration plants of the above three types in six countries comprising Bolivia, Guyana, Honduras, Jamaica, Nicaragua and Haiti [20, 23].

Several Caribbean countries comprising Barbados, St. Vincent, Grenadines, St. Lucia and Jamaica assisted by several international organisations such as the Food and Agricultural Organisation (FAO), Caribbean Development Bank (CDB), German Appropriate Technology Exchange (GATE) and the OLADE are engaged in promoting biomethanation [32]. The CDB and GATE introduced Regional Biogas Extension Programme (RBEP) for the Carribbean countries during 1983.

Besides being involved in promotion, RBEP also carried out R & D activities to develop plants best suited for the Caribbean region. F&AO during 1983 launched the Regional Biogas Cooperation Programme (RBCP) in the Caribbean countries, with a view to assist, among others, low-income farmers in owning their plants. RBCP and FAO organised a regional workshop on Biogas Technology Development in Jamaica in October, 1983. The workshop following deliberations recommended to create an information network for monitoring progress, develop strategies, expand training and extension services, and render follow up services for biogas plants. Strategy to use wastes varies from country to country depending on availability. Thus whereas Barbados prefers to use bagasse as feed, St. Vincent is more interested to use *arrowroot* starch for biomethanation. Poultry litter followed by goat and sheep droppings are commonly used feed materials. Estimates of livestock resources in Caribbean countries are given in Table 2.5 which can help to serve as an indicator of biogas potential.

Table 2.5 Estimates of Livestock Population in Different Countries

Livestock	*Population in Thousands in Different Countries*			
	Barbados	*St. Vincent Grenadines*	*St. Lucia*	*Jamaica*
Poultry	422	150	85	4200
Sheep	51	13	11	N.A.
Pig	38	6	10	255
Cattle	19	8	16	300
Goats	28	4	9	380

2.4 Biomethanation in Several European Countries

Pattern of biomethanation in European countries is different as compared to developing countries. This is largely due to lack of free availability of animal wastes with farmers as it is in countries like India and China. Consequently, family plants are a rare phenomenon in European countries. It is unlikely that low capacity family units will proliferate in Europe. In these nations, more complex tailor-made designs suited for community applications are in existence. Plants vary from small installations to complex units comprising gas purification, power generation, gas compression and water purification facilities. Biogas plants in several cases are part of total systems for sewage works, beet factories, alcohol distilleries, cattle feedlots or milking parlours. However, many of these plants were set up initially as waste-treatment rather than energy recovery units [8].

Table 2.6 Distribution of 501 Biogas Plants in Ten European Countries

Country	*Type of Wastes Used as Domestic Feedstock*			*Total*
	Agricultural	*Domestic Residues (Landfills)*	*Industrial*	
Belgium	21		9	30
Denmark	23		1	24
Germany	78	10	12	100
France	57	5	15	77
Greece	4			4
Ireland	5		2	7
Italy	62	1	12	75
The Netherlands	22		7	29
United Kingdom	21	9	17	47
Switzerland	108			108
Total	401	25	75	501

M. G. Demuynck, *et al.*, [8] while working at the University of Louvain in Belgium during 1982 surveyed working of plants in Belgium, Denmark, Germany (erstwhile FRG), France, Greece, Ireland, Italy, the Netherlands, United Kingdom and Switzerland. Whereas in 80 per cent of biogas plants agricultural wastes were used as feed materials, in case of 20 per cent of them industrial wastes were used (Table 2.6). Per cent distribution of 501 biogas plants in 10 European countries according to the type of plant and wastes used is given in Table 2.7. Certain difficulties faced by agricultural wastes fed plants were identified. It was found that the turn-key constructed plants were more expensive; there was no financial backing in some countries; there was lack of appropriate designs available to suit certain feed materials like bedding manure, etc. Similarly some difficulties were identified for plants that worked on industrial wastes. It was found that biological treatment of waste-waters was a more dominant objective than energy recovery. Even some highly efficient plant designs did not evoke much wide acceptance. It was also reported that legislation for water and air pollution control in several countries was not strictly enforced with the result that industries were not constrained to treat their waste-waters satisfactorily. As for engineering, operation and monitoring of plants, it was reported that most of them in their active span of working faced at least two major problems such as temporary shut-downs or souring of digester slurry; and four minor problems which however did not disrupt plant operations. Common engineering and maintenance problems faced were gas leakage from digester, failure of feed pumping device, inefficiency of mixing systems and problems with the electricity generation unit. Main operation problems faced were:

settlement of feedstock during storage, scum formation, pipelines blockage and gas-flow measurement with meters [8].

Table 2.7 Per cent Distribution of 501 Biogas Plants in Ten European Countries According to Plant Type and Wastes Used.

Type of Plant	*Type of wastes*		
	Liquid Agricultural	*Solid Agricultural*	*Industrial*
Completely mixed	68	9	21
Plug flow	18	25	
Partially mixed	2		
Two step	4	2	1
Batch	4	59	1
Anaerobic-filter	2	2	7
Accumulation	1		34
Upflow without carrier	1	2	22
IRIS system			10
Total	100	100	100

2.5 Involvement of the UN and Other International Agencies in Biogas Programme

Various United Nations agencies such as the Economic and Social Commission for Asia and the Pacafic (ESCAP), Food and Agricultural Organisation (FAO), United Nations Children's Fund (UNICEF), United Nations Development Programme (UNDP), United Nations Environment Programme (UNEP), United Nations Educational, Scientific and Cultural Organisation (UNESCO), United Nations Industrial Development Organisation (UNIDO), United Nations University (UNU) and World Bank (WB) are involved in some way or the other with development of biogas technology [6]. Important activities pursued by the UN and other international agencies in these endeavours are summarised in Table 2.8. Involvement of the ESCAP with these endeavours is particularly noteworthy. It organised several seminars and workshops on biogas technology starting from those held in New Delhi and Manila in 1975, in Suva in 1977 and in Bangkok in 1978. The FAO also worked in part for developing suitable processes to help use plant sludge as animal feed, and produce vitamin B_{12} from sludge for medicinal use by humans.

It also promotes use of plant sludge as enriched fertiliser. UNICEF views biomethanation as a means for improving rural hygiene including children health. UNDP provides financial and technical assistance to member countries for promoting biomethanation. UNDP in association with the UNEP organised an international workshop on biogas at Chengdu (China) during 1979. UNDP also assisted in setting up a research and development centre on biogas technology at Chengdu. UNESCO is mainly concerned with educational and cultural issues involved in biomethanation including alleviation of socio-economic barriers. UNEP views biogas programme as a means for pollution control and environmental improvement in rural areas. During 1981 it organised an International Conference on New and Renewable Sources of Energy at

Nairobi (Kenya), at which late Prime Minister of India, Mrs Indira Gandhi, delivered her keynote address. Like the UNDP, UNIDO also provides financial and technical assistance for promoting biomethanation in developing countries. It organises information dissemination and awareness campaigns. UNU in Tokyo (Japan) is involved with research and development activities on biogas. World Bank with its headquarters in Washington generates reports and provides funding for biogas programme. Constraints faced by countries at government level, implementing agencies involved and beneficiaries are taken into account for estimating magnitude of funding needed [6].

Table 2.8 Involvement of UN and other International Agencies in Promotion and Advancement of Biogas Systems

Agency	Activity
Economic and Social Commission for Assia and the Pacific (ESCAP)	Projects on biogas and integrated energy systems. Convening workshops. In past it organised workshops at Manila, New Delhi, Bangakok, Suva, etc.
Food and Agriculture Organisation (FAO)	Documentation concerning use of agro-industrial wastes, publication of directory of agencies involved in biomethanation. Organising workshops, study tours and field projects on organic recycling including biogas generation in countries like Afghanistan, China, Nepal, Papua New Guinea and Indonesia.
United Nations Development Programme (UNDP)	Implementation of biogas projects in Lesotho, Philippines and Tanzania.
United Nations Environment Programme (UNEP)	Implementation of rural energy programmes. Advancement of biotechnology, and photosynthesis.
United Nations Educational Scientific and Cultural Organization (UNESCO)	Promotion of basic and applied microbiological research including organising training programmes.
United Nations Children's Fund (UNICEF)	Promotion of biomethanation as a means for improving rural hygiene.
United Nations Industrial Development Organisation (UNIDO)	Promotion of biogas technology as part of integrated rural development programme.
United Nations Institute for Training and Research (UNITAR)	Imparting specialised training.
United Nations University (UNU)	Research on biomethanation for developing renewable energy sources and fertilisers for higher agricultural yield, and conserving natural resources.
World Health Organisation (WHO)	Issuance of technical bulletins in furtherance of sanitation and composting technologies.
Other Agencies	
Commonwealth Science Council (CSC), London.	Promotion of rural development and alternative energy programmes.
Inter-American Dev. Bank (IDB)	Promotion and funding of renewable energy programmes for fuel productioon from agricultural wastes in Central America.
Inter. Dev. Res. Centre (IDRC), Ottawa (Canada).	Funding of research projects including socio-economic evaluation of biogas systems.

Source : da Silva/UNESCO (1979), Quoted by : Ariane Van Buren, Transferring Biogas Technology to the Rural Area of the Third World, International Institute for Environmental and Development, London, 1980.

Various other international agencies such as Commonwealth Science Council (CSC), US Agency for International Development (USAID), International Development Research Centre (IDRC) in Ottawa (Canada), and Organisation of Economic Cooperation and Development (OECD) are also involved with biomethanation. During 1980, CSC in association with the KVIC organised a workshop on biogas technology which was attended by country representatives from Bangladesh, Fiji, Malaysia, Papua New Guinea, Sri Lanka, Tanzania and Jamaica. The workshop was organised as part of the CSC's Asia Pacific Rural Technology Programme. It provided an opportunity to review biogas technology and to promote its transfer to the Asia Pacific region. In recognition of very valuable and useful role played by the KVIC in promotion of biogas programme in India, CSC nominated it as the coordinator for biogas programme for Commonwealth countries in Asia and the Pacific region. USAID in particular has a strong legislative mandate to promote renewable sources of energy in developing countries. IDRC, a public corporation created by the Canadian Parliament in 1970, supports R & D activities relevant to the needs of developing countries. During 1978, IDRC published a monograph entitled 'Biogas Technology in the Third World: A Multi-disciplinary Review'. As early as 1969, OECD reported that its members already had more than 300 projects underway related to biogas and other renewable energy sources in sixty countries [28,34].

2.6 Biomethanation in Bangladesh

Bangladesh, a north eastern country in South Asia with an area of 55,598 sq. miles, has implemented its biogas programme with considerable success. Agencies involved in this effort are Ministry of Energy, Bangladesh Academy of Rural Development, Council of Scientific and Industrial Research, Bangladesh University of Engineering and Technology (Dacca) and Bangladesh Agricultural University (Mymensingh). Commonly used feed materials are cattle wastes, crop residues, water-hyacinth and algae. Biomethanation involves three distinct activities: setting up of experimental plants, establishing demonstration plants and family size units. In Bangladesh, plants of all popular designs, comprising fixed dome, movable drum and flexible bag type are promoted. In these efforts expertise of Bangladesh and India are mutually shared. In 1980, with the help of Bangladesh Agricultural University, one laboratory research plant and four demonstration plants were set up at different places in Mymensingh district. One research and two demontration plants were installed at the initiative of the Bangladesh Council of Scientific and Industrial Research. Two demonstration and one research unit were installed at the initiative of the Bangladesh University of Engineering and Technology. Environmental Pollution Control Board and the Bangladesh Academy for Rural Development installed one plant each. As in other countries, biomethanation is more exploited by rich farmers who own 10 or more heads of cattle. According to the findings of a rural energy survey, percentage of households who possess 10 or more heads of cattle is 8 per cent and they represent 29 per cent of total cattle population. This shows that only rich can afford to own biogas plants. According to the results of a survey conducted in 1977, a total of 17×10^6 lbs of dry animal wastes (188.7×10^6 lbs of fresh dung) is generated annually which can produce 133×10^9 ft^3 of biogas (equivalent to 1.64×10^6 tons of kerosene per year or 3.28×10^6 tons of bituminous coal per annum. Biogas potential from cattle wastes is tremendous considering the fact that during 1977 total cattle population in Bangladesh was over 29.33×10^6 cattle heads. Water-hyacinth whose growth is quite prolific in sprawling rivers and ponds of Bangladesh, is another potential feed material for biogas production. Bangladesh Bank and Bangladesh Small and Cottage Industries Corporation are also engaged in setting up biogas plants in several regions. In Bangladesh a family size biogas plant costs between Tk 10,000 to 25,000. High capital cost and lack of sufficient organic wastes are regarded as major difficulties in large scale popularisation of biogas plants in Bangladesh [11, 24].

2.7 Biomethanation in Brazil

Advancement and promotion of renewable energy systems including biogas received considerable fillip in Brazil during seventies and eighties. In this nation as typical in most developing countries there is a emphasis on both development and promotion of biomethanation. A number of agencies are involved in these efforts. [For more information reader may refer to directory of agencies given at the end of this book]. Coordenadoria de Energia has prepared well-documented manuals which provide technical guidelines for installation and operation of biogas plants. Such agencies have also set up demonstration units in far-flung areas of the country.

2.8 Biomethanation in Canada

Developments in Canada are more by way of developing the science further, rather than exploitation of biogas potential. This trend is in common with other developed countries but differs from developing countries where there is more emphasis on exploitation than technology advancement. Several institutions and organisations in Canada have done pioneering work in this field which include: Department of Agricultural Engineering at the University of Manitoba (Winnipeg) and the International Development Research Centre (Ottawa). Much of the work at the University of Manitoba was carried out by Prof. H. M. Lapp who wrote several comprehensive state-of-art reports on the subject besides conducting and directing research. Possibility of using biogas on running diesel engines and a transport pick-up were also examined. Universities of Guelph and Waterloo also conducted research on biomethanation. The monograph entitled 'Biogas Technology in Third World: A Multi-Disciplinary Approach', prepared by the IDRC has become a standard well-documented reference document. It was written by a multi-disciplinary team comprising A. Barnett, L. Pyle and S. K. Subramanian. National Research Council of Canada sponsored several studies for using food processing wastes as feed materials and produced numerous reports [28].

2.9 Biomethanation in P.R. China

Biomethanation in P.R. China received considerable impetus during popular movement '*Great Leap Forward*' launched in fifties. Energy crisis of seventies which rocked the world economy gave a renewed momentum to this programme. The programme received a fillip with the advent of a new national slogan '*New Big Leap*' ushered during mid-seventies. As a result between April 1975 and June 1978, number of biogas plants in China increased from less than half a million to around 7 million. Today China is credited with having maximum number of individual biogas plants. Chinese plant designs are simple in construction and belong to 'do-it-yourself' category. Although biogas potential is spread all over the country, areas south of Huang Ho and east of Tibet are more active. As reported by V. Smil [26], in areas south of Huang Ho, there existed a livestock population of more than 30 million buffaloes, 20 million large domestic animals, 75 per cent of total pig population, 75 per cent of total chicken and more than 50 per cent of rural population. Fully grown animals accounted for nearly 20 per cent and two year old animals 10 per cent of the total faeces generated (volatile solids). It was estimated that 50 per cent of available wastes could be used for biomethanation. In addition, about 70 per cent of crop residues and vegetable wastes could be used for biomethanation. According to some estimates, about 75 per cent of available crop residues/vegetable wastes could be used as animal feed and half of the remaining twenty five per cent biomethanation (Table 2.9 and 2.10). Sichuan became the first province in China in taking lead in biomethanation although later on the programme was extended to other provinces like Guangdong, Shanghai Shi, Zhejiang, Liaoning and Jiangsu with equal success. Most plants can be divided into two categories, namely, those in capacity range of 8–10 m^3 and those having 100 m^3 or more. Whereas gas from low capacity plants is mainly used for lighting and cooking, that from bigger plants is used to generate electricity, power agricultural machinery and pump water for irrigation [7]. A sizeable number of urban plants are owned and run by wine and spirit distilleries, and waste disposal and night-soil treatment units. Thus biogas plants in China are used to generate clean fuel as well as act waste treatment units [4,26]. With more than 40 per cent of world pig population in China, pig manure provides vast and valuable feed materials for biomethanation. Other commonly used feed materials are straw with fibres, rice, chaff, grass, leaves, kitchen garbage, and night-soil. Family plants in rural areas can be constructed with an outlay nearly one third of average per capita annual income [12].

One of the reasons for phenomenal success of biogas programme in China is creation of efficient rural organisation. There are three main levels of rural organisation: commune (sometimes having over 50,000 people), production brigade (having rural inhabitants between 1000 to 2500) and production teams (having 100 to 200 members). Team is a basic unit in commune and most important level in rural organisation. In China all four, namely, commune, production brigade, team and the individual household participate in plant construction with their involvement depending on plant capacity. Labour is provided by commune members; funding for large plants by production brigade and team and for small plants by owners themselves. Brigade is also concerned with workers training in plant construction. Before launching a biogas project, a brigade sends its members to another brigade already experienced in plant construction for training. Members following return to their brigade train other members in plant construction, operation and maintenance [26].

Table 2.9 Estimates of Biogas Potential in P.R. China

Source of Wastes	*Estimated Population*	*Contribution Towards Generating Volatile Solids and Biogas*		
		Volatile Solids (10^6 *T/Year)*	*Biogas (10^9 m^3 / Year)*	*Hard Coal Equivalent (10^6 T/Year).*
Large domestic animals	50 millions	21.3	10.6	8.78
Pigs	167 millions	21.3	10.6	8.78
Chicken	880 millions	1.6	0.8	0.66
Rural population (humans)	510 millions	14.3	7.2	5.96
Natural Phytomass	19.8 millions tons	15.8	7.9	6.55
Crop residues	52.5 millions tons	42.0	21.0	17.40
Total		116.3	58.1	48.13

Note: 1. Biogas production rate is assumed as 0.5 m^3/kg. VS day.
2. Calorific value of hard coal is assumed as 5800 kcal/kg.

Source : Smil, V, China's Energy: Achievements, Problems and Prospects, Praeger Publishers, NY, 1976, p. 107.

Table 2.10 Estimates of Biogas Potential from Crop Residues in P.R. China

Crop	*Annual Production (10^6 T) Residues*	*Multiplier Factor for Estimating Crop Residues (10^6 T)*	*Annual Crop Residue Production*
1. Rice	105.2	1.65	173.6
2. Wheat	31.0	2.57	79.7
3. Corn	29.0	4.00	116.0
4. Millet and Kaoling	22.0	2.80	61.6
5. Barley	19.0	2.30	43.7
6. Oats	2.5	2.30	5.8
7. Sweet Potatoes	106.6	0.30	32.0
8. White Potatoes	34.8	0.30	10.4
9. Soyabeans	11.6	1.90	22.0
10. Brown beans	3.3	—	—
11. Peas	3.3	—	—
12. Beans	1.4	2.30	3.2
1.3 Castor beans	0.1	—	—
14. Groundnut	2.5	3.00	7.5
15. Cotton lint	1.7	2.98	5.0
16. Rapeseed	1.0	—	—
17. Linseed	0.5	2.30	4.4

18. Sesame seed	0.4	—	—
Total	375.9		580.3

Source : Smil, V, China's Energy: Achievements, Problems and Prospects, Praeger, NY, 1976. Part of the above data was reportedly derived from the FAO, Production Year Book, Rome, 1973 and 'O.L. Dawson, Communist China's Agriculture, Praeger, NY, 1970.'

China has organised a number of national and international conferences on biogas technology. One such conference was held at Mianyang in April, 1975, which promoted 'do-it-yourself' approach in plant construction and operation. In furtherance of conference recommendations, state and district level offices were set up in Sichuan which were later extended to other states. Following this conference second one was again held at Mianyang in 1978 to review progress between 1975 to 1978 and to evolve long term strategy. The conference not only laid emphasis on building more family units but also promoted community plants. Another conference in the series was held in 1979 at which role of biomethanation in solving rural energy problems was emphasised. Later on, in keeping with the conference recommendations, a financial assistance package for owners was proposed. A number of taskforces for design, development and standardisation of plants were set up. In promoting biomethanation whereas national and provincial level concern was directed to avoid regional imbalances in energy supply, district and commune level concern was mainly to identify energy-deficient areas where biogas plants could be set up. It was decided to set up some 40 million plants for benefit of 70 million families. During 1979, Chinese experts participated in the International Workshop on Biogas Technology at Bremen (erstwhile Federal Republic of Germany). During the same year, UNDP and UNEP jointly organised a seminar on biogas technology at Chengdu. It, among others, stressed the need for initiating R & D projects. As a follow-up, several research centres such as Chengdu Institute of Biology (Sichuan), Szechuan Provincial Office of Biogas Development (Chengdu), and Guangzhou Institute of Energy Sources (Guangzhou) were set up [5,3,1].

Advent of biomethanation brought lot of changes in China's economic and social life. Following growing usage of gas, consumption of coal, kerosene and fuelwood came down [15]. The programme also brought improvements in environmental and hygienic conditions in rural areas (Table 2.11).

Chronology of developments in biomethanation in P.R. China is fairly old. It developed rapidly from 1920 to 1937 but slowed down afterwards till almost 1958. Status-quo prevailed until 1968 when it again started expanding fast. By the end of 1984, there had been more than four and half million operating plants in rural areas. They brought both quantitative and qualitative change in rural energy scene. Over the years supply base of biogas has been widened from household to rural industry, rural to urban areas, and individual units to community plants. Rapid progress of biomethanation in P.R. China has proved that digester construction technology is crucially important in success of biogas programme. Impressive strides have been made in digester design, materials and construction technology [40].

Mr. Lo Gua-rui, Xinzhu county, Taiwan, is believed to be inventor and pioneer of biogas digesters in China. Way back in 1920, Mr. Lo Guo-rui designed and built the first Chinese biogas digester which was named 'Guo-rui Natural Gas Generator of China.' In China based on extensive research, field tests, and experience with existing designs, a number of fixed dome plant designs of cyclindrical, spherical and ellipsoidal shape have been developed and standardised over the years. Household plants are very popular across the length and breadth of the country with proportion of operating plants over 95 per cent. Cylindrical shaped plants constitute more than 90 per cent of the total constructed digesters. Design based on Chinese fixed dome digesters have gained wide popularity in several developing countries including India [40].

Table 2.11 Improvement in Hygienic Conditions and Reduction in Incidence of Diseases in Several Areas of P.R. China Following Biomethanation.

Area	Type of Disease	*Number of Diseases Before Biomethanation*		*Number of Diseases After Biomethanation*	
		Among Humans	*Among Cattle*	*Among Humans*	*Among Cattle*
Sichuan Province	*Shistosomiasis*	26,000	2650	2400	350
Yongxing P.C., 2-PB	*Shistosomiasis*	143	—	2	—
Yongxing P.C., 7-PB	*Shistosomiasis*	134	—	46	—
Yongxing P.C., 7-PB	*Shistosomiasis*	70	5	1	1
5-PT	*Hookworm*	45	—	0	—
	Nematode	50	—	0	—
	Malaria	50	—	0	—
Puming P.C.	*Shistosomiasis*	1200	—	0	—
Puming P.C., 4-PB,	*Shistosomiasis*	12	—	0	—
4-PT	*Hookworm*	60% of Population	—	10% of Population	—
	Mawworm	80% of Population	—	10% of Population	—
	Diarrhoea	5	—	0	—
	Hepatitis	3	—	0	—
Deyang Horticultural	*Shistosomiasis*	20	—	0	—
Farm	*Hepatitis*	6	—	0	—
	Mawworm	20% of Population	—	0	—
Baichigan P.B.	*Tapeworm*	10	—	0	—
Lichun P.C. and Helu county					
Shuzuang P.B.	*Diarrhoea*	15	—	0	—
Tongxian county					

Source : Hilliges, G., et al., Biogas in Peoples Republic of China, A Project Report Based on Exchange of Experiences in the Field of Biogas among P. R. China, India and F. R. Germany (erstwhile), Bremen, March, 1981.

Although China is regarded as the birth place of fixed dome plants, the country has also seriously experimented with floating drum type plants. The first cylindrical floating-gasholder plant was designed by Mr. Tian Li-Fang in 1930.

Recently in 1980 a biogas plant with separate floating gasholder was designed and successfully built by the Biogas Extension Office at Huangzhou. In 1983 Wuhan Energy Institute designed and built a floating gasholder type digester made of synthetic fibre glass which resembled an Indian floating type digester. Bioenergy Laboratory of Shanghai Science and Technology Association in eighties developed a flexible floating gasholder. In 1984 a floating ferrocement gasholder was successfully developed by the Chengdu Biogas Scientific Research Institute of the Ministry of Agriculture, Animal Husbandry and Fisheries. For handling large volumes of feed materials conveniently, semi-plastic and complete red mud plastic digesters were also developed in seventies [40]. RMP digesters are especially appropriate for large and medium-sized plants where biogas is also used as an engine-fuel.

Plants have also been set up in China for treating distillery wastes. In 1967, Distillery of Langyang (Helan province), successfully built two 2000 m^3 biogas plants to treat distillery wastes using sweet potato as raw material and achieving daily gas production of 800-1000 m^3. In China a number of plants have also been set up to treat municipal refuse, night-soil, slaughter-house wastes and industrial organic wastewaters [40]. In recent years due to ever growing demand for energy and stringent emission control for environmental protection, number of large and medium-sized plants have gone up phenomenally.

Chengdu Biogas Scientific Research Institute carried out year long experiments for studying biogas yield with different feed materials under varying conditions [40]. The experiments have shown that at average ambient temperature of 16°C an 8 m^3 digester can produce 548m^3 of gas per year, with a average gas production per fermenting volume as 0.235 m^3/m^3·d, and gas production rate per fermenting slurry as 0.325m^3/kg. TS and 0.567 m^3/kg. VS.

2.10 Biomethanation in Germany

The erstwhile Federal Republic of Germany (FRG) set up a number of large capacity biogas plants. Many of them were set up in fifties when biogas was relatively less known and popular in the world. Several agencies are involved in its advancement and promotion in Germany. Notable, among them, are the Bremen Overseas Research and Development Association (BORDA) in Bremen and the German Appropriate Technology Exchange (GATE) in Eschborn. [For more information, reader may refer to the directory of agencies involved given at the end of this book]. Biogas in Germany mainly used for generating power as part of decentralised energy systems. Ministry of Agriculture assisted by several independent agencies is responsible for biogas-related projects in the country [30].

2.11 Biomethanation in Fiji

Biomethanation in Fiji, a country of over 300 islands, dates back to 1969 when sixteen plants in the capacity range 1.5 to 7m^3/day were set up. Earlier plants were of rectangular brick-steel lid type and later ones were mostly made of corrugated iron sheets which were low cost and easy to operate. Following shortage of fossil-fuel reserves, Fiji government promoted new and renewable sources of energy including biogas. It introduced several tax concessions mostly during 1976 to encourage conservation of coal, oil and electricity and promote wider use of these new sources of energy. Biogas in Fiji is mostly used for cooking and lighting. Most of the existing plants can be categorised into three different sizes, four different designs and three different types. At many places bag type digesters made of Neoprene rubber and based on Taiwanese design are also in use. A summary of cost comparisons of different capacity plants [14] is given in Table 2.12. A number of beneficiaries also use kerosene as a supplementary fuel. Initially plants were mainly viewed as a device for treatment of piggery wastes but later on as source of energy as well. In Fiji integrated energy systems comprising livestock, acquaculture and agriculture were also set up which were found to be particularly appropriate for small and large farms. The overall plan of the integrated systems covered algae basins, fish ponds and gardens which supplied fuel, feed and fertilisers. Three plastic-lined basins were built to receive effluent from digester for growing algae as animal feed. Resulting effluent from algae basin was discharged into a deep pond for fish and duck-farming and final effluent utilised both as fertiliser and for irrigation.

Table 2.12 Comparative Cost of Select Digester Designs in Fiji

Type and Capacity	Number of Pigs	Biogas Output (ft^3)	Cost (US dollars)
300-gallon capacity digester in concrete and steel	30	300	600 (1969)
1000-gallon capacity digester in concrete and steel	100	1000	200 (1971)
1000-gallon capacity digester in Neoprene rubber	100	1000	500 (1975)
1000-gallon capacity digester all in concrete	100	1000	1500 (1978)

Source : Goodman, L. J. and Love, R.N. (eds.), Biomass Energy Projects: Planning and Management, Pergamon Press, NY, 1981, p. 133.

2.12 Biomethanation in Indonesia

Biogas related activities in Indonesia are fairly wide-spread and actively pursued. Several institutions in Jawa such as University of Pajajaran, Institute of Technology (Bandung), University of Gohah Mada (Jogja), University of Brawijaya (East Jawa), Industrial Research Institute (Jakarta), Padjadjaran Univ. (Bandung) and Yayasan Dian Desa Yogyakarta (Jakarta) carried out substantive research on biogas technology. A number of voluntary agencies are involved in promoting biomethanation in Bali where Christian Life Foundation built up several family- sized plants in past. Biogas plants in Indonesia have been set up with different objectives. For instance plants at Bali were set up keeping hygienic and environmental considerations in view to utilize pig faeces which are available in plenty. On the other hand plants at educational institutions in

Jawa were set up for demonstration-cum-research purposes. Studies were carried out in past for determining alternative feed materials and for developing cheaper plant design. Institutional plants are also used for demonstration purposes for popularising biogas programme. Development Technology Centre (DTC) at the Institute of Technology (Bandung) set up three types of digesters at a nearby dairy farm. One of them was of 330 litre capacity, vertical type and comprising 55 gallon drums; second one was of 370 litre capacity, horizontal displacement type and made up of drums; and third one was of 3 m^3 capacity and made of concrete. For popularising biomethanation in Indonesia, DTC also published a booklet entitled 'Biogas Technology'. Being predominatly a Muslim nation, use of pig faeces as feed material despite vast availability are generally viewed as social stigma due to religious sentiments.

2.13 Biomethanation in Iran

Centre of Endogenous Development Studies is promoting biomethanation in Iran. Being an oil-exporting nation there is no overriding need as such for exploiting renewable sources of energy, nevertheless there is a growing awareness that tapping of such sources may become necessary in future. Being a clean fuel development and utilisation of biogas is regarded as an important goal. A number of movable drum-type biogas plants patterned after the Indian design were set up in the Niazabad town in seventies. Comprehensive plans have been drawn for setting up more family-sized and community plants in other parts of Iran [30]

2.14 Biomethanation in Jamaica

Biogas programme in Jamaica was launched in early eighties. Under this programme nine family size plants of Chinese, Guatemalan and Mexican designs were installed in 1982. In addition, few other medium sized plants of 40 m^3 biogas/day capacity were installed during 1982-83. These plants were mostly owned by institutions and cooperative farms. Government of Jamaica set up a research centre to carry out R & D activities on biomethanation. Scientific Research Council is also involved in promoting biomethanation in Jamaica [24, 36].

2.15 Biomethanation in Japan

Japan like several western countries is more involved in developing biomethanation rather than exploiting it. Initially, units were set up in Japan for pollution and environment control rather than energy recovery from waste materials. Several institutions like National Institute of Animal Industry (Chiba), and Fermentation Research Institute (Anage) carried out studies in past on high temperature anaerobic fermentation in *thermophilic* range. Several distilleries in Japan are equipped with biogas plants.

Viewed in the perspective that three-fourth of energy needs are met by imports, biomethanation assumes significance as it provides gas, and fertiliser and protects environment. As estimated by Kiyowori Haga, *et al.*, 5.6×10^7 tons of animal faeces are generated every year in Japan. Assuming that it contains at least 10 per cent of organic matter and assuming that 300 m^3 of biogas is produced per ton of organic matter, 1.68 x 10^9 m^3 of biogas can be produced every year in Japan. This in terms of heat value is equivalent to 1.01 x 10^{13} kcal per year with 6000 kcal/m^3 as the heat value of biogas. With the norm of 10^7 kcal as equivalent to 1 kilolitre of petroleum, it is equivalent to 1.01 x 10^6 kilolitres of petroleum products produced annually. As per the present consumption rates, although biogas can meet no more than 0.27 per cent of energy demand in Japan which is estimated as 3.8 x 10^{15} kcal per year, it nevertheless represents 18 per cent demand of petroleum products in agriculture sector estimated as 5.5 x 10^6 kilolitres per year. Furthermore, assuming that an average Japanese family consumes 70 m^3 of fuel gas per month, biogas can substitute fuel needs of over 2 million families. Technique of biomethanation in Japan is fairly old. Developments date back to 1926 when several working models that used kitchen wastes as feed materials were developed. Small biogas units were used in Tohaku region for many years. Japan is one country in the region where several of its plants works in *thermophilic* range utilising a wide variety of organic wastes [30].

2.16 Biomethanation in South Korea

South Korea is a pioneer country in biomethanation where as of 1975, there existed over 29,000 biogas plants. Office of Rural Development (ORD) in Suweon provides necessary technical and financial support in setting up plants. In 1977, Rural Energy Resources Research Division was created as a separate division within the ORD for giving a greater thrust to biogas programme and performing R & D activities. The government provides necessary assistance in setting up plants at whose initiative, a 155 m^3 capacity community biogas plant was set up at Livestock Experiment Station in Suweon. Rural Guidance Bureau and the Institute of Agricultural Engineering and Utilisation have carried out studies under ORD supervision for making plant operation more efficient and economical. In order to keep cost low, labour is normally provided by beneficiaries themselves. As the country experiences severe cold during winter when temperature are as low as (–)17°C, biogas yield is very low during November to March. For minimising heat losses and conserving digester heat during winter, gasholders are covered with agricultural wastes like straw. Plants thus work more efficiently in southern than northern parts [28, 30].

2.17 Biomethanation in Malaysia

Like other Asian countries, Malaysia is also not behind others in promoting biomethantion. Several wastes like tapioca residues, rubber effluent,, palm oil sludge which the country produces in plenty, are used as feed materials. Based on 1976 livestock census, gas to be extent of 1.6 x 10^9 ft^3 per annum can be produced from traditional livestock wastes. As for biogas potential from palm oil sludge, peninsular Malaysia produced 53.75 million tons of palm oil sludge in 1976 which if anaerobically fermented could produce 2.6 x 10^9 ft^3 of gas annually [30]. However, there is a considerable gap between the potential for biogas plants and actual numbers that exist. A large number of biogas plants in Malaysia are concentrated in states of Ketah, Kelantan and Selangar. Several educational institutions are engaged in carrying out studies on biomethanation. For example, National University of Malaysia studied the potential of using tapioca wastes as feed.

2.18 Biomethanation in Nepal

Nepal has evinced keen interest in promoting biomethanation. Biogas programme in Nepal as elsewhere received considerable fillip following energy crisis of seventies. According to some estimates, around 27.7 million tonnes of cattle dung is generated every year and if some 40 per cent is anaerobically digested, resulting biogas can meet fuel needs of 4,65,000 households or about 2.55 million people [28]. Analysis of commercial and non-commercial energy consumption in Nepal during 1974-75 indicated that domestic sector was responsible for as high as 93 per cent of total energy consumption out of which fuelwood alone accounted for 87 per cent of total energy consumption. Apart from meeting energy needs, biomethanation is also viewed as a solution to growing deforestation problem and as a source of enriched fertiliser.

In the entire Himalayan Kingdom of Nepal, as per revised estimates, some 62,000 tonnes of dung is collected and burnt annually which burns 4,600 tonnes of nitrogen and 3,200 tonnes of phosphorus. As per estimates made by Agricultural Input Corporation, total consumption of chemical fertiliser in Nepal during 1975-76 was 18,600 tonnes of nitrogen, 4,050 tonnes of phosphorus and 2,260 tonnes of potassium. Plant sludge can meet large proportion of country's fertiliser needs besides generating enormous amount of clean fuel. Biogas potential from animal dung is estimated as 613 million m^3 with an additional 60 million m^3 from co-digestion with night-soil [36]. Wastes from cattle including buffaloes, and night-soil are main feed materials. Number of households possessing cattle including buffaloes was estimated as 2 million in 1991. With an assumption that 65 per cent of them possessing enough cattle to own a plant, there is a potential of 1.3 million biogas plants in Nepal [43]. The average beneficiary possesses 5.5 members per family and 4.7 cattle or buffaloes. A comparison of potential and existing number of plant shows that there is vast untapped potential. Average daily gas demand per household is 1.8 m^3 while the potential is estimated as 1.40 m^3 per day [39].

One of the earliest biogas plants in Nepal was built in 1963 at the St. Xavier School campus at Godavari (Kathmandu) at the initiative of B. R. Soubolle, a leading exponent of biogas programme in that country. In 1972, a 200 ft^3 (6 m^3) capacity household plant was set up in Kathmandu. In 1975, Energy Research and Development Group set up a committee to promote biomethanation. Division of Soil Science and Agricultural Chemistry at Khumaltar (Kathmandu), Development and Consulting Services at Butwal and the Nepalgunj Agricultural Station worked jointly and in individual capacities in providing training and extension services. Department of Agriculture and American Peace Corps also jointly worked for a biogas research and extension project entitled 'Energy

Needs in Food System'. Under this project two biogas plants one at Lalitpur and another one at a refugee camp near Trisuli were set up. Besides these agencies, Agricultural Tools Factory, *Baluja Yantra Shala* and the Biratnagar Workshops provided requisite technical support in setting up plants [30]. Realising the importance of biogas in meeting growing energy needs, Department of Agriculture in his Majesty's Govt. of Nepal developed a five year plan for plant installation under which during seventies some 1000 biogas plants were set up. [21]. The biogas programme received special attention during 1975-76 which was declared as the Agricultural Year by the Nepalese government during which as many as 196 plants were set up. Till July 1993, over 13,000 plants have been constructed in 75 districts of the country mainly by the *Gobar Gas Company* with loans provided by the Agricultural Development Bank of Nepal. *Nepal Gobar Gas Pvt. Ltd.* (Kathmandu), *Kisan Gobar Gas Udyog* (Nuwakot), *Nepal Gobar Gas Vistar and Vikas Co. Pvt. Ltd.* (Dumre), *Rashtriya Gobar Gas Nirman and Sewa Ltd.* (Bharatpur) and *Baral Gobar Gas Sewa Kendra* (Pokhra) are some of the major companies engaged in setting up biogas plants (Table 2.13A).

Dome type biogas plant based on Chinese design is well-suited for Nepalese conditions. Technical performance of this plant is rated high. A recent survey showed that 85 to 90 per cent of plants built are operational. Most of the existing plants are owned by wealthy farmers who have high risk-taking and investment capabilities. Small farmers who own only few cattle and need a small biogas plant, encounter economic difficulties in setting up a plant. Areawise most of the biogas plants have been installed in the *Terai* region (lowlands) of Nepal. The high transportation costs of the construction materials make biogas less attractive to farmers in the hills [42]. The government provided 25 per cent subsidy on investment costs during 1988-89 and 1989-90 resulting in increase in plants installed from 676 in 1987-88 to 1100 in 1989-90. In wake of vast untapped potential of biomethanation, the GGC and ADB (Asian Development Bank) have sought the help of Netherland Development Organisation (SNV-Nepal) to support implementation of biogas programme in Nepal. GGC, ADB and SNV- Nepal have evolved a '*Biogas Support Programme (BSP)*' to be collectively implemented over a period of five years (1992-97). Along with private sector participation, BSP aims to increase number of plants by 20,000 and enhance attractiveness of the programme to small farmers and those residing in hills by additional subsidies. An outlay of Rs. 20 crores (about $ 5 million) is earmarked for BSP out of which 81 per cent is allocated for subsidy on investment costs, 1 per cent towards study on privatisation, 3 per cent on institutional strengthening, 10 per cent on programme management by the SNV— Nepal, and 5 per cent for miscellaneous expenditure. BSP is expected to result in a saving of 49 million kg of fuelwood per annum, availability of 365 thousand tons of enriched fertiliser and improved hygienic conditions for 20,000 families [42].

Biomethanation has gone a long way in reducing workload of women in Nepal. According to the result of a study carried out by M. M. van Valiet and W. J. van Nes [44], availability of biogas reduces the workload of a family from a minimum of 2 to maximum of 7 hours per family per day. These findings relate to the study carried out in Rupandehi which is lowland district near the Indian border. Women in Nepal spend considerable time in collection of fuelwood and fodder. Prior to plant installation 6.6 hours per day were spent on fuelwood collection This was done by female family members (31 per cent), servants (31 per cent), all family members (23 per cent) and male-female family members (15 per cent). The average time needed to cook and clean utensils following cooking decreased by 50 per cent from 4.4 hours per day before installation to 2.2 hours after installation. Cooking with biogas was found to be more convenient and clean compared to fuelwood.

Table 2.13 Comparative Price of Different Capacity Biogas Plants Manufactured by Leading Nepalese Firms (1992)

(Nepalese Rupees)

Name of the firm	*4 m^3*	*6 m^3*	*8 m^3*	*10 m^3*	*15 m^3*	*20 m^3*
Gobar Gas Co.	14712	16986	20236	23558	30124	36620
Kisan Gobar Gas	14030	16474	19974	24394	29885	35050
Nepal Gobar Gas Vistar	14612	16036	20037	23358	30024	36420
Rashtriya Gobar Gas	11936	14239	17532	20215	25971	30987
Baral Gobar Gas	—	—	—	17761	18525	19553

Source : Devkota, G. P., Privatisation in Gobar Installation, Biogas Forum, 1992.

A national level Biogas Information Seminar was organised by ADB/N, GATE, UNICEF and SNV-Nepal/BSP at Kathmandu on July 29 and 30, 1994. It was inaugurated by Mr. Girija Prasad Koirala, Prime Minister of Nepal, and was attended by more than 125 participants representing national level planners, bureaucrats, administrators, institutional heads, donor agencies, private enterprises, NGOs and beneficiaries. It recommended that more of such seminars be organised to disseminate information to a larger section of population. Biogas programmes should be integrated with the livestock development programmes. Private enterpreneurs and NGOs should be trained in biogas technology. Biogas should be included as a subject in school curriculum. A blueprint for comprehensive biogas policy should be evolved. There should be a single coordinating agency for biogas promotion in a country although institutions and organisations can be several. Interest on loan assistance for setting up biogas plant should not be more than 10 per cent. Procedures for securing loans should be made simpler. Alternative feed materials such as municipal solid wastes and industrial wastes should be increasingly used. Research should be carried out to develop cheaper and cost-effective biogas plants [45].

2.19 Biomethanation in Pakistan

The advent of biomethanation in Pakistan is relatively of recent origin although some plants were set up in seventies. During the year 1980, Directorate General of Energy Resources (DGER), Govt. of Pakistan, launched a 10-year demonstration-cum-development programme for biogas technology. The programme was later implemented and monitored by the Directorate General of New and Renewable Energy Resources (DGNRER). The programme was launched with a view to provide clean cooking fuel in rural areas as a substitute for conventional fuels like kerosene. During 1980, it was proposed to set up 15,000 biogas plants in Pakistan and cover some 2000 villages under demonstration programme. DGNRER also tried to popularise a particular plant model named ER 150 after duly approving its design. A number of ER 150 models which can produce 4 m^3 of gas per day have been installed in rural and urban areas. Pakistan Council of Appropriate Technology (PCAT) undertook propagation of biogas programme in 1976 and installed 21 Chinese type biogas plants to assess workability of these plants under different climatic conditions prevailing in the country [41]. After identifying some technical problems such as gas leakage, movable drum type plants similar to Indian design were also set up but with some modifications to bring down installation cost without seriously affecting efficiency. PCAT initially set up 746 plants of 3-7 m^3 size range in various parts of the country. Out of 746 plants, 100 plants were installed by providing 25-70 per cent subsidy. Another agency by name 'Appropriate Technology Development Organisation' was engaged in past in design and construction of 10 m^3 capacity fixed dome type plants based on Chinese design. Until March 1991, over 4000 biogas plants have been installed by the DGNRER costing Rs. 33.34 million [4]. Most of the these plants in size range 5-7 m^3 were installed during the period 1981-85. In eighties, two community plants of movable drum type were also installed one each in Karachi and Lahore.

One reason for limited number of plants in Pakistan can be attributed to easy availability of LPG, electricity and fuelwood in countryside and inadequacy of promotional efforts. Although no comprehensive socio-economic survey has been reported, based on rough assessments it is believed that not less than 70 per cent of the available plants are functional.

DGNRER encourages plant construction on a 'do-it-yourself basis' [41]. It has also developed a manual for the benefit of prospective beneficiaries. DGNRER also evolved guidelines for plant construction on turn-key basis. Some of the demonstration units were set up at military dairy-farms, universities and integrated rural development centres with government bearing the major cost. Pakistan is also credited with having developed its own design of a biogas lamp. On the whole Pakistan has made an impressive progress in biomethanation. In view of the country's commitment to socio-economic development, rapid industrialisation and improvement in living standards in rural areas, government agencies in Pakistan are committed to pursue biogas programme vigourously.

2.20 Biomethanation in Philippines

After India and China, Philippines is the third major country in Asia-Pacific Region where biogas programme has been pursued with spectacular success. Office of Energy Affairs (OEA) through its Non-Conventional Resources Division (NCRD) is promoting non-conventional technologies including biogas. The OEA-NCRD combine has assisted numerous projects in past for promoting biomethanation. In 1990 the OEA introduced Financing Energy Services for

Small Scale Energy Users (FINESSE) which was aided by the World Bank and Dutch government. FINESSE aims to promote technologically efficient and cost-effective energy systems by funding small and medium size beneficiaries [37]. Other important agencies associated with advancement of biogas technology are—Non-Conventional Energy Resources Development Programme sponsored by the Ministry of Energy, Bureau of Animal Industry, National Housing Authority, University of Philippines at Los Banos, National Institute of Science and Technology (NIST) and the Liberty Flour Mills (LFM). Biomethanation is viewed as eco-friendly appropriate technology particularly for agricultural farms where pollution abatement measures are generally not followed [33]. Bureau of Animal Industry statistics indicated that some 13 million metric tonnes of animal wastes were generated in Philippines during 1975, capable to produce biogas equivalent to 3 million barrels of oil produced per annum. In view of biogas technology being techno-economically competitive and having vast potential, biomethanation is an important feature of energy policy in Philippines [28, 33].

Way back in 1975, an International Workshop on Biogas Technology and Utilisation was organised in Manila, which was first of the two workshops organised by the Economic and Social Commission for Asia and the Pacific (ESCAP). For taking stock of global developments another workshop was held in Manila on July 22, 1976 which was jointly organised by the Division of Chemical and Pharmaceutical Sciences of the NIST and the LFM. It discussed issues relating to biogas generation and utilisation in global perspectives with special reference to developments in Asian countries. Biogas technology is marketed commercially in Philippines. There are more than ten firms which market small to large-scale biogas systems. These firms are adequately equipped to design, fabricate, train and offer after-sales-services to end-users [37]. The commonly marketed biogas systems are floating and fixed dome types. The floating dome type plants based on Indian design are popular in large farms.

One of the earliest large scale biogas plant was set up at Maya Farms at Angono at Antipolo Hills in Rizal province forty miles south of Manila. The unit originally consisted of 24 digesters batch-fed on 45 days cycle which utilised wastes from an adjoining piggery farm maintaining over twenty thousand pigs. Later on five more digesters of the Indian and Taiwanese designs were added. Biogas from the plant is used for meeting fuel meeds of canteen, a meat processing plant and soap cannery unit. The Maya Farm is truly an integrated crop-livestock-fish-meat processing and canning unit which meets energy needs of the entire farm. It also published a monograph entitled '*Biogas and Waste Recycling: Philippines Experience*' for benefit of others planning similar plants. The Maya Farms in past also provided technical assistance to other piggeries for setting up similar plants.

Besides large units, a number of small to medium-sized plants were also set up at different locations, many of them with the assistance of National Science Development Board (NSDB) and Economic Development Foundation. Luzon State University developed a new design and construction technology for biogas plants [28, 33]. The Central Azucarera de Tarlac obtained biogas from distillery wastes and filter-cake which could be used to reactivate spent carbon in boilers

[33]. University of Philippines at Los Banos set up an integrated energy system at its campus which utilised both gas and sludge for cooking, refrigeration, fish-rearing and rice-cultivation. Biogas unit consisted of two plants connected in series that utilised pig faeces from nearby farms as feed [33]. Performance audit of biogas plants in past indicated that nearly 60 per cent of small to medium scale systems are no longer in use [37]. Lack of adequate maintenance support was identified as the main reason for decline in small-scale systems.

2.21 Biomethanation in Sri Lanka

Introduction of biomethanation in Sri Lanka dates back to early seventies when a 2.8 m^3 demonstration plant was set up by the Industrial Development Board (IDB). Sri Lanka with an agricultural economy and a land area of 6.5 million hectares has a population of over 15 million people about seventy per cent of whom live in rural areas. With assistance from the UNEP and Oklahoma State University (USA), the government set up a rural energy system for meeting energy needs of a small village community comprising 50 to 200 families. In mid-seventies experiments were carried out by the Department of Civil Engineering at University of Peradeniya for using water-hyacinth and seaweeds for biomethanation. Studies were also carried out for minimising cost of fixed dome and movable drum type plants. IDB developed a low-cost dome type biogas plant '*Lakegen*' with no movable parts. It consists of two underground brick static tanks at different elevations for achieving higher gas pressure. Vide its energy policy, the government in past encouraged setting up integrated energy systems in rural areas comprising solar, wind and biogas plants [33].

2.22 Biomethanation in Taiwan

Biomethanation is fairly popular in Taiwan where it started in seventies. Availability of large recurrent quantities of hog faeces provided necessary incentive for pursuing this programme in this country where five to six million hogs are raised annually. Available hog faeces if fully utilised for biomethanation can produce 5×10^8 m^3 of gas per year with heat value of 4.78×10^{12} kcal. In seventies, there were some twenty hog farms in Taiwan having more than ten thousand heads each. Over 7500 family plants existed in 1975 which utilised several designs. This country introduced flexible biogas plants made of rubber. Initially the bag was made of laminated neoprene which was later on replaced by a new material developed by the Union Industrial Research Institute at Hsinchu. This material named as '*Red Mud Plastic*' lasts from 10 to 20 years. Apart from family units, the government has promoted integrated energy systems which combines biogas plants with algal and fish-cultivation [28,30].

2.23 Biomethanation in Tanzania

Biomethanation has been carried out with considerable success in Tanzania. Small Industries Development Organisation (SIDO) in past has been responsible for setting up large number of floating drum plants of Indian design. It has also been engaged in providing training and extension services. Tanzania with a rural-based economy and large cattle population is ideally suited for pursuing a biogas programme. Arusha Appropriate Technology Project (AATP) was started in 1977 of which biomethanation was a major programme. Under the AATP, a low-cost design of biogas plant was developed whose gasholder comprised discarded oil drums interconnected by lead or plastic pipes. The AATP set up 11 plants in 1977, 41 in 1978 and 52 in 1979 in Arusha and other regions. In 1970 when there was no energy crisis, 150 floating drum type plants had already existed. In 1980, however, 50 per cent of them were no longer in operation mainly due to corrosion problems. *Centre for Agricultural Mechanisation and Rural Technology (CAMARTEC), Arusha, introduced Biogas Extension Service (BES) in 1983 under the International Biogas Extension Programme of the German Appropriate Technology Exchange (GATE).* At present several institutions and private entrepreneurs are associated with biogas extension [38]. Due to their low-costs, fixed dome plants are becoming more popular these days than moving drum type. By 1985 end, 40 fixed dome type plants were built in Arusha region. However, 80 per cent of them did not function efficiently following gas leakage through cracks in domes. Problem of gas leakage was solved by providing a stay ring at the annular contact between digester and gasholder [35]. The CAMARTEC also supervises the lamps and burners manufacture in Tanzania. The Centre has conducted programmes on biogas usage as engine fuel in past which were attended by participants from several countries including Kenya, Burundi and Malawi [35]. Following vigourous promotion by the SIDO, CAMARTEC and other agencies, plants tally crossed 500 mark in 1992. Bioslurry is commonly distributed under natural gravity [38].

2.24 Biomethanation in Thailand

Introduction of biomethanation in Thailand dates back to early sixties when it was conceived more as a means for waste treatment than energy recovery. Following energy crisis in seventies, emphasis shifted from waste treatment to energy recovery. Several agencies are involved in promotion and advancement of biomethanation in Thailand. Prominent, among them, are Technical Division of the National Energy Administration, Thailand Institute of Scientific and Technological Research (Bangkok), Energy Division of the Asian Institute of Technology (Bangkok), Agricultural Engineering Division of the Kasetsart University (Bangkok) and Applied Scientific Research Corporation (Bangkok). Sizeable number of plants installed in Thailand are of movable drum type based on the Indian design. Under the Fifth Economic and Social Development Plan (1982-86) considerable outlays were allocated for promotion of biomethanation. In past the Government of Thailand provided several incentives for popularising family, institutional, and community plants in the country [30].

2.25 Biomethanation in Turkey

Biomethanation in Turkey made headway from 1980 onwards only. Fifty per cent of population living in more than 35,000 villages depend on agriculture and animal husbandry. In 1980, UNICEF signed an agreement with the Turkish Government to provide technical and financial support for conducting research on biogas in the country [29]. This work was coordinated by State Planning Office, Ministries of Agriculture and Forestry, Village Affairs and Cooperatives, and Energy and Natural Resources; and the Mineral Research and Exploration Institute. Soil and Water Research Institute (Topraksu) under the Ministry of Village Affairs and Cooperatives has been most actively involved with design and manufacture of prototype digesters and experimental investigations. Beginning 1982, Topraksu has installed over 1000 biogas plants having 6, 8, 12, and 50 m^3 capacity in rural areas. The government also provides a credit of up to an amount equivalent to 1600 US dollars with an annual loan interest of 16 per cent payable by beneficiaries. Two plants based on Chinese and Indian design with some modifications were constructed and tested by the Topraksu. However, more plants are of fixed dome type based on Chinese design. Even though the total number of animals in Turkey is very large, more than half of rural families do not possess sufficient animals to generate feed which is one of the constraining factors in its promotion. Collection of animals wastes can be easier if animals are kept in barns for larger periods in a year. In Turkey, however, animal husbandry is traditionally done by free grazing. Shift in focus from free grazing to confinement is difficult due to economic and social reasons. Biogas is more suited for meeting water heating and cooking needs and not for power generation as rural electrification is already nearly complete. There is, however, lack of awareness towards hygienic and ecological benefits of biogas [27].

2.26 Biomethanation in UK

Like the pattern observed in most developed nations, biogas in the UK is produced by large installations. There are a number of installations but practically none of them are used for meeting cooking and lighting needs of individual households. There are several firms that design and manufacture large capacity biogas plants. Notable among them are: Anaerobic Digestion and Biomass Advancement Company at Marborough, British Anaerobic and Biomass Association Limited at Wiltshire and Lockstocke Development Company at West Sussex. [For more information reader may refer to the directory of agencies given at the end of this book]. In addition, several agencies such as Anaerobic Digester Research and Development Unit at Polytechnic of Wales, Department of Microbiology at the Rowettt Research Institute at Aberdeen (Scotland) and Warren Spring Laboratory (WSL) at Stevenage are engaged in carrying out research in different areas of biomethanation. Department of Energy is the nodal agency for sponsoring research on biogas technology in the country [14, 30].

Livestock farms produce approximately 1.8 million tonnes of poultry wastes and some 14 million tonnes of livestock slurry from pig and cattle are produced in the UK each year. These wastes are traditionally used as fertiliser after landfilling which in many cases may cause environmental problems such as odour and water pollution. Anaerobic digestion which is eco-friendly is better

Table 2.15 Some Important Biogas Installations in the USA

S. No.	*Plant Location*	*Source of Feedstock*	*Type and Capacity*	*Type of Digester*	*Utilisation of Gas*	*Plant Construction Agency*
1.	Guymon, OK (Two units)	Cattle	Full Scale 50,000 gal.	Continuous feed, complete mix, mesophilic	In-house applications through pipeline	Thermonetics Calorific Recovery
2.	Ripon, WI	Poultry	Full Scale 50,000, gal	-do-	House heating	Wayne Gibbons
3.	Mt. Pleasant, McCable Farm, IA	Swine	Full Scale 50,000 gal.	Gas mixing	Boiler heating	Harold Mc Cable
4.	State Univ., Ames, IA	Beef		Complete mix, mesophilic	Experimental Purposes	University of Iowa
5.	Edwall, OK	Cattle	Pilot Scale 1800 gal.	Continuous feed	-do-	Earth Cyclers
6.	Aurora, OR	Cattle	Pilot Scale	Modular	-do-	David House
7.	Hartford, NY	Cattle	Pilot Scale	Plug-flow	-do-	Bill Jewell, Cornell
8.	Hartford, NY	Cattle	Full Scale	Conventional	Space Heating	Bill Jewell, Cornell
9.	Hartford, NY	Cattle	Full Scale	Plug-flow	-do-	Bill Jewell, Cornell
10.	Dewey, IL	Poultry	Full Scale 25,000 chickens (15×10^6 BTU/day)	Plug-flow	Running Diesel Engine	Bill Jewell, Cornell
11..	Lancaster, PA	Poultry	Full Scale 75,000 chickens (15×10^6 BTU/day)	Plug-flow	Electricity Generation and Crop Drying	Energy Recovery
12.	Wyoming, MN Verlo Larson	Swine	Full Scale 10,000 gal.	Continuous feed, completely mixed	Digester Heating	Phil Goodrich, Univ. of MN

(*Contd.*)

Table 2.15 *(Contd.)*

S. No.	*Plant Location*	*Source of Feedstock*	*Type and Capacity*	*Type of Digester*	*Utilisation of Gas*	*Plant Construction Agency*
13.	Jefferson, WI Habeck Farm	Cattle	Full Scale 32,000 gal.	Batch fed completely mixed vertical tank	Digester heating	A.O. Smith Corp.
14.	North Glenn Co. Karl's Dairy Farm	Cattle	Full Scale 400 cows	Plug-flow	Electricity generation	Energy Harvest, Schaeffer & Roland, Chicago.
15.	Monfort, Co.	Dirt lot waste	Pilot plant 110 lbs/day	Continuous feed, completely mixed	Experimental purposes	Hamilton standard
16.	Bartow, FL Kaplan Feedlot	Cattle	Full Scale 570,000 gal.	Continuous	House heating	Hamilton Standard
17.	Arvada, Co.	Cattle	Pilot Plant 200 gal.	Plug-flow, Mosophilic	Digester heating	Biogas of Colorado
18.	Lamar, Co.	Cattle	Full Scale 6000 gal.	Plug-flow, mobile unit, mesophilic	Digester heating	Biogas of Colorado
19.	West Union, IA	Chicken	Full Scale 160,000 layers	Gas mixed mesophilic		
20.	Greeley, Co.	Beef	Full Scale (2.5×10^6 BTU/day)	Plug-flow	House heating	Energy Recovery Inc., Aravda.
21.	Henderson, MN	Hog	Full Scale (5×10^6 BTU/day)	Plug-flow	Electricity generation	Energy Recovery Inc., Aravada.
22.	Custer, MI	Cattle	Full Scale 350 cattle	Plug-flow, bag type	30 kw electricity generation	Energy Harvest Shaeffer & Roland
23.	Rice Lake, MI	Dairy Cow	Full Scale 110 cows	Plug-flow	12 kw electricity generation	Energy Harvest Shaeffer & Roland

(Contd.)

Table 2.15 *(Contd.)*

S. No.	*Plant Location*	*Source of Feedstock*	*Type and Capacity*	*Type of Digester*	*Utilisation of Gas*	*Plant Construction Agency*
24.	Monroe, WA	Dairy cow	Full Scale 350 cows	Complete mix system, gas recirculation	Electricity generation	Ecotope Group
25.	Clay centre, NE	Cattle	Pilot Scale	Semi-continuous	Experimental purposes	Hamilton Standard
26.	University Park, PA	Dairy	100^3	Mixed 2 chamber Mesophilic	Digester heating	Penn. State Univ.
27.	Stillwater, MN	Poultry, Human, Vegetable	Homestead size	Plug flow mesophilic	Cooking and heating	Rutan Research, Al, Rutan
28.	Cardiff, CA	Sewage Sludge and Water Hyacinth	Pilot Scale 200 gal.	Plug-flow	Experimental purposes	County owned water treatment facility

Source: Abeles, T.P., *et al.*, Energy and Economic Assessment of Anaerobic Digesters and Biofuels for Rural Waste Management, Report No. EPA-600/7-78-174, U.S. Environmental Protection Agency, Cincinnati. Ohio 45268, Dec., 1978, p. 157-160.

option in many cases. One difficulty with this approach is that of resource location which are widely dispersed for most individual farms to economically exploit. Thus cooperative ventures which result in greater economies-of-scale can be more attractive alternative. Potential of power generation from farm wastes is considerable in UK (Table 2.14). It includes livestock population, wastes generated, and potential of biogas generation [46]. There is considerable scope for improving economics of biogas generation from mixtures of farm wastes, industrial wastes, municipal wastes and sewage sludge through centralised anaerobic digestion (CAD) systems. These schemes could serve either a single large farm or several farms typically within a radius of about 10 km for high moisture content feedstocks (6-10 per cent dry matter); and high dry matter feedstocks such as agro-industrial waste and poultry litter. Landfill costs in the UK are still relatively low for disposal of industrial organic wastes but may increase in future as more stringent disposal legislation is imposed. This and implementation of environmental protection legislation to farming operations may provide fillip to CAD in the UK. CAD is well established in Denmark where Danish government imposes very high taxes on domestic fossil fuels to promote NCES. Future success of CAD in the UK shall depend on resource mapping of both farmyard manures and agro-industrial wastes; and economic and financial analysis to establish economic benefits of CAD. These measures shall help to develop appropriate algorithms to model how much of the national resource could be gainfully exploited. Interest in generating energy from farm wastes is anticipated to grow further following planned reduction in technical and institutional barriers and growing concern for environmental issues.

Table 2.14 Biomethanation Potential in UK

Resource	*Population*	*Dry weight of waste (tDS/d)*	*Potential methane yield (m^3/d)*	*Potential electrical energy (MWe)*	*Potential annual output (GWeh/yr)*
Cattle	12200000	28500	5700000	713	6242
Pigs	7900000	400	800000	99	876
Poultry	124000000	5000	1000000	125	1095
Straw		19000	1000000	138	1205
Total		53500	8600000	1075	9418

Note : tDS/d = Tonnes dry solids per day

Source : Dagnall, S., UK Strategy for Centralised Anaerobic Digestion, Bioresource Technology, 52, 276 (1995).

2.27 Biomethanation in USA

Biomethanation in the United States is more pursued by way of advancement rather than exploitation. Animal faeces, the main feed material for these plants, are not that freely and widely available with farmers as in developing nations. It is unlikely that individual family size plants on the pattern of those in China or India will proliferate in the USA. Large installations farms, cattle feedlots, food processing units and several dairy farms characterise the pattern of biomethanation in this country. Most of the installations originally existed mainly as a device for waste treatment which following energy crisis in seventies were also viewed as a source of energy recovery. Biomethanation potential is enormous in the US although only a small part is being exploited. According to some estimates nearly 255 x 10^6 tonnes of animal faeces are generated per annum which if anaerobically fermented can produce gaseous fuel equivalent to 274 trillion BTUs of natural gas. Besides animal wastes, other feed materials capable to generate biogas include 430 x 10^6 tonnes of crop residues per annum, 123 x 10^6 tonnes of urban refuse and 13 x 10^6 tonnes of municipal sewerage. Some major organic wastes are from cattle, beef, swine, poultry and humans. In addition, municipal refuse, seaweeds, water-hyacinth and fruit and vegetable wastes are other commonly used feed materials. Biogas in the US is mainly used for power generation, heating and experimental purposes and rarely as a cooking and lighting fuel. This country is in forefront in harnessing gas from landfill sites. There are a number of operating units in California, New Jersey, Washington DC, New York and Chicago all of which have full-fledged engineering facilities for gas recovery. Landfill gas is mainly used for power generation and meeting fuel needs of small-scale industries and utilities [1, 30].

A number of institutions are involved in carrying out studies particularly relating to kinetic and thermal aspects of biomethanation. These include : Department of Agriculture at the Cornell University (New York), Department of Civil Engineering at the University of Illinois, Imperial Valley Biogas Project (California), Dyanatech R & D Co. in Cambridge (Massachussets), Department of Microbiology at the Colorado State University at Fort Collins (Colorado) and Municipal Environmental Research Loboratory in Cincinnati (Ohio) [1]. Biomethanation, particularly for meeting individual household needs may not play a significant role in the USA in the manner it did in India and China. One reason is that wastes are too widely scattered for economics to be attractive although biomethanation makes sense for helping solve disposal problems of municipal and industrial wastes.

2.28 Concluding Remarks

Biomethanation has been pursued with fair to a very high degree of success in several nations. Progress made in countries like China, India, Nepal, Philippines and Tanzania is particularly impressive. Notable success has been achieved by those countries that have an effective implementation strategy. India's '*National Programmes on Biogas Development*' based on turnkey approach or Nepal's '*Biogas Support Programme*' or Tanzania's '*Biogas Extension Service*' are specific examples which show how phenomenal success can be achieved with effective implementation strategies. There are also examples of nations where biogas programme was started with lot of initial enthusiasm but could not be sustained for long either for want of political will or lack of effective implementation strategy. In some countries it could not make headway as economic-viability of biogas plants remained in doubt and direct consumption of organic wastes preferred over biomethanation [5, 22]. Based on previous studies, G. Leach [19] reviewed the state-of-art of biomethanation in South

Asian countries comprising India, Pakistan, Bangladesh and Sri Lanka and concluded that success of biogas programme is linked to income, household size, animals possessed, prevalent price of other fuels and implementation strategy employed. In general, the poorer the country and the less availability of conventional fuels, the greater will be the dependence on biofuels. On the whole biomethanation has made significant progress world over which has caught the imagination of people. If biomethanation can achieve phenomenal success in two large Asian countries, India and China, there is no reason why it cannot make similar success in other countries.

REFERENCES

1. Abeles, T. P., *et al., Energy and Economic Assessment of Anaerobic and Biofuels for Rural Waste Management*. Report No. EPA-600/7-78-174, US Environmental Protection Agency, Cincinnati, Ohio, Dec., 1978, p. 157-160.
2. Bembalkar, N. L., Energy Crisis; Indian Response, *Science Express (Indian Express)*, Visakhapatnam, Feb., 26, 1991.
3. BORDA, *Biogas: A Manual for the Realisation of Biogas Programmes*, Bremen Overseas Research and Development Association, Bremen, Federal Republic of Germany (erstwhile), May, 1979.
4. BORDA, *Biogas in the Peoples Republic of China*, Project Report, Federal Republic of Germany (erstwhile), March, 1981.
5. Brown, N., Biogas Systems in Development, *Appropriate Technology*, **14(3)**, 5-7 (1987).
6. Buren, A. V., *Tranferring Biogas Technology to the Rural Areas of the Third World*, International Institute for Environment and Development, London, 1980.
7. Dawson, O. L., *Communist China's Agriculture*, Praeger, New York, 1970.
8. Demuynek, M. G., *et al.*, Assessment Study on Biogas Plant in Europe, in 'A-Strub (ed.), Energy from Biomass, *Proc. 2nd EC Conference*, Applied Science Publishers, London, 1982', p. 620-625.
9. *Economic Times*, Renewable Sources of Energy (Infrastructure), Economic Survey 1990-91, New Delhi, July 21, 1991, p. 14.
10. Eusebio, J. A. and B. I. Rabino, Research on Biogas in Developing Countries, *Compost Science*, Mar.-Apr., 24-27 (1980).
11. Eusuf, M., Development of Biogas in Bangladesh: Technological Aspects, *Khadi Gramodyog*, **27(3)**, 182-184 (1980)
12. FAO, *China, Azolla Propogation and Small Scale Biogas Technology*, FAO Soils Bulletin No. 41, Food and Agriculture Organisation, Rome, 1979.
13. Fernandes, W., Where Biogas Plants are Built by Volunteers, Special Supplement on Science and Technology, *Times of India*, Aug. 15, 1986.
14. Goodman, L. J., and R. N. Love (eds.), *Biogas Energy Projects: Planning and Management*, Pergamon Press, NY, 1981, p. 133.
15. Hilliges, G., *et al.,. Biogas in Peoples Republic of China*, A Project Report based on Exchange of Experience in the Field of Biogas among China, India and F. R. of Germany (erstwhile), Bremen, March, 1981.
16. *The Hindu*, Role of Renewable Sources of Energy, Visakhapatnam, Dec. 27, 1990.
17. *The Hindu*, Renewable Energy Sources: Need for Greater Priority, *The Hindu Survey of Indian Industry*, 1990; Feb., 1991, p. 177-179.
18. *The Hindu*, Biomas; A Potential Energy Source, Visakhapatnam, May 14, 1991, p. 17.
19. Leach, G., Household Energy in South Asia, *Biomass*, **12(3)**, 155-184 (1987).

20. Lee, K. C., Biogas Programme in the Latin American Region and in Jamaica, *Khadi Gramodyog,*. **27(3)**, 185-87 (1980).
21. *Newsletter Biogas*, Biogas Plant in Nepal: Past and Present, Summer, No. 1, 1978.
22. Murthy, R. K., The Energy Crunch: An Appraisal, *The Hindu*, June 18, 1991, p. 17.
23. Pimental, D. and M. Pimental, *Food and Energy and Society*, Edward Arnold Publishers, London, 1979.
24. Rao, J. K., R & D in Biogas Technology, *The Hindu*, Hyderabad, Sept. 4, 1984.
25. Reddy, A. K. N., Energy Planning : Time for Tough Decisions, *The Hindu*, Visakhapatnam, Nov. 23, 1990.
26. Simil, V., *China's Energy : Achievement, Problems and Prospects*, Praeger Publishers, NY, 1976, p. 107.
27. SPO, *Turkish Government, UNICEF Joint Biogas Project : Design and Construction of Small and Medium Size Biogas Digesters in Rural Areas (in Turkish)*, State Planning Office, No. 7, Ankara, Turkey, 1984.
28. Subramanian, S. K., *Biogas Systems in Asia*, Management Development Institute, New Delhi, 1978.
29. Tasdemiroglu, E., Review of Biogas Technology in Turkey, *Biomass*, 0144-4565, 137-148 (1989).
30. TERI, *Directory of Sources of Information on Biogas Technology*, Tata Energy Research Institute, Mumbai, 1981.
31. TERI, Biogas Situation in Pakistan *Biogas News*, No. 2, Tata Energy Documentation and Information Centre, TERI, Mumbai, June, 1984.
32. TERI, Caribean Biogas Programme, *Biogas News* No. 2, Tata Energy Documentation and Information Centre, TERI, Mumbai, June, 1984.
33. Ventura, P. F., Status of Biomethanation Development in the Philippines, *Jour. Applied Microbiology Biotechnology,* **4(1),** 151-158 (1988).
34. WAPT, *A Report Prepared by the World Agricultural Production and Trade*, NY, 1968.
35. Alinea, E. K., Status of Biogas Technology at CAMARTEC-Arusha, *Biogas Forum*, **I(52),** 14-17 (1993)
36. Devkota, G. P., Privatisation in Gobar Gas Installation, *Biogas Forum*, **II(53)**, 4-7 (1993).
37. Herula, C. S., Biogas in the Philippines, *Biogas Forum*, **IV(51)**, 7-8 (1992).
38. Kellner, C., Biogas Dissemination : The Tanzanian Experience, *Biogas Forum*, **IV(47)**, 4-9 (1991)
39. Leermakers, M. and W. J. van Nes, Extension of Biogas in Nepal, *Biogas Forum* **I(56)**, 9-14 (1994)
40. Qiang, C. G., *et al.*, Overview of Biodigestion Development in China, *Biogas Forum*, **I** No. **48**, 7-19 (1992).
41. Siraj, W. U., Biogas Programme in Pakistan, *Biogas Forum*, **II (49)**, 10-13 (1992).
42. Van Nes, W. J., The Biogas Support Programme in Nepal, *Biogas Forum*, **II (53)**, 4-7 (1993).
43. Van Nes, W. J., and J. Lam, Desired Future Structure of Biogas Sector in Nepal, *Biogas Forum*, **I** (No. **56**), 4-8 (1994).
44. Van Vliet, M. M. and van Nes, W. J., Effect of Biogas on the Workload of Women in Rupandehi-district in Nepal, *Biogas Forum*, **II (53)**, 12-16 (1993).
45. BNRM, Suggestions by Nepali Biogas Workshops, Biogas and Natural Resources Management Report No. 46, *Biogas Forum*, **III** No. **58**, 20-21 (1994).
46. Dagnall, S., UK Strategy for Centralised Anaerobic Digestion, *Bioresource Technology*, **52**, 276 (1995).

Chapter 3

BIOMETHANATION IN INDIA : ACHIEVEMENTS AND PROSPECTS

3.1 Introduction

Biomethanation has made rapid strides in India starting from sixties onwards as evident by growing number of plants installed. Work pertaining to the development of biogas technology started in beginning of the 20th century though biomethanation as a programme was launched only in sixties. During sixties and seventies, Khadi and Village Industries Commission, Union Ministry of Agriculture and Rural Reconstruction, State Agro-Industries Corporations and State *Khadi* Boards were some of the main agencies that worked for promotion of biogas programme in the country. However, in eighties biomethanation received support not merely as a rural industry programme but more as a part of new and renewable sources of energy (NRSE) which the government vigorously promoted. In 1981 the government set up Commission for Additional Sources of Energy (CASE) and the Department of Non-Conventional Energy Sources (DNES) a year later. DNES was entrusted with biogas programme and emerged as the principal nodal agency for its promotion and coordination in the country. In subsequent years practically every state set up its own agency for popularising non-conventional energy sources, which supplemented the efforts of the DNES. Following promotion by the DNES, biogas programme gathered a rapid momentum when plants were set up in lakhs every year which continues unabated even today. By 1995-96 a total of 23.60 lakh plants were set up under the National Project for Biogas Development (NPBD) which the DNES introduced in Sept. 1982 [4]. NPBD was a sequel to the '*All-India Coordinated Project for Biogas Technology*' which the government launched in 1975-76. Tata Energy Research Institute (TERI) which was initially established in Mumbai and later shifted to New Delhi in mid-eighties undertook several projects which helped a great deal to promote biomethanation, It prepared numerous reports, documents and reference material based on extensive field and deskwork. In this chapter it has been endeavoured to summarise historical developments, progress achieved and important milestones of biogas programme.

3.2 Historical Background

Developments in the field of biomethanation started quite early in India. A sewage treatment plant was set up at Dadar (Mumbai) in thirties which also produced sewage gas comprising methane and carbon dioxide besides waste treatment. Way back in 1920, Prof. N. V. Joshi of the Indian Agricultural Research Institute (IARI), New Delhi along with his co-workers conducted experiments to produce biogas through anaerobic fermentation of waste paper, banana skin and other city wastes. Anaerobic fermentation picked up further momentum when Dr S. V. Desai of the IARI visited the Sewage Treatment Plant at Dadar in 1937 for studying the process. In 1945, S. V. Desai and S. C. Biswas of the IARI produced biogas from anaerobic fermentation of cattle wastes which became an important milestone in this direction. Enthused by the growing interest in the process, Prof. N. V. Joshi developed a small batch type biogas plant in 1946. This plant which was made at a cost of Rs. 1800 to produce 20 ft^3 of gas per day mainly comprised a digester and a separate gasholder. Being one of the very early designs, it however could not continue for long due to several limitations. Two plants made to this design were set up in Pune during late forties but unfortunately could bot function following bursting of digesters. Another plant as per this design was set up in Chennai but that too met the same fate. It was later on analysed that foaming of slurry was main factor for bursting of digesters. However, these designs acted as landmark for further development. Almost contemporary of these developments, Y. N. Kotwal and his associates conduced studies at the Sewage Plant at Dadar. They found that when urine was added to the slurry, it fermented more rapidly leading to higher gas yield. It was also noted that when larger quantity of urine was added, fermentation was more rapid but caused excessive formation of foams within digester. Motivated by earlier endeavours, J. J. Patel at the Agricultural College (Pune) developed a new design in 1951 by combining digester and gasholder in a single unit. This plant costing Rs. 1800 and designed to produce 200 ft^3 of gas per day was named as '*Gramalakshmi* Gas Plant'. Distinctive features of this plant covered integration of gasholder and digester in one unit, underground location of gasholder, continuous daily feeding of slurry, consistency in the level of digester contents, automatic overflow of the digester slurry, and facility of rotating gasholder to break scum formation. In 1952-53 the IARI further modified its design, and built up two plants side in its premises by side in which gasholders were provided as covers over digesters [14, 15].

Satish Chandra Dasgupta and Swami Vishwakarmanand during 1952 tried to improve construction technology of biogas plants further. They developed a plant design free of any masonary work. Their design included half split bambooes to support digesters, three to four supports along with pulleys and counter weights to guide gasholders and use of mild steel in gasholder construction. Later on C. N. Acharya, a very distinguished scientist of the IARI, visited this plant to improve it further. S. C. Dasgupta and C. N. Acharya found that reduced pressure on biogas produced within gasholder tended to increase gas yield. For using this fact in plant operation it was proposed that gasholder should be kept at negative pressure (minus half inch water column) when not in use. Adjacent to plant some eighty bricks were kept in three or four buckets for puting on gasholder top in stages for building requisite pressure according to the need. Conversely, with emptying of gasholder following gas utilisation, bricks were removed in stages for releasing necessary pressure.

Enthused by the spree of charges in plant design, J. J. Patel in 1952 incorporated further modifications in his design if the '*Gramlakshmi* Gas Plant.' The modified design included four chambers in the digester and an annular water chamber around the digester top for accommodating gasholder which was guided by four supports. This design though free from several earlier problems was also not very successful due to two main limitations. First, masonary surface of inner wall of annular water chamber of the gasholder tended to absorb gas which lowered gas yield. Secondly, there was no facility for mashing of semi-dry animal wastes and crop-residues before feeding to the plant which caused blockages in slurry from one chamber to another. To overcome these difficulties another improved version of the '*Gramlakshmi* Gas plant' was developed by J. J. Patel in 1954. This modified design had two chambers, one to act as primary and the other as secondary digester with gasholder to have rotary as well as vertical movements. Its digester made of brickwork was kept underground and fitted with two pipes one for feeding and the other for taking out the digested slurry. Gasholder was designed in a manner so as to ensure uniformity of its pressure at 3 to 4 inch water column thus dispensing with need for any counter weights for regulation. It was designed for a retention period of about 50 days. This modified design of the '*Gramlaxmi* Gas Plant' became the major landmark for biomethanation in India as also several other countries. This modified design was later on adopted by the KVIC for its countrywide popularisation. Apart from the KVIC which acted as nodal agency for biogas programme during sixties and seventies, another Lucknow-based Planning Research and Action Division (PRAD) also adopted it for promotion.

Mr. Jashbhai J. Patel, the principal originator of the design of the '*Gramlakshmi* Gas Plant' virtually emerged as the doyen of 'Biogas Programme in India'. For his distinguished contributions many called him '*The Father of Biogas*' [26]. While reviewing the history of biomethanation in India, it will not be incongruous to recall his brief life-sketch. Born on February 22, 1909 in a tiny sleepy village, Valvod in Gujarat, Mr. Patel did his graduation in agriculture from the College of Agriculture at Pune in 1931. During his long association with dairy and agriculture, he was deeply involved in development of a suitable system for fermentation of cattle dung for producing biogas. With his untiring efforts, he developed three models of biogas plants during the period 1949-53 which he called as *Gramlakshmi*, I, II, and III. The *Gramlakshmi* III met with unqualified success and was adopted by the KVIC for providing non-conventional energy in rural areas. Jashbhai was appointed the first Director of the newly created Biogas Department under the KVIC which he served till his retirement in 1965. After his retirement, he worked as an honorary Advisor to the KVIC for biogas programme for several years. He addressed the International Biogas Conference held at Pune in 1990. He died on May 20, 1992. With his demise, the entire biogas world has lost a *Gramlakshmi* and a true champion of the biogas programme [26].

Apart from popularising *Gramlakshmi* III design, PRAD also modified the design initially developed by S.C. Dasgupta and Swami Vishwakarmanand. The guiding spirit at the PRAD in promoting biomethanation was Mr. Rambux Singh who like Jasbhai Patel has emerged as another legend. At the time the PRAD involved itself with biomethantion, accessories like burners and lamps were not of satisfactory design. Burners were made of empty cigarette and shoe-polish tins with their burning efficiency not more than 30 per cent. KVIC and PRAD initially tried to promote plants of the IARI design. Their promotion however posed problems as costs were high and there were some manufacturing problems too. Biogas plants made to the IARI design with attached pulley and counterweights created alignment problems causing tilting of gasholder sideways which caused uneven and interrupted gas supply. KVIC, then switched over to popularise plants which had no pulley and counterweights attached. Gas pressure was instead sought to be built up with weight of gasholders. Biogas burners popularised by the KVIC were made of cast iron having about 60 per cent efficiency. KVIC also carried out R & D activities pertaining to biogas technology. As a case-in-point, its R & D wing developed a design of horizontal plants which are appropriate for areas where excavation involving hard rocks or high water-table is an expensive proposition [13, 15].

Based on Chinese design and field trials carried out at the Gobar Gas Research Station, Ajitmal (Etawah), PRAD developed a fixed dome type plant which was named as the '*Janata* Biogas Plant (JBP)'. Action for Food Production (AFPRO), a Delhi based non-governmental organisation, developed a fixed dome type plant in eighties which has been named as '*Deenbandhu* Biogas Plant' [17]. It has a curved bottom and a hemispherical top without any cylindrical portion in between. Swami Vimuktanand at Ramakrishna Mission (Calcutta) developed a design popularly called '*Belur Math Model*'. This design has a separate two piece gasholder and a separate digester. Jyoti Solar Energy Institute at Vallabh Vidyanagar (Gujarat) developed a plant design which is particularly appropriate for digesting agricultural waste materials and crop residues. Very recently the PRAD (Lucknow) has developed a fixed dome type design which is even much cheaper than *Deenbandhu* model. MNES has accorded approval to this Janata II model for promotion in the country.

3.3 Reference to Biogas in Some Important Seminars and Conferences held

A number of seminars and conferences on biomethanation have been organised in past in India and abroad. One important workshop on biogas technology and utilisation was held in New Delhi from July 28 to Aug. 2, 1975 which was organised by the ESCAP and inaugurated by Mrs. Indira Gandhi, the late Prime Minister of India. In her inaugural address, Mrs. Indira Gandhi, underscored the need for promoting biomethanation which is particularly appropriate for developing countries [5]. This Workshop was attended by twenty five participants from thirteen developing countries, a large number of observers from India as well as representatives from several international organisations. The Workshop made a number of recommendations for promotion of biogas programme at national, regional and global levels. After the New Delhi Workshop, ESCAP organised a second workshop at Manila (Philippines) from

Oct. 13 to 18, 1975 in association with the National Science Development Board (NSDB). This Workshop was attended by twenty participants from twelve countries, a number of observers and representatives from the UNICEF, WHO and UNDP. It was inaugurated by Mr. Abid Hussain, the then Chief of the ESCAP Division of Industry, Housing and Technology and presently, Vice-Chairman of Rajiv Gandhi Foundation. In his welcome address, he emphasised many advantages of developing an integrated farming system involving production of energy and fertiliser, cultivation of food and agriculture crops, and a simultaneous growth of algae and fish [6]. The Workshop discussed fermentation of organic wastes, production and use of biogas, biomass-culture for integrated farming system, and socio-economic aspects of biogas programme.

Mr. R. Venkataraman, the then President of India, in his first Dr. Y. Nayudamma Memorial Lecture delivered at the Administrative Staff College of India, Hyderabad, on June 22, 1986 in memory of the eminent leather technologist who died in the 'Kanishka' air crash this day in 1985, emphasised the importance of biogas in country's energy supply. He advised [24] :

> 'There is a need for scientists to identify requirement of the people, isolate technology component and communicate it to the multi-disciplinary scientific teams to resolve their problems rapidly. In view of the country's vast population, there is a need to exploit solar energy, windpower and biogas on a massive scale for providing appropriate technology and simple devices along with needed training to the rural poor'.

In a full-page supplement 'India Leads in the Global Renewable Energy Movement' brought out in the leading English dailies on July 13, 1994, coinciding with the visit of US Energy Secretary Ms. Hazel O'Leary, Mr. P. V. Narasimha Rao, the then Prime Minister of India, observed that the energy from the sun, wind, biomass and water can reduce the pressure on conventional sources which are depleting very fast. This energy is renewable, unlimited and pollution-free. He noted :

> 'Non-conventional energy is fast catching the imagination of the people in our country. Efficient and economical means to harness such new energy sources are being developed. The experience so far has been encouraging. India is poised to assume a leading role among the countries which are rich in such resources.'

Following interactions between Mr. P. V. Narasimha Rao, Prime Minister of India and Ms. Hazel O' Leary, US Energy Secretary, in New Delhi on July 12, 1994, India and the United States have decided to constitute a joint sub-commission to monitor the cooperation between the two in energy sector. The visiting delegation led by Ms O'Leary also held discussions with Mr. L. M. Menezes, the then Secretary, Ministry of Non Conventional Energy Sources and visited Solar Energy Centre, Gurgaon. Possibilities of cooperation between India and the USA in the fields of solar, wind and biomass energy were explored. It was agreed to exchange scientific information and test data in respect of solar, biomass and other fields between the Indian Solar Energy Centre and the National Renewable Energy Laboratory (USA).

Mr. P. V. Narasimha Rao, the then Prime Minister, while addressing a conference of state ministers incharge of non-conventional sources of energy in New Delhi on Oct. 28, 1992, observed that the country's ambitious programme to generate more electricity could face serious problems due to their adverse impact on environment. In this context he noted that the alternative was to exploit adundant sources of renewable energy such as biogas, solar energy, wind-power and wave-energy. The shortage of power is a dampner to the government plans to woo more foreign investors into the country, and renewable energy sources if adequately harnessed can overcome these problems substantially. In his message to the special supplement, 'Non-conventional Sources of Energy: A Decade of Achievement', *Indian Express,* Oct. 28, 1992, Mr. Rao noted:

> 'Energy from the sun, wind and biomass can alleviate the pressure on traditional sources that have been considerably depleted. Practical and economical means to harness such new energy sources need to be developed. India's initiative in this regard must succeed.'

Mrs. Indira Gandhi, the late Prime Minister of India, during her regime laid considerable emphasis on development of new and renewable sources of energy. For giving a fillip to these sources her government established the *Commission for Additional Sources of Energy (CASE).* At the time of its inauguration she observed [8] :

> 'A crucial factor for technological development is energy. Other countries are experimenting in a big way uses of renewable sources. In the long run hydel power, solar and bioenergy will be the principal forms of energy used in our country. The tapping of solar energy through three known methods, namely, photosynthesis, photovoltaic systems and photothermal applications can meet our energy needs. We are also using nuclear energy for this and other developmental purposes. Immediately, however, we must be very thrifty in our use not only of indigenous and imported petroleum products but even of coal. We have decided to establish a 'Commission for Additional Sources of Energy' to advise us on energy policy and to draw up and implement programmes on alternative sources. This will give fresh impetus to innovative thinking. Programmes such as a tree for every child, nutrition gardens, forest farming, raising of energy plantations, organic recycling through conservation and use of all waste products and harnessing of renewable forms of energy based on sunlight, wind and *biogas* are all potentially valuable mass movements, provided they are formulated thoughtfully and implemented with the involvement of local population.

Mrs. Indira Gandhi in her key-note address at the UN Conference on '*New and Renewable Sources of Energy*' held in Nairobi, Kenya, during Aug. 10-20, 1981 observed [9] :

> 'In contemporary India, as in some other parts of the world, every experience ranges from most primary and uncomplicated form as manual labour and animal power to the most advanced use of nuclear and solar energy technology. Impressive quantities of commercial energy are regularly produced from sources like coal, lignite, oil and hydro-electric potential of waterways. At the same time there is also an extensive use of familiar non-commercial sources like fuelwood, agricultural wastes and animal dung. We are trying to harness vast potential of sun, wind and water. Solar energy is already being used for cooking, for heating water and pumping it for irrigation purposes, for crop-drying, etc. Another promising item in new energy programme is *biogas* which is mainly produced from agricultural wastes and uses items such as water-hyacinth that has for years been blocking waterways and clogging fields. Throughout the world, measures must be taken to make increasing use of renewable energy sources such as hydropower, large-scale programme of reforestation and energy plantation, *biogas* production, use of solar energy for low and medium temperature requirements like water-heating and of wind and solar energy for pumping.'

Mrs. Indira Gandhi, in her inagural address to the World Energy Conference held in New Delhi during Sept. 18-26, 1983 emphasised the significance of biogas based decentralised energy systems for a country like India. She observed [10]:

> 'All known strategies of development and of raising the living standard of the poor are energy-intensive. So long as they are dependent on centralised energy systems, people's need are not likely to be met in full measure. Decentralised systems are necessary to promote regional self-reliance and help further utilisation of materials such as animal and plant wastes which are available in villages. These processes could be managed and maintained even by those who do not have such education. Community biogas plants based on cowdung deprive those who used to freely gather such fuel for their own use. What was non-commercial has become commercial. Economic growth means finding remedies to social and economic problems that keep arising.'

A national workshop on 'Policy Issues in the Biogas programme' was organised by the AKRSP and sponsored by the *Aga Khan Foundation and European Commission* in New Delhi during August 21-22, 1995. The workshop noted that biogas helps to keep a smoke-free kitchen, reduces cooking time, and dispenses with fuelwood collection and preparation of dung-cakes. Present day emphasis on natural resources management in wake of dwindling forests and wildlife has caused renewed interest in biogas technology. The workshop attended by representatives from MNES, Ministry of Forests and Environment,

AKRSP, TERI, and several NGOs deliberated on potential and pitfalls of biomethanation programme in India. The workshop highlighted little known benefits of biogas technology and the misplaced negative perception of biogas. The workshop evaluated alternative strategies for popularising biogas among 70 per cent of India's household who need sustainable, eco-friendly and low-cost fuel and fertilisers. The workshop felt that an awareness of benefits that biogas can bring to rural women should be fostered with greater vigour. It should be promoted that biogas can substantially minimise drudgery of rural women and save them from risks involved in fuelwood collection. The workshop recommended that the NPBD undertake publicity campaigns assisted by voluntary organisations particularly those working for women welfare to increase awareness of these benefits of women. Multi-faceted benefits of biogas programmes such as eco-friendly, forest-conservation and health-improvement need to be promoted. The workshop also recommended that the government should continue to support the programme by way of subsidy. In the context of government's emphasis on market-driven economy, reducing or withdrawing subsidies shall affect the capability of marginal and landless farmers in setting up plants; adversely affect forest resource management, and compound economic problems of already over-burdened women. The workshop recommended a more carefully planned market-driven economy in which the strategy can be gradually introduced with circumspection and caution. The participants stressed that at least for five years the subsidy should not fall below its current level. During this period, orgaisations can experiment with different marketing and production strategies to develop an effective national programme.

Mr. B. R. Prabhakara, the then Secretary, MNES, while addressing the National Workshop on '*Policy Issues in the Biogas Programme*' organised by the AKRSP in New Delhi during August 21-22, 1995, provided an overview of implementation of the NPBD. MNES considers biogas as well as improved cookstoves among the most important non-conventional energy sources available to rural households. For funding biogas programme, although essentially an appropriate-technology based programme, possibility of external funding from international organisations should be explored.

MNES's principal strategies to promote biogas technology involve funding training, extension work, and cash subsidies. MNES releases money to state nodal departments and a few select NGOs. MNES has a goal of gradually reducing and finally eliminating subsidies. Although MNES is under pressure to eliminate subsidies, it also feels that in the absence of alternative mode of support, state support in the form of cash subsidy is desirable. At the same time, MNES should evaluate the effectiveness of the subsidy based on experience of field extension workers and other concerned staff. The subsidy serves a crucial purpose as an incentive for men to invest in biogas technology, as women are the major beneficiaries of the programme while economic decisions are taken by men. The workshop was also addressed by Prof. T. K. Moulik, a renowned expert in economic and implementation aspects of biogas programme; Mr. Barry Underwood, Chief Executive (AKRSP); and Mr. V. K. Sardana, Chief Executive Officer (Aga Khan Foundation).

In a full-page MNES supplement in leading English dailies on July 13,1994 on *'Global Renewable Movement'*, Mr S. Krishna Kumar, the then Minister of State for NCES, in his message observed that the MNES had launched ambitious programmes across the entire spectrum of renewable energy to take India to leading position. He said :

> 'The enormous potential of India in wind, small-hydro, *biogas*, solar, ocean, geothermal and other forms of renewable energy is now sought to be harnessed for national development with special emphasis on reaching renewable energy to rural areas for upliftment of the life of the people. With entrepreneur development and market orientation, with international assistance and wide participation of industry and the public, renewable energy in India is developing into a national mass movement.'

Mr. Krishan Kumar, in another message to a previous MNES supplement in *Hindustan Times* on June 26, 1993, stressed the need for private sector participation in implementing NCES projects. He added :

> 'Energy availability in rural areas cannot be isolated from the basic needs of sustenance like food, shelter, drinking water and clean environment. In view of finiteness of fossil fuels and increasing environmental concerns, central government is giving impetus for wide propagation of non-conventional energy systems for use in rural areas. These include improved *chulhas* and *biogas* plants for cooking energy, solar PV systems for lighting and water-pumping, and biomass-based systems for varied applications. We look forward to non-governmental organisations and rural energy entrepreneurs actively participating in his great effort.'

Mr. S. Krishna Kumar in an interview published in a recent issue of *Prospects* (1995), an Asian journal devoted to renewable energy, noted (p. 24):

> '*Biogas* and improved fuel-efficient cooking stoves are among the government's traditional programmes. By making energy from wastes resources readily available in the villages, we can save on fuelwood and environmental degradation, reduce petroleum imports and substitute commercial electricity. The raw material for *biogas* is easily available, the technology is cheap, the equipment rudimentary and the cost is reasonable. The government aims to bring about a widespread acceptance of *biogas* in rural India over the next six years. In Indian conditions and those in other developing countries, development of rural energy resources is vital.'

Mr. K. C. Pant, Chairman, Tenth Finance Commission and former Union Minister for Energy and Chairman of the erstwhile Advisory Board on Energy, while addressing a seminar on Energy Management organised by the HRD Foundation and the IREDA (Indian Renewable Energy Development Agency) in New Delhi on Dec. 21-22, 1992 emphasised the need for harnessing renewable energy sources. He stressed the need to tap unlimited energy from the sun, wind, sea and biological wastes along with expansion in conventional energy supplies.

Dr R. K. Pachauri, Director, TERI, and Advisor to the UNDP, in his foreword to the TERI *Manual for Extension Workers for Spherical Biogas Plants* (1988), noted that biogas has energed as a significant resource with tremendous potential to benefit a large number of individuals and communities. For programme success it is desirable to look into some larger issues of biogas technology. According to Dr Pachauri, these issues involve questions of participation by local communities in biogas programme, development of appropriate skills at the local level, setting up the infrastructure for successful implementation of programmes, and opportunities for deriving commercial benefits for those who would be implementing these programmes. In his foreword to another TERI document '*Manual for Plant Owners of Biogas Plants* (1988)', he noted :

> 'Given the importance of biogas technology in the developing countries, dissemination of knowledge for establishing biogas plants and the benefits that can be derived from them assumes importance. The experience of India in this field is particularly relevant to the spread of knowhow in other countries of the world.'

According to G. S. Golueke, renowned biogas expert who worked at the Sanitary Engineering Research Laboratory, University of California, Berkeley (USA), biogas recovery from wastes through anaerobic fermentation is a very appropriate technology for a developing country like India. He wrote [11] :

> 'Anaerobic digestion holds greatest promise in energy recovery from animal manures because it seems almost to be the most economical and yet environmentally satisfactory method of treating such wastes especially those from confined rearing operations. Manures are readily decomposable and in a state readily amenable to being slurried. The amount of gas produced per unit amount of manure is significantly influenced by the food intake of the animals. Thus gas production per pound of manure from animals raised in the USA is greater than that from animals raised in *India*.'

Commonwealth Science Council and the KVIC jointly organised an international workshop on biogas in Mumbai during Sept. 8 to 10, 1980 to find ways and means for promoting biogas programme in Commonwealth countries. In his inaugural address, Mr R. V. Swaminathan, the then Union Minister of State for Agriculture and Rural Reconstruction, emphasised the importance of biogas as a supplementary fuel. He observed [22]:

> 'It is estimated that over 30 per cent of 980 million tonnes of cattle dung produced in India is burnt for fuel which could otherwise be used as organic manure. In terms of nutrients, it is equivalent to a third of India's chemical fertiliser used. If the entire quantity of 980 million tonnes of cattle dung

> produced is put through gas plants, it can produce enough gas for household needs for 487 million people and production of country's organic manure will increase by 350 million tonnes. There are other waste materials like poultry-droppings, crop-residues and night-soil which can be digested in a biogas plant. It is thus necessary to develop appropriate technologies through which all the organic waste materials can be gainfully utilised for energy production.'

According to B. R. Soubolle, a noted biogas expert, biomethanation is very appropriate decentralised process for energy generation particularly in developing countries. He wrote [21]:

> 'People living in remote areas of South-East Asia, or other tropical or sub-tropical countries, where electricity is not available and fuel is hard to get, have a very cheap, abundant and efficient fuel in the gas produced from ordinary cowdung. This gas (marsh gas or methane) is generated with the greatest of ease simply by letting a mixture of cowdung and water, called slurry, ferment in a well-like pit without exposure to air. The gas rises to the surface and collect in a drum, whence it is piped to the kitchen store. A farmer with a couple of bulls or buffaloes for ploughing and one or two cows for milk gets enough dung every day to produce sufficient gas for all the cooking needs of a village family of six. The cooking is clean and hygienic, the pots do not get black, there is no smoke or smell, and the gas is non-toxic. And after extracting the gas to cook his food, and to light his house at night, the farmer still has all the dung left, well-fermented and rotted, to fertilise his fields.'

Dr. M. S. Swaminathan, India's legend in agriculture and a former Member of the Planning Commission, laid considerable emphasis on promotion and advancement of biogas technology in India. While addressing a seminar on theme '*Indian Agriculture at the Cross-roads*', he observed [6] :

> 'There is concomitant danger of increasing carbon dioxide concentration in the atmosphere due to the continuous use of fossil fuels and it is reported that emissions of this gas have grown from 2,000 million to 18,000 million tonnes per year during the period 1900 to 1975-76. Increase in the concentration of carbon dioxide could lead to increased temperature and thereby exert serious effects on the climate. The transport of petroleum and its products leads to considerable pollution of oceanic waters which may in the long run reduce the biological productivity of oceans and the quality of fish. Solar energy coupled with *biogas* technology can provide a considerable proportion of the energy needed for enhanced productivity. The *biogas* technology based on animal refuse and cellulosic wastes is attracting world-wide

> attention. It is important that our efforts in combining technology and social action to make *gobar* gas utilisation a mass movement are not behind those of several other countries who took the technology from us'.

An International Conference on '*Biogas: Technologies and Implementation Strategies*' was held at Pune (India) on Jan. 12, 1990. At this Conference, Government of India's commitment to creation and widespread utilisation of new and renewable sources of energy was highlighted. It was pointed out that India was poised for a bigger and wider role in the field of biogas. However, a need for mobilising additional funds for starting and sustaining biogas programme was felt. This conference was jointly organised by the BORDA (Bremen Overseas Research and Development Association) and UNDARP (United Socio Economic Development and Research Programme), a Pune-based NGO, actively supported by the erstwhile DNES (now MNES) and German Federal Ministry for Economic Cooperation under the aegis of German Appropriate Technology Exchange (GATE). The Conference which was third in series with earlier two held at Bremen was, among others, addressed by Dr. Maheshvar Dayal, the then Secretary Incharge, DNES, and Mr Repnik and Dr. U. Reeps, representing GATE and BORDA, respectively. It was attended by delegates from over thirty countries which brought together rich and diverse experience on a common forum. Conference papers were categorised in five groups — technology, social issues, finance, implementation and evaluation [28]. Following extensive deliberations it was unanimously agreed that anaerobic fermentation helps conserve resources, protects environment, and leads to sustainable development. Small digesters in rural areas and large ones for treatment of municipal and industrial wastes should be popularised in urban areas.

3.4 Estimates of Biogas Potential in India

Biogas potential in a country can be assessed from the availability of livestock and biomass. Estimates of India's livestock population and norms of cattle and buffalo ownership according to land holdings and family size are given in Table 3.1 and 3.2, respectively [1, 19, 20]. S. Neelakantan [18] while working at the National Dairy Research Institute at Karnal (Haryana) made statewise estimates of biogas potential. As worked out by him, estimates of wastes availability in different states of India are summarised in Table 3.3. Estimates of biogas potential based on wastes availability and mean ambient temperature are summarised in Table 3.4. It can be seen from these estimates that total wet dung output of 960.863 million tonnes comprises 192.172 million tonnes of dry matter and 180.642 million tonnes of organic matter. Out of 180.642 million tonnes of organic matter, cattle dung accounts for 136.816 and buffalo dung 43.826 million tonnes. As for the organic matter utilisation, a portion of 124.643 million tonnes is used as manure and remaining 52.385 million tonnes is burnt annually as dung cakes.

Table 3.1 Broad Estimates of India's Livestock Population

Category	Activity Break-up	*Population in millions*
Cattle	Males exclusively used for work	69.10
	Males partly used for work	2.26
	Females partly used for work	1.98
Buffaloes	Males exclusively used for work	6.98
	Males partly used for work	0.62
	Females partly used for work	0.39
	Total work animal population up to 1965-66	81.41
	Estimate of work-animal population up to 1980-81	87.26

Note: (Based on growth rate of 4.4 per cent animal population per decade). Assuming that 0.5 h. p. is a available per animal, total power available from work animals = 43.63 million h. p.

Source: Directorate of Economics and Statistics, Indian Agriculture in Brief, Ministry of Agriculture, New Delhi, 1974.

Table 3.2 Norms of Cattle and Buffalo Ownership According to Land-Holdings and Family Size in India

Operational Holdings (Aceas)	*Average Family Size*	*Rural House hold*	*Cattle*		*Buffalo*		*Total*	
			Adult	*Young*	*Adult*	*Young*	*Adult*	*Young*
Up to–0.49	7.23	26054	0.32	0.17	0.11	0.07	0.43	0.24
0.50–0.99	4.59	4355	0.99	0.30	0.25	0.11	1.24	0.41
1.00–2.49	4.77	11326	1.55	0.41	0.36	0.16	1.91	0.57
2.50–4.99	5.27	11717	2,31	0.64	0.49	0.21	2.80	0.85
5.00–7.49	5.85	6537	3.12	0.86	0.69	0.33	3.81	1.19
7.50–9.99	6.13	3483	3.44	1.09	0.82	0.45	4.26	1.54
10.00–12.49	6.54	2492	3.79	1.04	0.04	0.50	4.73	1.54
12.50–14.99	6.70	1438	4.27	1.35	1.05	0.58	5.33	1.93
15.00–19.99	6.91	1832	4.60	1.40	1.21	0.66	5.81	2.06
20.00–24.99	7.40	1088	5.01	1.50	1.23	0.68	6.24	2.18
25.00–29.99	7.34	629	5.36	1.66	1.49	0.80	6.85	2.46
30.00–49.99	7.84	1047	13.24	1.80	2.24	1.04	15.48	2.84
50.00 & above	8.73	468	9.46	2.73	3.60	1.57	12.06	4.30
All Size	5.02	72466	1.84	0.56	0.45	0.26	2.29	0.79

Sources : 1. NSS, National Sample Survey, 17th Round, 1961-62, New Delhi.
2. Parikh, K. S., Energy, Second India Studies, Macmillan Co of India Ltd., Delhi, 1976, p. 66-98.

Note : It is to be seen that households possessing seven or more acres of land have sufficient number of animals to provide needed feedstock for operating their own family size plants.

Table 3.3 Availability of Animal Wastes in Different States of India

Sl. No.	State or Union Territory	Avg. Dung Produced by an Animal per Day (kg)		Animal Dung Produced Annually (million tonnes)		
		Cattle	Buffalo	Cattle	Buffalo	Total
1.	Andhra Pradesh	11.41	10.20	51.410	25.293	76.703
2.	Assam	10.52	13.32	25.195	2.817	28.012
3.	Bihar	8.50	13.91	47.022	18.550	65.572
4.	Gujarat	10.77	9.59	25.722	10.911	36.713
5.	Haryana	17.54	16.95	14.259	11.968	26.227
6.	Himachal Pradesh	18.56	19.44	12.814	2.947	15.762
7.	Jammu & Kashmir	10.51	9.54	6.873	1.491	8.364
8.	Madhya Pradesh	6.40	8.96	57.556	18.338	75.894
9.	Karnataka	11.59	10.09	40.996	10.849	51.815
10.	Kerala	3.62	4.96	3.799	0.852	4.631
11.	Maharashtra	16.78	16.68	90.019	18.518	108.537
12.	Nagaland	8.94	12.32	0.249	0.019	0.268
13.	Orissa	10.61	12.73	41.168	5.897	47.065
14.	Punjab	18.58	14.66	21.437	19.616	41.053
15.	Rajasthan	12.30	12.71	58.911	19.580	78.491
16.	Tamil Nadu	14.37	13.62	56.957	13.543	70.500
17.	Uttar Pradesh	13.20	10.99	175.887	45.819	171.706
18.	West Bengal	9.53	12.15	43.748	4.623	48.371
19.	Delhi	17.38	20.24	0.479	0.764	1.243
20.	Chandigarh	15.32	20.55	0.044	0.124	0.168
21.	Pondicherry	13.09	13.11	0.342	0.051	0.393
22.	Manipur	9.27	12.13	0.976	0.210	1.186
23.	Tripura	8.69	12.92	1.866-	0.234	2.100
24.	Andaman, Nicobar & Laccadive Islands	3.83	4.82	0.070	0.020	0.090
	All India Average	11.3	11.60			
	All India Totals			727.749	233.114	960.883

Source : Neelakantan, S., Assessment of the Availability of Cowdung for Potential Gobar Gas Production in India, Agricultural Situation in India, 30(9), 633-636 (1975).

Table 3.4 Estimates of Biogas Potential in Different States of India

Sl. No.	State or Union Territory	Monthly Mean Temperature (°C)		Biogas Production Rate m³/ton of wet dung	Annual Dung Availability (Million tonnes)	Annual Biogas Potential (Million m³)
		Min.	Max.			
1.	Andhra Pradesh	22.8	33.7	75.0	76.703	5752.8000
2.	Assam	16.8	29.4	64.0	28.012	1792.5120
3.	Bihar	17.0	31.7	68.0	65.572	4458.6240
4.	Gujarat	20.7	32.6	71.0	36.713	2606.2680
5.	Haryana	13.5	34.2	66.3	26.227	1739.1816

6.	Himachal Pradesh	10.7	28.7	52.3	15.762	824.6664
7.	Jammu & Kashmir	3.2	21.0	23.2	8.364	194.0448
8.	Madhya Pradesh	17.7	33.3	67.2	75.894	5100.4800
9.	Karnataka	21.9	29.4	69.3	51.815	3590.8488
10.	Kerala	26.6	28.8	74.0	4.631	342.7680
11.	Maharashtra	21.2	30.8	68.7	108.537	7456.6980
12.	Nagaland	16.8	29.4	64.0	0.628	16.8960
13.	Orissa	20.6	31.3	72.8	47.065	3426.2592
14.	Punjab	13.5	34.2	66.3	41.053	2721.7476
15.	Rajasthan	15.8	33.7	69.6	78.491	5463.0432
16.	Tamil Nadu	24.7	31.6	77.3	70.500	5449.6500
17.	Uttar Pradesh	15.1	32.9	61.1	171.706	10491.3588
18.	West Bengal	18.1	29.7	66.3	48.371	3337.6680
19.	Delhi	13.5	34.2	66.3	1.243	82.7484
20.	Chandigarh	13.5	34.2	66.3	0.168	11.1384
21.	Pondicherry	24.7	31.6	77.3	0.393	30.6108
22.	Manipur	16.8	29.4	64.0	1.186	76.0320
23.	Tripura	16.8	29.4	64.0	2.100	134.4000
24.	Andaman, Nicobar & Laccadive Islands	28.9	32.3	82.7	0.090	7.4688

Source : Neelakantan, S., Assessment of the Availability of Cowdung for Potential Gobar Gas Production in India, Agricultural Situation in India, 30(9), 633-636 (1975)

Whereas S. Neelakantan [18] has placed the estimates of dung availability at 960.863 million tonnes, estimates made by others vary. This is bound to differ as animal population keeps changing and different research organisations and individuals adopt different assumptions and norms as the basis. For instance, the latest livestock census indicates that India has 270 million cattle which surely must not have been the basis for the estimates made by S. Neelakantan. However, there is no agreed consensus on the cattle population. One recent survey puts the figure close to 180 million, a striking difference of 90 million [27]. When the estimates about animal populations differ so widely, dung estimates and proportions available for biomethanation shall be accordingly different. As an example of variation in dung estimates, Department of Science and Technology (DST) puts it at 865 million tonnes per annum (MTPA), National Council of Applied Economic Research (NCAER) at 1335 MTPA and Planning Commission at 1200 MTPA. By different researchers dung potential has thus been estimated varying from 700 MTPA to as high as 1400 MTPA. It is to be realised that dung by its vary nature is difficult to measure. Unlike kerosene or cooking gas, it is not a single homogeneous commodity having uniform measurable properties [27]. Nevertheless despite these variations there are standard statical procedures available to estimate dung supply with fair degree of accuracy.

Considering that 21.47 lakh biogas plants constructed up to 1994-95 cover barely over 1.5 per cent of total rural households, there is vast untapped potential for installing many times more plants [30, p. 30]. With an assumed norm that adult bovine generates 10 kg of animal wastes per day 75 per cent of which is collected, 262 million adult bovines generate nearly 2 million tonnes of animal wastes per day. With the norm that 25 kg of animal wastes can produce 1 m^3 of biogas, it is possible to feed some 40 million biogas plants of 2 m^3 capacity daily which are 25 times the plants (16 lakhs) installed up to 1991-92 [30, p. 30]. Biogas produced if full potential is tapped can meet fuel needs of 120 million rural households in India [29, 31].

Estimates for biogas potential in general do not take into account magnitude of animal wastes scattered on grazing lands, roadsides and elsewhere. In this context it is relevant to note that pattern of dung disposal is also influenced by seasonal factors. As reported by S. Neelakantan [18], during winter season 58 per cent of dung dumped in manure pits and 40 per cent is used in the form of dung cakes; and during summer 76 and 32 per cent and during rainy season 85 and 13 per cent, respectively. Nearly 2 per cent of available dung is used in all seasons for other household purposes like periodic floor and wall coating with dung slurry. Statistics showing trends in pattern of dung disposal are shown in Table 3.5.

Table 3.5 Pattern in Animal Wastes Disposal in Different States of India

Sl. No.	*State*	*Dung Produced by Bovine per day (kg)*	*Disposal Pattern*		
			Per cent Dumped in Manure Pits	*Per cent Used as Dung Cakes*	*Other Uses (per cent)*
1.	Andhra Pradesh	3.6	89	7	4
2.	Bihar	3.6	40	59	1
3.	Gujarat	3.7	76	24	Negligible
4.	Haryana	6.8	51	47	2
5.	Himachal Pradesh	5.3	98	2	Negligible
6.	Karnataka	2.9	88	9	3
7.	Kerala	2.6	93	4	3
8.	Madhya Pradesh	2.6	70	29	1
9.	Maharashtra	3.2	86	11	3
10.	Punjab	7.0	52	47	1
11.	Rajasthan	5.7	67	30	3
12.	Tamil Nadu	5.8	17	19	4
13.	Uttar Pradesh	4.7	48	48	4
	All India Average		69	29	2

Source : Neelakantan, S., Assessment of the Availability of Cowdung for Potential Gobar Gas Production in India, Agricultural Situation in India, 30(9), 633-636 (1975).

3.5 Progress of Biomethanation in India

Biomethanation has made phenomenal strides in India starting from sixties onwards as reflected by plethora of plants installed in recent years. Statewise progress of plants installed during seventies and up to the end of eighties is separately summarised in Table 3.6 and 3.7, respectively. Financial gains of biomethanation during sixties and seventies are given in Table 3.7. In 1975 the government launched *'All-India Coordinated Project of Biogas Technology'* under which by the end of the sixth plan, nearly half a million plants were set up. During the sixth plan (1980-85) an outlay of Rs. 50 crores was allocated for the biogas programme and a target of 4.5 lakh family sized biogas plants was set up which was however later curtailed to 3.5 lakh plants. A total of 2,93,277 plants were set up by February, 1985 and the year 1984-85 ended with a shortfall of only 3 per cent [2]. During 1981-82, 1982-83, 1983-84 and 1984-85 targetted and actual installed number of plants were : 35,000 and 25,369; 75,000 and 57,498; 75,000 and 92,590; and 150,000 and 180,430, respectively.

Table 3.6 Statewise Progress of Biogas Programme in India During Seventies

Sl. No.	*Region*	*Cumulative Number of Biogas Plants as on March 31 of Different Years*			
		1972-73	*1973-74*	*1974-75*	*1978-79*
1.	Andhra Pradesh	36	534	816	2406
2.	Assam	17	27	23	72
3.	Bihar	85	100	143	5374
4.	Gujarat	2694	2994	3416	7510
5.	Haryana	175	180	2181	10512
6.	Himachal Pradesh	6	7	7	15
7.	Jammu & Kashmir			2	97
8.	Karnataka	234	524	1054	6842
9.	Kerala	93	108	306	1062
10.	Madhya Pradesh	191	106	417	1551
11.	Maharashtra	1350	1530	2018	7651
12.	Manipur				24
13.	Nagaland				6
14.	Orissa	15	17	51	512
15.	Punjab	91	99	198	5136
16.	Rajasthan	40	45	111	377
17.	Tamil Nadu	198	248	520	4683
18.	Tripura			1	73
19.	Uttar Pradesh	416	470	805	13560
20.	West Bengal	22	22	51	1723
	Union Territories				
21.	Chandigarh				51
22.	Dadra & Nagar Haveli				4
23.	Delhi	3	3	6	45
24.	Goa, Daman & Diu	7	7	7	105
25.	Pondicherry	4	4	6	33
	Total	6002	6975	12139	69424

During the sixth plan period the government set up the National Biogas Development Board (NBDB) with a view to advance grants, subsidies and loans to organisations associated with the implementation of biogas programme. These

Table 3.7 Number of Biogas Plants Installed in Different States and Union Territories (U.T.) of India During the Period 1974-81 to 1989-90

States/Union Territory	*1974-81*	*1981-82*	*1982-83*	*1983-84*	*1984-85*	*1985-86*	*1986-87*	*1987-88*	*1988-89*	*1989-90*	*Total*
Andhra Pradesh	2854	548	3143	7446	18525	19120	19181	9800	8378	3111	92106
Arunachal Pradesh	0	0	0	0	0	0	0	0	24	2	26
Assam	75	21	17	206	681	457	1245	2107	2015	1808	8632
Bihar	9826	2103	6012	5999	5634	9744	9642	8350	6379	4726	68415
Goa	0	27	123	191	189	156	143	110	231	279	1449
Gujarat	9185	1793	4876	6502	9451	13563	14534	11697	14227	16305	102093
Haryana	10277	2	2297	2549	2986	2266	2449	1580	1919	2031	28406
Himachal Pradesh	0	0	250	770	2510	2650	2850	3201	3505	5087	20823
Jammu and Kashmir	0	3	1	78	44	155	132	132	106	59	710
Karnataka	7799	126	3437	6895	9113	8361	11324	9457	9380	7932	73824
Kerala	1587	262	5260	1282	4236	3937	3424	2937	3148	3722	25055
Madhya Pradesh	3625	259	4772	5523	4405	5668	4539	2976	4759	4400	40953
Maharashtra	11933	2883	8029	22882	52546	60344	59553	60170	54191	50104	382635
Manipur	0	0	3	14	4	0	0	16	112	190	339
Meghalaya	0	1	9	9	52	0	0	6	30	60	167
Mizoram	0	4	8	2	0	115	152	90	114	106	591
Nagaland	0	16	6	18	9	3	16	5	51	0	124
Orissa	622	337	1211	1427	2914	5429	4432	6094	9377	17184	49029
Punjab	5612	505	1161	1360	1943	1807	2478	1409	2060	2077	20414
Rajasthan	409	1111	2407	2178	8078	5398	4371	3840	3738	3734	35273
Sikkim	0	10	8	13	0	0	25	55	90	163	364
Tamil Nadu	6565	670	5019	7231	18340	19514	22079	23128	16488	15044	134078
Tripura	0	0	0	0	0	0	0	0	39	75	114

Table 3.7 (*Contd.*)

Uttai Pradesh	27883	11018	15317	15170	27698	28096	31495	19198	19261	13553	208689
West Bengal	2413	162	1316	1425	3007	3268	6515	7184	7200	10398	42888
Andaman & Nicobar	0	0	2	1	0	10	40	19	15	11	98
Chandigarh	0	0	22	24	6	5	5	5	5	5	77
Dadra & Nagar Haveli	470	18	27	27	0	11	15	45	11	10	603
Daman & Diu	0	0	0	0	0	0	0	0	0	0	0
Delhi	51	9	9	35	135	80	112	100	63	35	629
Pondicherry	0	0	87	95	95	65	30	19	35	21	447
Lakshadweep	0	0	0	0	0	0	0	0	0	0	0
Total	101213	21888	60089	89352	172601	190222	200833	173659	166953	162241	1339051

Source :
1. Khandelwal, K.C., Biogas for Better Life, *Changing Villages*. 6(4), 274-283 (1994).
2. Khandelwal K.C. and Mahdi, S.S., *Biogas Technology: A Practical Handbook*, Tata McGraw-Hill Publishing Company Limited, New Delhi, 1986.
3. Ramana, P.V., Biogas Programme in India, *TIDE*, 1(3), 3 (1991).

organisations included KVIC, State Khadi and Village Industries Boards, State Agro-Industries Corporations, Dairy Development Corporations and several voluntary agencies. NBDB also extended technical and managerial support including supply of raw materials. The governing body of the NBDB comprised representatives of the Union Ministries of Agriculture and Rural Reconstruction, Industry, Finance, and Science and Technology, Planning Commission, National Dairy Development Board, KVIC, Tata Energy Research Institute, Rural Electrification Corporation, PRAD (Lucknow), State Agro-Industrial Corporations, voluntary organisations, and select fertiliser producers and educational institutions. It suggested launching of National Project for Biogas Development for providing a concerted thrust to biomethanation. DNES under the Union Ministry of Energy was entrusted to implement the NPBD in September, 1982. DNES introduced several measures for the success of the NPBD which included upward revision in the central subsidy for plant construction, availability of more effective post-installation follow-up services, creation of large cadre of trained manpower in the field, and supply of raw materials [2]. Comparative cumulative progress of plant installation before and after launching of the NPBD is evident from the data summarised in Table 3.7.

Table 3.8 Achievements of Biogas Programme in India During Sixties and Seventies.

Year	*No. of Biogas Plants Added*	*Total Biogas Plants at the Year End*	*Value of Production (Rs. in lakhs)*			*Disbursements (Rs. in lakhs)*		
			Biogas	*Manure*	*Total*	*Grant*	*Loan*	*Total*
1962-63	315	315	1.29	0.86	2.15	0.87	0.87	2.14
1963-64	203	518	2.13	1.42	3.55	1.21	1.79	3.00
1964-65	230	748	3.09	2.06	5.15	0.75	1.10	1.85
1965-66	204	952	3.93	2.62	6.55	2.03	2.96	4.99
1966-67	313	1265	5.35	3.48	8.83	1.58	3.38	4.96
1967-68	436	1701	6.95	4.67	11.62	3.94	5.64	9.58
1968-69	664	2365	9.48	6.65	16.13	5.41	15.45	20.86
1969-70	720	3085	13.17	8.48	21.65	6.13	29.93	36.06
1970-71	811	3896	15.97	10.71	26.68	9.76	29.92	39.68
1971-72	1041	4937	21.25	20.78	42.03	7.27	32.32	39.59
1972-73	1065	6002	26.61	25.26	51.87	9.17	31.20	40.37
1973-74	856	6858	100.11	70.43	170.54	11.82	24.09	33.91
1974-75	6650	13508	197.21	138.70	335.91	36.07	3.18	39.25
1975-76	13476	26964	369.95	301.14	671.09	44.62	1.03	45.65
1976-77	18595	45579	624.89	508.66	1133.55	92.49	3.32	95.81
1977-78	14340	59919	821.50	668.70	1490.20	103.11	1.25	104.36
1978-79	9505	69424	1130.92	774.77	1905.69	153.76	11.13	162.89
1979-80	10689	80113	1247.04	854.28	2101.32	161.36	0.92	162.28
Total	80113					606.73	199.85	805.58

Source : KVIC, Gobar Gas on the March, A Report, Khadi and Village Industries Commission, Mumbai, March, 1979, p. 11. Entries for 1978-79 and 1979-80 were separately provided to the author by Mr. H.R. Srinivasan, the then Director (GGS), KVIC., Mumbai vide a private communication on Nov. 15, 1980.

Table 3. 8A Progress of Biogas Plants in Different States During the Period 1989-90 to 1992-93

State/Union Territory	*As on*	*Additions*			*As on*
	Mar. 13, 1990	*1990-91*	*1991-92*	*1992-93*	*Mar. 31, 1993*
Andhra Pradesh	89327	8578	13777	14269	125951
Arunachal Pradesh	24	28	37		89
Assam	8537	1191	986	810	11544
Bihar	38553	4313	5397	3502	71765
Goa	1448	216	203	157	2024
Gujarat	92908	26537	33086	35551	188082
Haryana	18129	1948	1997	2028	24102
Himachal Pradesh	20822	3668	3510	3401	31401
Jammu and Kashmir	708	185	62	40	995
Karnataka	65968	6586	8600	15763	96917
Kerala	23471	3700	4436	2882	34489
Madhya Pradesh	37332	3637	4135	8035	52139
Maharashtra	370662	50384	52085	28465	500596
Manipur	339	77	116	150	682
Meghalaya	167	52	50	60	329
Mizoram	591	120	78	120	909
Nagaland	124				124
Orissa	48407	13022	8386	12521	82336
Punjab	14802	2393	2197	2144	21536
Rajasthan	34864	3518	4169	3139	45690
Sikkim	364	175	275	214	1028
Tamil Nadu	127096	9983	9184	8112	154375
Tripura	114	50	110	65	339
Uttar Pradesh	180806	17063	12414	10447	22073
West Bengal	40474	8702	7121	5942	62239
A & N Islands	98			10	108
Chandigarh	77	1	4		82
Dadra & Nagar Haveli	133	10	6	5	154
Daman and Diu					
Delhi	578	22	17	4	621
Lakshdweep					
Pondicherry	447	25	18	25	515
Other				30242	30242
Total	1237394	166184	171456	188103	1763133

Source : 1. MNES, Various reports including Annual Report : 1992-93.
2. TERI, Various reports.

During the seventh plan (1985-90) it was planned to set up 15 lakh biogas plants with their yearwise break-up as 2.5, 2.75, 3, 3.25 and 3.5 lakh, respectively. Annual targets were revised following mid-year review. For instance during the first four years of the seventh plan (1985-86 to 1988-89) actual installed and revised planned figures in lakhs (within brackets), respectively were: 1.9 (1.5), 2.0 (1.5), 1.7 (1.2) and 1.7 (1.5). The progress of biogas programme had been particularly noteworthy following launching of the NPBD in 1981. During initial years of the NPBD focus of the programme was limited to 100 odd districts. Prior to the launching of the NPBD with KVIC efforts till 1980-81, little over 1 lakh plants were set up. During the sixth plan (1980-85) the plant tally rose to over 4 lakh by March 31, 1985. Impressive success scored during the sixth plan led to fixing higher targets for the seventh plan (1985-90) and extending the programme to all the districts. By the end of the seventh plan following all-out efforts by the MNES and concerned agencies, a total of 12.37 lakh biogas plants were set up with 85 per cent of them reported as functional [32]. These plants saved as estimated 4.53 million tonnes of fuelwood annually valued at Rs 181 crores besides generating 21.76 million tonnes of enriched fertiliser valued at Rs 180 crore per annum. Under the NPBD, 17.63 lakh plants were installed till 1992-93 end which have saved an estimated 5.5 million tonnes of fuelwood annually besides generating over 25 million tonnes of enriched fertiliser (Table 3.8A). As a whole the NPBD contributes a saving of over Rs. 420 crores per annum in terms of reduced fuelwood consumption and fertiliser substitution. With the growing success of the NPBD targets for 1993-94 were revised to 2 lakh plants from 1.35 lakh originally planned and total eighth plan target was revised to 12.5 lakh plants instead of 10 lakh plants. Revision in targets could be possible due to spectacular success achieved through turn-key approach and NGOs participation. As a result of all-out efforts number of plants installed by 1995-96 end has reached 23.6 lakh mark and the tally is expected to rise phenomenally in future (Fig. 3.1 and 3.2). Twenty three and a half lakh plants are estimated to generate gas equivalent of over 75 lakh tonnes of fuelwood valued at Rs 375 crores per annum, and 345 lakh tonnes of enriched organic manure per annum. Till 1994-95 only 18 per cent potential has been exploited (Table 3.8B).

Table 3.8B Number of Biogas Plants Installed in India under National Project on Biogas Development

Sl. No.	*State/UTs*	*Estimated Potential*	*Plants Installed till March 31, 1995*	*Potential Exploited (per cent) till March 31, 1995*	*Plants Added during 1995-96*	*Plants Set up till March 31, 1996*
1.	Andhra Pradesh	1065600	173388	16.2	20632	194020
2.	Arunachal Prad.	7500	137	1.8	15	152
3.	Assam	307700	13690	4.5	365	14055
4.	Bihar	939900	81047	8.6	892	81939
5.	Goa	8000	58327	29.6	35	58362
6.	Gujarat	554000	248381	46.6	19038	267419
7.	Haryana	300000	27335	9.5	1773	28108
8.	Himachal Prad.	125600	35647	28.4	1231	36878
9.	Jammu & Kashmir	128500	1116	0.9	20	1136
10.	Karnataka	680000	121002	23.6	16721	137723
11.	Kerala	150500	41433	28.0	1250	42683
12.	Madhya Pradesh	1491200	89022	6.0	18220	107244
13	Maharashtra	897000	556693	62.1	12362	569055
14.	Manipur	38700	877	2.3	142	1019
15.	Meghalaya	24000	379	1.6	—	379
16.	Mizoram	2200	1109	50.4	82	1191

17.	Nagaland	6700	294	4.4	182	476
18.	Orissa	605500	106774	17.6	12024	118798
19.	Punjab	411600	27992	6.8	3464	31456
20.	Rajasthan	915300	55096	6.0	4047	59143
21.	Sikkim	7300	1382	19.0	224	1606
22.	Tamil Nadu	615800	176271	28.6	4029	180300
23.	Tripura	28500	476	1.7	51	527
24.	Uttar Pradesh	2021000	251276	12.7	10616	261892
25.	West Bengal	695000	81151	11.7	6620	87771
26.	A & N	2200	113	5.1	5	118
27.	Chandigarh	1400	87	6.2	5	92
28.	D & N Haveli	2000	159	4.8	3	162
29.	Daman & Diu	100	—	—	—	—
30.	Delhi	12900	654	5.1	3	657
31.	Pondicherry	4300	530	12.3	—	530
	Total	12050000	2147386	18.1	171305	2358691

Source : MNES, Annual Reports 1994-95 and 1995-96, New Delhi.

3.6 Advent of Community Plants

During the sixth plan the government gave a new direction to biogas programme when it encouraged setting-up community plants. This resulted in transforming the biogas programme from a mere household affair to a commercially viable national endeavour. During 1981-82, the government initiated action for setting up 21 community plants as demonstration-cum-experimental units in select parts of the country. These community plants were proposed at Veeran Juniyapuram and Karimnagar in Andhra Pradesh, Barakur in Udipi Taluk in Karnataka, Gandhigram in Tamil Nadu, Shapura in Madhya Pradesh, Devgarh in Rajasthan, Dhaniv and Walhe in Maharashtra, Hissar in Haryana, Masudpur in the Union Territory of Delhi, Hooghli in West Bengal; Gujaria, Kashipur, Jainpur, Pantnagar and Doiwala in Uttar Pradesh; Patna in Bihar, Janagaria in Himachal Pradesh, Anand in Gujarat, Ludhiana in Punjab and an unnamed site in Kerala. In view of financial difficulties faced by farmers in owning family-size plants, considerable outlays were earmarked for community and institutional biogas plants during the seventh and eighth plans which could meet the need of 25 families or so at a time. During the seventh plan (1985-90), while setting an ambitious target of 15 lakh family size plants, it was also proposed to set up 1200 community or institutional biogas plants though, in reality, achievements were much lower [23].

Rural Electrification Corporation set up first ever community plant in the country at Kodumunja village in Andhra Pradesh in 1977 which was followed by the one at *Fateh-Singh-Ka-Purwa* at Etah (U.P.) which was set up by the PRAD. Installation of community plants received significant fillip in the country when the DNES launched '*All India Coordinated Project for Demonstration-cum-Experimental Community/Institutional Biogas Plants*' with a target of 100 and 450 plants, respectively during the sixth and seventh plans [25]. Initially,

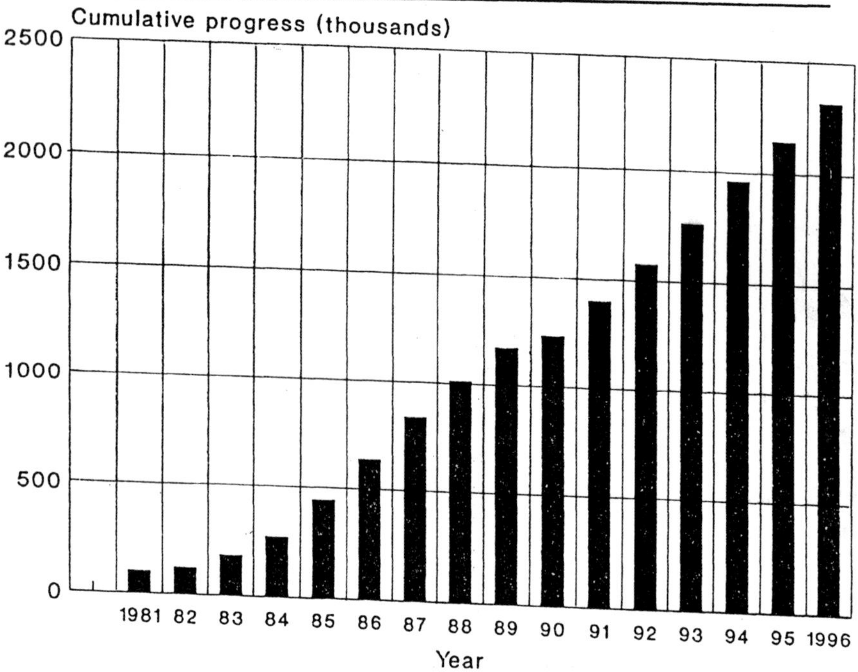

Fig. 3.1 Cumulative progress of plants installed during 1980-81 to 1995-96

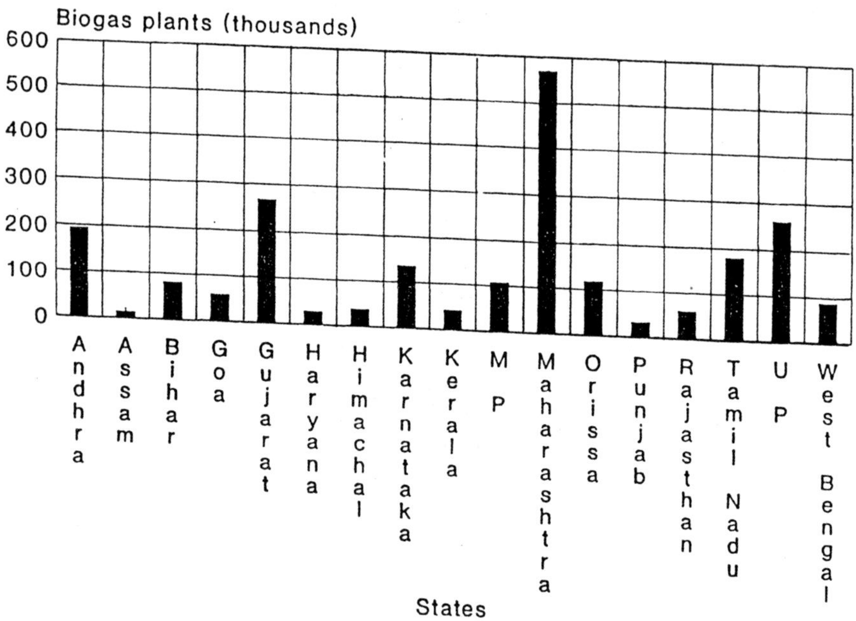

Fig. 3.2 Statewise distribution of biogas plants as on March 31, 1996.

erstwhile DNES fixed 100 per cent subsidy for community and 75 per cent for institutional plants which was later on reduced to 90 and 70 per cent during 1985-86. Number of community, instructional and night-soil based plants has increased from 264 in 1986-87 to 585 in 1989-90 and 1009 in 1992-93 (Table 3.9 and 3.10). It is to be pointed out that though pace of setting up community and institutional plants has not been as fast as family type plants, nevertheless the concept has caught the imagination and may surge forward in coming years.

Table 3.9 Number of Community and Institutional Biogas Plants Installed in India upto Dec. 31, 1988.

States	*Cum. Plants as on March 31, 1987*		*Plants Installed Between Apr. 2, 1987 and Dec. 31, 1988*		*Cum. Plants As on Dec. 31, 1988*	
	CBP	IBP	CBP	IBP	CBP	IBP
Andhra Pradesh	21	24	1	4	22	28
Assam	—	1	—	—	—	1
Bihar	6	—	—	—	6	—
Delhi	1	—	—	—	1	—
Gujarat	16	6	1	5	17	11
Haryana	2	4	1	1	3	5
Himachal Pradesh	—	1	—	—	—	1
Karnataka	21	24	1	4	22	28
Kerala	—	2	—	—	—	2
Madhya Pradesh	17	17	—	2	17	19
Maharashtra	2	26	8	—	10	26
Orissa	4	3	—	1	4	4
Pondicherry	—	1	—	—	—	1
Punjab	28	4	13	3	41	17
Rajasthan	5	17	—	—	5	17
Tamil Nadu	2	2	—	6	2	8
Uttar Pradesh	32	16	—	8	32	24
West Bengal	2	—	—	—	2	—

Source : MINES & TERI, various reports.
Note : CBP: Community Biogas Plant
IBP: Institutional Biogas Plant.

Table 3.10 Statewise Distribution of Community Plants as on March 31, 1990.

State	*Community Plants*	*Institutional Plants*	*Night-Soil Based Plants*	*Total*
Andhra Pradesh	21	42	3	66
Gujarat	23	31	—	54
Madhya Pradesh	41	22	—	63
Maharashtra	36	29	4	69
Punjab	69	11	—	80
Rajasthan	5	29	—	34
Tamil Nadu	4	27	—	31
Uttar Pradesh	35	28	—	63
Others	20	24	—	44
Total	254	233	7	494

Source: MINES & TERI, various reports.

3.7 Functional Status of Biogas Plants

One of the responsibilities of the MNES is to organise evaluation of performance of biogas plants. For evaluating the functional status of biogas plant during 1985-90 it involved the National Council of Applied Economic Research (NCAER). The NCAER evaluated the status with the help of two separate surveys undertaken in 1985-86 and 1987-88. The first survey covered the study of 10,883 plants in 12 states while the second involved 19,842 plants in 23 states (Table 3.11 and Fig. 3.3). Based on these two surveys about 88 per cent plants were reported functional in 1985-86 and 84 per cent in 1987-88. The two surveys nearly covered 3600 villages in 251 districts. It was also found that fixed dome *Deenbandhu* models were very popular and were being preferred over other designs. The survey also revealed that nearly two-third of all plants were owned by small and marginal farmers many of them belonging to the scheduled castes and scheduled tribes communities. An important finding was that marginal farmers owned 38 per cent of all the plants [32, 33].

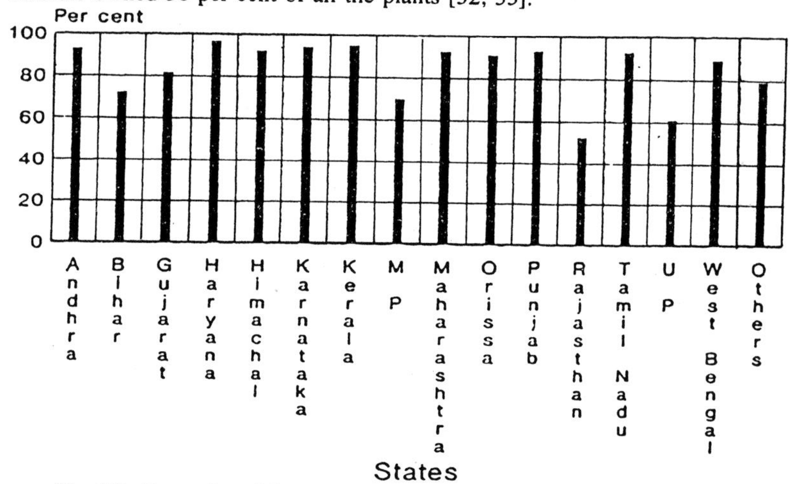

Fig. 3.3 : Proportion of functional plants in different states during 1887-88.

M. S. Pathania and K. N. S. Sharma [34] carried out a study to analyse the reasons for non-functioning of biogas plants. A sample of 50 non-functional plant owners comprising 29 *Janata* type and 21 KVIC type plants was selected on random basis. Constraints categorised as economic, sociological, operational and health, labour and input-related were identified through a structured questionnaire which were largely similar for both KVIC and *Janata* designs except two specific reasons in each case. In case of Janata plants, 86 per cent reported 'cracks in walls' as the main cause for non-functioning and 83 per cent attributed it to the 'leakage in domes'. As for KVIC type plants, 90.5 per cent identified 'corrosion in tank' as the main constraint whereas for 71 per cent 'destruction of dam' was the main problem. Some major reasons for non-functioning of both Janata and KVIC designs were accumulation of water in gas outlet, gas leakage from pipeline, problems of waste-collection and effluent disposal, use of sub-standard material in plant construction, and low gas yield during winters. Major economic problems identified were non-availability of loan for repair and maintenance, high installation costs and lack of any direct income. Sociological problems included hesitation in mixing cow-dung, emptying of slurry-tank by family members and psychological dislike for gas coming from animal wastes. Some non-functioning plant owners had a feeling that biogas plant spreads unpleasant odour and increases number of flies and mosquitoes. Input-related constraints covered 'non-availability of spares' and 'scarcity of water in summer' as one of the reasons for non-functioning. Lack of availability of skilled or cheap labour also contributed to the non-functioning of plants. According to some owners of non-functioning plants, biogas-cooked food is not good for health and gives bad smell of gas. None of the respondents however reported any specific health problems from biogas. For removing their misplaced suggestions, plant owners need be trained. Extension workers should visit beneficiaries personally to dispel misconceptions and allay their doubts.

Table 3.11 Functional Status of Biogas Plants (In per cent) in Select States

States	*Functional Plants*		*Non-Functional Plants Due to Operational Problems*		*Non-Functional Plants Due to Structural Defects*	
	1985-86	*1987-88*	*1985-86*	*1987-88*	*1985-86*	*1987-88*
Andhra Pradesh	93.6	92.9	4.6	5.3	1.8	1.8
Bihar	82.1	72.4	2.6	23.8	15.3	3.8
Gujarat	NA	81.7	NA	11.2	NA	7.1
Haryana	79.4	96.7	13.0	1.7	0	1.6
Himachal Pradesh	100.0	92.2	0	1.7	0	1.6
Karnataka	86.1	94.1	6.6	4.0	7.3	1.9
Kerala	95.7	95.4	3.3	3.4	1.0	1.2
Madhya Pradesh	64.3	70.1	4.2	24.9	31.5	5.0
Orissa	93.9	91.1	2.2	5.3	3.9	3.6
Punjab	96.6	92.9	1.7	5.2	1.7	1.9
Rajasthan	NA	52.1	NA	37.9	NA	10.0
Tamil Nadu	88.1	92.8	6.5	4.9	5.4	2.3
Uttar Pradesh	77.0	60.6	5.2	27.6	17.8	11.8
West Bengal	NA	89.2	NA	5.6	NA	5.2
Others	NA	78.8	NA	20.0	NA	1.2

Source : Ramana, P.V., Biogas Programme in India, TIDE, 1991, p. 4 (based on a DNES Study).

3.8 Concluding Remarks

Progress of biomethanation has been fairly impressive in India and the country is only next to China in this regard. Government lent adequate support from time to time by launching schemes like *All-India Coordinated Project for Biogas Technology* (1976) and the *National Project for Biogas Development* (1982). It also provided substantial subsidy in past for setting up individual family-size and community biogas plants. Biomethanation has made rapid strides in India and is poised for further progress. Economies-of-scale is the only factor which comes in its way and that may also vanish with growing advances in technology and continuous R & D efforts. The energy-starved age in which we live today demands that renewable sources of energy like organic wastes should be fully exploited for biomethanation and other forms of energy. Biogas which is obtained from these wastes provides a clean cooking and lighting fuel and improves hygienic conditions. Biogas utilisation is one of the fields in which India can proudly claim to be ahead of the developed world. During 1980 Commonwealth Science Council (CSC) in recognition of the useful role played by the KVIC in promotion of biogas programme in India, appointed it as the coordinator for biogas programmes in Commonwealth countries of Asia and the Pacific Region as a whole [16]. India's endeavours in the field of biomethanation were also appreciated by the Second World Congress on Energy and Environment held in New Delhi during Nov. 7-9, 1986. Apart from coordinating biogas programmes of the Commonwealth countries, India also shared experiences with other nations such as Philippines, Thailand, Indonesia, Brazil, Iran, Iraq, Srilanka, Nepal, Bangladesh, Tanzania, *Botswana*, Somalia and Uganda. Exchange of experiences in past has helped a great deal to solve common problems and serve as a continuing basis for international cooperation and partnership.

REFERENCES

1. Directorate of Economics and Statistics, *Indian Agriculture in Brief*, Ministry of Agriculture, New Delhi, 1974.
2. DNES, *A Personal Communication*, Department of Non-Conventional Energy Sources, New Delhi, Feb. 16, 1989.
3. *Economic Times*, Energy Minister Reiterates Commitment to Alternative Energy Sources, New Delhi, Jan. 13, 1990, p. 6.
4. *Economic Times*, Economic Survey : 1990-91, New Delhi, July 21, 1991, p. 14.
5. ESCAP, *Report of the Workshop on Biogas Technology and Utilisation*, Economic and Social Commission for Asia and the Pacific, New Delhi, July, 1975.
6. ESCAP, *Report of the Manila Workshop on Biogas Technology and Utilisation*, Economic and Social Commission for Asia and the Pacific, March 24-April 2, 1976.
7. *Financial Express*, Wider Use of Solar Heating Systems Favoured, Madras, Aug. 11, 1991, p. 13.
8. Gandhi, I., *'Renewable Energy in Action'*, A Report, Commission for Additional Sources of Energy (CASE), New Delhi, 1981.
9. Gandhi, I., Key-note Address at the UN Conference on New and Renewable Sources of Energy, Nairobi, Kenya, Aug. 10-20, 1981, *Economic Times*, Aug. 11, 1981.
10. Gandhi, I., Inaugural Address at the World Energy Conference, New Delhi, Sept. 18-26, 1983, *Economic Times*, Sept. 19, 1983.
11. Golueke, C. G., and P. H. Mc Gauhey, Waste Materials, in *'Annual Review of Energy*, Vol. 1, Palo Alto, California, 1976', p. 157-277.
12. *Hindustan Times*, Biogas Plant Target Exceeded, New Delhi, 1990.
13. Khandelwal, K. C., Biogas for Better Life, *Changing Villages*, **6(4)**, 274-283. (1984)
14. Khandelwal, K. C. and S. S. Mahdi, *Biogas Technology: A Practical Handbook*, Tata Mc Graw-Hill Publishing Co. Ltd., New Delhi, 1986.
15. KVIC, *Gobar Gas on the March*, A Report, Mumbai, March, 1979, p. 11.

16. Moulik, T. K, India's Biogas Programme : Last 10 years Experience, *Economic Times*, Aug. 20, 1986, p. 5.
17. Myles, R. and A, Dhussa, Comparison of Performances of Janata and Deenbandhu Biogas Plants, *paper presented at the National Conventions on Solar Energy*, Dec. 1-4, 1987.
18. Neelkantan, S, Assessment of the Availability of Cowdung for Potential Gobar Gas Production in India, *Agricultural Situation in India*, **30(9)**, 633-636 (1975).
19. NSS, *National Sample Survey*, 17th Round, 1961-62, New Delhi.
20. Parikh, K. S., Energy, *Second India Studies*, Macmillan Co. of India Ltd, Delhi, 1976, p. 66-98.
21. Soubolle, B. R., *Fuel Gas From Cowdung*, Bulletin published by *Sahayogi Prakashan*, Tripureshwar, Kathmandu, Nepal, 1976.
22. Swaminathan, R. V., Development of Alternative Energy Sources, *Khadi Gramodyog*, **27(3)**, Dec., 147-148 (1980).
23. TERI, *Energy Data Directory and Yearbook*, New Delhi, 1988, p. 239.
24. *Times of India*, A News Item, June 23, 1986.
25. Ramana, P. V., Biogas Programme in India, *TIDE*, **1(3)**, 1991, p. 4.
26. BORDA, In Memory of Shri Jashbhai J. Patel, *Biogas Forum*, 26 (1993).
27. Sharma, S., Policy of Dung Needed, *The Pioneer*, New Delhi, June 23, 1994.
28. Sasse, L., Brief Report on International Conference on Biogas Technology and Implementation Strategies, Pune, *Biogas Forum*, **I** (No **4**), 18-19 (1990).
29. TERI, *Managing the Biogas Programme (Sponsored by the MNES)*, New Delhi, 1994.
30. TERI, *Biogas Technology (Sponsored by the MNES)*, New Delhi, 1994.
31. TERI, *Biogas : A Source of Rural Employment (Sponsored by the MNES)*, New Delhi, 1994.
32. TERI, *TEDDY : 1995-96*, New Delhi, 1996.
33. Venkataramana, P., National Project on Biogas Development, *Urja Bharati*, **3(3)**, 33-35 (1993).
34. Pathania, M. S. and Sharma, K. N. S., An Analysis of Constraints for Non-Functional Biogas Plants in Haryana, *Khadi Gramodyog*, **41(4)**, 175-180 (1995).

Chapter 4

IMPLEMENTATION STRATEGIES AND ECONOMIC ASPECTS OF BIOMETHANATION

4. 1 Introduction

With the introduction of biogas programme in sixties a number of implementation strategies have been tried in India. The programme has been implemented by various agencies under different ministries using varied approaches. The programme in its present from as the 'National Programme of Biogas Development' is being coordinated by the MNES. Turnkey approach of programme implementation is by far the most successful strategy developed so far which has paid rich dividends. This approach is largely responsible for spurt in number of plants installed in recent years. MNES is supported in its task by numerous state and district level agencies including autonomous organisations and institutions. In wake of globalisation and new economic policy, MNES in recent years has encouraged private entrepreneurs and NGOs role in promoting renewable energy technologies. For implementing new economic policies effectively, MNES has evolved innovative strategies and action plans, restructured different programme divisions based on technologies and user applications, brought related programmes from other ministries to its fold, prepared plans for phased withdrawal of subsidies for certain key technologies with corresponding replacement by soft-loan option. MNES also sponsors renewable energy research projects at universities, research centres and NGOs which have well-equipped research facilities.

As for biogas plant economics, construction and operating costs of a plant depend on its size. The bigger the size, the lower are costs per unit of gas yield. In India plants usually cost more than what a low income group beneficiary can afford. This is for this reason the government has been providing subsidy which under new economic policy is however being phased put. Movable drum type plants on account of steel parts cost more than fixed dome type plants. In a floating drum type plant, gasholder alone accounts for half of the total cost. For keeping the costs of fixed type plants further low, it is advisable to use broken stones, self-made bricks, hydrated lime and locally available binders, sand and gravel. Overall construction and operation costs on per household basis are lower for a community plant than a family unit. If gas from a community plant is used in a centralised community kitchen without connecting households through pipelines, set-up costs can be kept low. Operating costs of community plants are more on account of added supervision and control. Costs and resulting profitability of a plant is affected by its average service life. Service life of a plant ranges between 25 to 40 years. Fixed dome plants have longer life than floating drum type plants. In a movable drum type plant, digester lasts for about 30 years, gasholder 20 years, and pipes and accessories for about 20 years. Operating costs per cubic metre of biogas is more for a floating drum type than a fixed dome unit. Among fixed dome models, *Deenbandhu* model is by far the cheapest. Biogas yield per unit digester volume is more for a movable drum type than fixed dome type plant. As for the relative fertiliser yield, no significant difference is discernible from the two types of plants.

4.2 Organisation for Implementing NPBD

NPBD is based on multi-model collective approach that involves all concerned state and district level agencies in programme implementation. For effective implementation at village level, MNES has assigned responsibilities with appropriate controls at national, state and district levels. KVIC which prior to takeover by the MNES was the principal agency for implementation, continues to discharge important responsibilities. At state level under the overall supervision of the MNES, NPBD is coordinated by a nodal agency such as Department of Agriculture or Department of Rural Development whereas actual execution is coordinated by implementing agencies such as state renewable energy agency or state agro-industries corporation. In addition, the programme is also implemented by autonomous organisations like KVIC, NDDB (National Dairy Development Board) or AIWC (All India Women's Conference). At the district level, the programme is implemented by either DRDA (District Rural Development Agency), BDO (Block Development Office), turnkey entrepreneurs, NGOs, *gram panchayats*, dairy cooperatives, etc. National level organisations such as the KVIC and NDDB implement the NPBD through network of their branches and cooperatives for which separate targets are set. Under the NPBD a state can have more than one implementing agencies—GAIC (Gujarat Agro-Industries Corporation), KVIC, NDDB, AKRSP (Aga Khan Rural Support Programme) [35, p. 14]. For monitoring and guiding state nodal agencies MNES has set up its offices in eight states located at Chandigarh, Bhopal, Guwahati, Hyderabad, Lucknow, Madras, Ahmedabad and Bhubaneswar. For providing training and technical support to plant builders, seventeen Regional Biogas Development and Training Centres (RBDTCs) have been set in different states (See appendix IV) which also assist regional MNES staff in their activities [35, 19].

4.3 Turnkey Approach for Implementing Biogas Programme

Turnkey approach has largely contributed towards phenomenal success of the NPBD. Turnkey workers (TKWs) and NGOs located in different states are capable of installing several thousand plants a year. TKWs virtually work from concept to commissioning, from motivating the beneficiary to constructing a plant with requisite quality and ensuring trouble-free plant operation for a stipulated period of three years. In fact the turnkey approach has paved the way for privatisation of plant construction. TKWs work at the village level and have interface both with the beneficiaries and the implementing agency [27]. Most TKWs generally build plants in cluster of villages which facilitates material transportation and movement. Materials are often purchased in bulk which saves cost and maintains quality [35, p. 27].

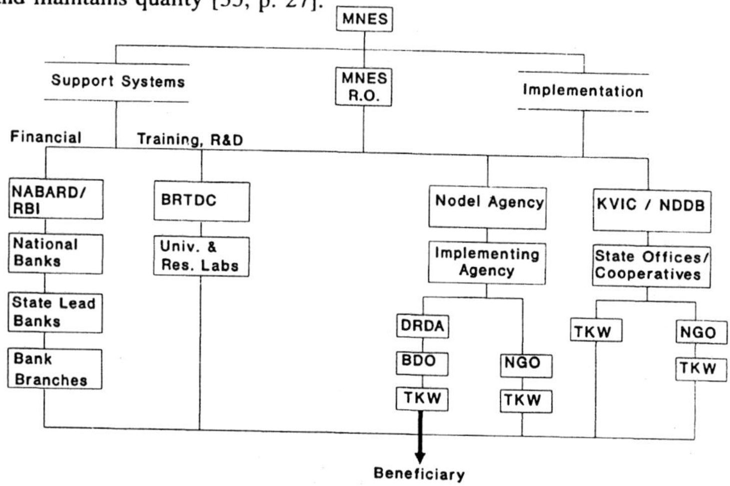

Fig. 4.1 Organisation for NPBD

Source: TERI, Biogas: A Source of Rural Employment, New Delhi, p. 15.

A typical turnkey worker, though no qualifications are prescribed, could be a mason involved in construction work in a cluster of villages, or a manufacturer of accessories for stoves and agricultural equipments. Unemployed or underemployed individuals in rural areas are also eligible to become TKWs. For enrollment as TKW or turnkey agency, applications need be addressed to district agencies such as district rural development officer or block development officer [35, p. 27].

Following enrollment, TKWs are assigned targets for building plants. Size of the target is determined on the basis of overall target for the area to which TKW belongs and the number of TKWs operating in that areas. Training of TKWs and masons who build the plant is a very important activity in the whole programme. TKWs and masons are trained at the RBDTC for which MNES sanctions Rs. 30,000 and Rs. 9,000 per training programme, respectively.

TKW's work begins with collecting some basic information from beneficiary such as number of cattle possessed, available space, income level, numbers of family members, etc. In case loan is needed, TKW helps the beneficiary in filling in the concerned forms and carrying them to the local lead bank for sanction. Once the loan is obtained construction is started. Construction work is carried out by trained masons under the supervision of TKW who himself may or may not be having masonry experience [35, p. 26].

Following work completion, a TKW is paid a fee of Rs. 500 per plant commissioned for which he has to offer a three year warranty for trouble-free operation. A limited payment is withheld which is released only after the warranty period is over. During three year warranty period a TKW is required to visit plant owner for routine inspection and repair a defect at his own cost if the beneficiary is not responsible for it. A TKW is entitled to a fee of Rs. 150 if he connects the inlet of the plant to an household lavatory as an incentive to promote night-soil based plants [35, p. 27]. Turnkey approach for plant construction has also helped in creating employment opportunities for rural poor either on full or part time basis. According to MNES estimates, NPBD had generated by 1993-94 over 80 million mandays of employment for rural workforce [37].

Several autonomous organisations and NGOs are also implementing the NPBD on turnkey basis. In the first category, a prominent organisation is KVIC which till the launching of the NPBD used to be the principal coordinating agency for programme implementation. However, even with the framework of the NPBD it continues to be a national level implementing agency for meeting plan targets assigned by the MNES with the only difference that it is one of the several organisations implementing the biogas programme. KVIC installs plants through turnkey approach by using the countrywide network of its offices with h adquarters at Mumbai. KVIC's turnkey approach is business-like characterised by high degree of autonomy to turnkey workers. KVIC is highly flexible in its approach considering that it also implements the NPBD through a network of turnkey workers who may not be its employees. TKWs also perform the role of motivator and facilitator or sometimes they engage rural youth at a small fee on per plant basis. KVIC head office in Mumbai splits the targets for different states in consultation with state units in keeping with their past performance [33, p. 28-31].

For plant maintenance the KVIC has introduced '*Post-Installation Routine Maintenance* (PIRM)' which needs particular elaboration. PIRM covers all the plants for a period of three years following the two year warranty period. This implies that each plant has a maintenance cover for a period of five years. It was introduced as a pilot scheme in 1989-90 in select districts of Maharashtra and Gujarat which have a high concentration of biogas plants [27]. The scheme is operated by a cadre of Trained Mechanics (TMs) who are taken from the same block or district where the scheme is under implementation or at best from nearby block or district. No particular qualifications are prescribed for the TM but before assigning a job they are imparted one to two weeks training depending upon the background. Each TM visits allotted plants in their area at least three times in a year with first visit during April-July, second during August-November and third during December-March period. Each beneficiary is provided a book to record the work done by him [27]. Payment for the repair done is made after checking from the certificate issued by the plant owner. Each TM is provided a tool-kit for repairing plants. [For more details, reader may contact KVIC, 3 Irla Road, Vile Parle (West), Mumbai 400056].

Several NGOs are also acting as turnkey organisations which include AFPRO in Delhi, Aga Khan Rural Support Programme (AKRSP) in Gujarat,

MYRADA in Karnataka, *Gram Siri* in Andhra Pradesh, and *Gram Vikas* in Orissa. These organisations act both as implementing and turnkey organisations. They receive the turnkey fee from the MNES which through their organisational channels is passed on to their supervisors and masons for plant construction. MNES follows multi-model and multi-agency approach in implementing the NPBD. In this context it will be appropriate to describe the implementation mechanism of the AFPRO and the GAIC, Ahmedabad [33, p. 21], as pertinent examples.

AFPRO, a national non-governmental voluntary organisation (NGVO) has played a pioneering role in implementing the NPBD through NGVO network. AFPRO is responsible for developing the *Deenbandhu* model, under the supervision of R. M. Myles, J. B. Singh, A. Dhussa and other distinguished professionals, which is so far the most successful and popular design. The AFPRO network is instrumental in setting up large number of fixed dome *Janata* and *Deenbandhu* biogas plants. By September, 1993, the network had installed more than 75000 biogas plants in different regions. To promote NPBD it has 65 NGVOs as Programme Implementing Partners (PIPs) who control over 90 extension centres in seventeen states. Many of these centres are located in remote, tribal, underdeveloped and hilly areas and provide both social and technical services in promoting biogas technology.

GAIC is one of the four implementing agencies in Gujarat other being KVIC, NDDB and AKRSP. GAIC has its own distribution and marketing network created to supply tractors, spare parts, fertiliser and farm implements. GAIC started installing biogas plants in 1976 much before the advent of the NPBD. After the NPBD was introduced, GAIC created an energy division with field level arrangements for programme implementation. GAICs activities that interface with the NPBD are as follows [33, p. 23] :

* Procurement, storage and distribution of construction material and other inputs.
* Collection and disbursement of subsidy among beneficiaries.
* Training of motivators, supervisors, masons and prospective owners.
* Arranging institutional finance for the beneficiary.
* Awareness campaign and promotion of biogas programme.
* Post-installation repair and maintenance of biogas plants.
* Coordination with programme implementing agencies at state level.

GAIC prefers cluster approach by constructing minimum six plants at a time in one village before moving to the next. At the block/village level, GAIC trains educated rural youth as '*Self Employed Biogas Supervisors* (SEBGS)' who supervise plant construction. After the motivator identifies a prospective beneficiary, he is introduced to the SEBGS. The SEBGS then fills up a form and sends it to the state nodal agency for approval. Following approval, site and size selection is made by the SEBGs in consultation with the beneficiary and the Agro Service Centre (ASC). Plant is constructed by a trained mason whereas the construction material is supplied by the local ASC which is adjusted against the subsidy drawn. SEBGS is paid a fee of Rs. 300 for every plant constructed. He is held responsible for trouble-free operation of plant for a period of three years. Following plant completion, ASC staff physically verifies and issues a completion certificate based on which balance subsidy amount is released. SEBGS visits a plant minimum once in six months during the three year warranty period to ensure trouble-free operation. A restriction is imposed on SEBGS that he cannot build more than 200 plants in a year for avoiding decline in quality [33, p. 22-28].

GAIC provides every beneficiary a booklet giving operation and maintenance instructions. He is also given two postcards addressed to the concerned SEBGS to contact him in the event of any complaint. Every GAIC constructed plant has a specific code engraved on it indicating the construction year, serial number of SEBGS and serial number of the subsidy register maintained by the ASC. This number on the postcard helps to locate the plant by the SEBGS after receiving the complaint. In many areas repair crew also visits plants on motor cycle for attending repair on priority. GAIC receives Rs. 175 as service charge for every plant constructed. The unemployed rural youth who acts as facilitator receives Rs. 50 as motivation fee [33, p. 27].

MNES has evolved a detailed feedback mechanism for continuous monitoring of the NPBD at field level. Under this mechanism, block level functionaries carry out physical inspection of plants set up under their control, district level functionaries 5-10 per cent of the plants, and state level functionaries 1-5 per cent of the plants [35, p. 20]. MNES also commissiones special plant performance evaluation projects to specialised agencies from time to time. Study awarded to the NCAER to evaluate functional status of biogas plants during the period 1985-90 is the case-in-point. Studies involving socio-economic evaluation of plants were in past assigned to the Indian Institute of Management. Ahmedabad and the Operations Research Group, Vadodra, MNES regional offices at Ahmedabad, Bhopal, Chandigarh, Delhi, Hyderabad, Jaipur, Lucknow, Nagpur and Chennai also monitor the progress of the NPBD under their respective jurisdictions, conduct technical inspections and monitor training and turnkey activities at large [34, p. 23].

Several agencies are engaged in implementing biogas programme in India including the AKRSP which launched its programme in 1986 in wake of Junagadh's ecological crisis. *Gir National Forest* was continually deforested by villages in search of fuelwood, and Junagadh's once productive soil was becoming increasingly saline following excessive use of fertilisers. Within this backdrop biogas was viewed as attractive alternative and complimentary to the AKRSP's wasteland development programme. AKRSP first started with KVIC's family-sized floating drum model followed by *Janata* model and finally *Deenbandhu* model [38].

AKRSP purchases materials directly from market instead from the GAIC. Material is first transported to villages from where it is distributed to individual beneficiaries. As the scale of the programme grew, AKRSP followed GAIC's example and set up block-level material distribution centres to enable beneficiaries to collect material at their own cost. AKRSP inspires and motivates villages to set up plants and also involves some key villagers as extension volunteers (EVs). They assist in material distribution, supervision and construction of plants and paid an incentive fee of Rs. 50 per plant. In Junagadh when the material distribution was shifted to blcok-level centres, role of local EV was accordingly diluted. With more villages to cover, an EV who was named as Master Extension Volunteer (MEV) operated in more than one village. An MEV receives an incentive fee of Rs. 150 per plant and is entrusted with the task of motivation, allocation of masons, material distribution at the cluster level and supervision. He also coordinates the activities of village EVs [38].

AKRSP also organises training of masons. Training of womenfolk is also an important activity of the AKRSP. Video-cassette in *Gujarati* on biogas usage and plant maintenance are shown to the women as part of training activity. In some cases awards are also given to owners for best plant maintenance and up keep [38].

4.4 Role of Cooperatives in Implementing Biogas Programme

Biogas plants in a region can also be set up by forming a cooperative society. It is to be noted that cooperative law aims at incorporating the principles of cooperation that distinguish it from industrial law. It is characterised by open and voluntary membership, democratic control, savings belonging to members, limited interest on capital, cooperative education and cooperation among cooperatives. A Mazumdar and N. K. Gopalkrishnan [28] initially proposed the idea of forming cooperatives for implementing biogas programme. Cooperatives can be formed at village, block and district levels and a national level organisation can coordinate and monitor their activities. One of the eligibility criterion for membership is that those keen live within the area of operation of society and are competent to contract. The responsibility for audit of cooperative institutions vests with the registrar of cooperative societies while in industrial organisations members appoint auditors on recommendations of Board of Directors in the annual general meeting. A cooperative functions like any other organisation except that its by-laws differ. Members of a biogas cooperative should know how to prepare feasibility reports, arrange loan from banks and subsidy from government agencies, and possess cadre of trained mechanics for setting-up plants and day-to-day maintenance and operation. It should also have facilities for training and development of individuals engaged in construction, operation and maintenance of biogas plants. A cooperative has an added advantage of being eligible to receive financial assistance from cooperative financial institutions. In the field of training and development, a cooperative institution can receive aid from the State Cooperative Union, District Cooperative Union and National Cooperative Union of India.

Apart from forming separate cooperatives, biogas programme can also be pursued by linking with the dairy and milk cooperatives. National Dairy Development Board (NDDB) which helps to establish village dairy cooperatives can work towards making biogas programme an integral part of such cooperatives. Dairy cooperatives can be particularly helpful in popularising biogas programme among milk producers. The integrated dairy-cum-biogas cooperative can ensure that each milkshed has at least one community and one demonstration-cum-instructional plant. Visit of farmers to demonstration units can go a long way to promote biomethanation [24, 25].

4.5 Fiscal Incentives for Installing Biogas Plants

In order to promote biomethanation in rural areas, government has been providing subsidy to counter socio-economic disparities existing in the country. This however has been sufficiently reduced over the years and is likely to be completely withdrawn in future in keeping with its new economic policy. Apart from the central subsidy several state governments such as Bihar and Gujarat provide additional subsidies and other financial incentives. The basic idea behind giving subsidy is to meet part expenditure of the plant cost by the government which a poor beneficiary cannot afford. In the initial years of the NPBD the subsidy amount was fixed which met 50 per cent of the total cost which at present is barely one-third of the plant cost (Table 4.1 and 4.2). Each beneficiary is eligible to avail the benefit of subsidy. Pattern of subsidy available accords due priority to the development of economically backward classes like scheduled castes/scheduled tribes, small and marginal farmers and landless labourers. Several hilly regions, where cost of transporting plant material is higher compared to more easily accessible hilly regions, can avail higher subsidies to

Table 4.1 Pattern of Subsidy Available in Different Regions Under the NPBD (In Rupees)

Plant Capacity (m^3/day)	*Region A*	*Region B*	*All others*
1	2800	2000	1500
2	3200	2400	1800
3 & 4	3500	2600	2000

Region A : NE region states excluding plains of Assam, Sikkim, J & K, HP and 8 hilly districts of UP (excluding *terai* regions of two hilly districts), A & N Islands and Lakshadweep.

Region B : SC/ST, desert districts, small and marginal farmers, landless labourers, plain areas of Assam, *terai* region of two hilly districts UP, western ghats and other notified hilly areas.

Source : TERI, Biogas: A Source of Rural Employment, New Delhi, 1994, p. 17.

Table 4.2 Pattern of Subsidy Available for Community, Institutional and Night-soil based Biogas Plants.

Plant Capacity (m^3/day)	*Community Biogas Plant*	*Institutional Plant*	*Private and Profit Making Institutions*	*Night-Soil based Biogas Plants*
8	—	—	—	22000
10	—	—	—	25000
15	44000	35000	22000	40000
20	56000	45000	28000	50000
25	70000	55000	35000	60000
35	100000	80000	52000	560000
45	150000	95000	64000	625000
60	170000	115000	76000	800000
85	200000	140000	94000	—

Source : MNES and TERI, Various Reports.

compensate for high transportation cost. In the initial years the subsidy was available to plants of 1 to 10 m^3 capacities which later was confined to plants of 1 to 4 m^3 capacities only with a view to achieve higher capacity utilisation and benefit weaker sections only [37].

MNES also provides bank loan facility to prospective owners to meet non-subsidy component of the plant cost. After complete withdrawal of subsidy component in future, bank loan facilities shall be the only source of finance available to the beneficiary. Every state has a lead bank with network of branches in all districts to coordinate activities relating to loan sanction. At national level overall coordination is done by the National Bank of Agriculture and Rural Development (NABARD) which provides automatic refinance facility to a commercial bank as a fixed percentage of the total loan amount sanctioned in a year [33, p. 10].

MNES is the principal nodal agency for target-setting at the national level. Before the commencement of financial year, MNES proposes statewise targets in keeping with the past and prospective performance. Based on these targets 50 per cent of the estimated outlay is released in advance to the state implementing agencies. Balance 50 per cent is released by the MNES after achieving 25 per cent of annual targets. Final adjustment in the amount is made by the MNES on the basis of feedback received from implementing agency at the year end [35, p. 16].

State nodal agency splits up the targets districtwise. Implementing agencies release funds for district level functionaries according to the targets. The funds advanced include subsidy amount, service charges, turnkey job fee and different promotional incentives including expenditure on training [35, p. 16].

Service charges to support their staff are paid to the state nodal agencies and implementing agencies on the basis of number of plants set up (Table 4.3). A promotional fee of Rs. 50 per cent plant supervised is paid to village level functionaries or approved workers of the KVIC who act as motivators and facilitators for setting up plants. MNES has also started a special scheme under which incentives are provided to use biogas slurry as fertiliser. A number of plots have been set up in agricultural fields to demonstrate the effectiveness of plant effluent as enriched fertiliser. As an incentive to promote biogas manure, MNES provides funds for setting up demonstration plots. As a result of these measures in several states, biogas manure is looked upon as a primary product from a plant with biogas as an output of secondary importance. After review of the NCAER study report on functional status of biogas plants, MNES in 1987 announced a special scheme under which owners of non-functional plants were offered a lump-sum amount between Rs. 750 to Rs. 1000 per plant for their repair [33, p. 14-17].

Special incentive of 50 per cent of the conversion cost with a ceiling of Rs. 2500 is available for popularising dual-fuel engines (80 per cent biogas and 20 per cent diesel) besides a flat subsidy of Rs. 2500 for setting up higher capacity plants of 6 to 10 m^3 capacity needed for dual-fuel operation. The incentive amount is aimed to meet expenditure towards conversion kit and 3 to 4 m^3 capacity rubberised nylon balloon needed for storage and transportation of biogas from the plant site to the dual-fuel engine site [35, p. 18].

MNES provides gold and silver shields to states and districts for outstanding achievements in different categories. This includes awards for maximum number of biogas plants and demonstration plots set up and training programmes organised. Awards are also given for organising special competitions for developing low-cost and more efficient models of biogas plants. Promotion and awareness campaign for biogas programme is done both by the MNES and state nodal agencies. This is done by advertisements through print-media, radio, television, fairs, exhibitions and mobile-vans. Funds for communication and publicity are provided by linking with the target covered which vary from Rs. 1 lakh to Rs. 5 lakh [33, p. 17].

Table 4.3 Norms for Payment of Service Charges

No. of Plants	Amount of Service Charges Payable
100 — 999	Rs. 35000 + Rs. 250 per plant in excess of 100 plants.
1000 — 3000	Rs. 5 Lakh + Rs. 150 per plant in excess of 1500 plants.
3001 — 7000	Rs. 7.5 Lakh + Rs. 100 per plant in excess of 3000 plants.
70001 — 20000	Rs. 11.5 Lakh + Rs. 75 per plant in excess of 7000 plants.
Above 20000	Rs. 21.5 Lakh + Rs. 50 per plant in excess of 20000 plants.

Source : TERI, Managing the Biogas Programme, New Delhi, 1994, p. 16.

4.6 Modalities for Obtaining Bank Loan

For securing financial assistance a prospective beneficiary should contact a biogas motivator or a local village functionary. The functionary should work out loan requirement based on assessment of the plant capacity and unit cost. A blank application collected from the nearest bank can be submitted through the block development officer or other notified authority, after filling in necessary details. The form usually contains following information [6].

- Number and age of cattle heads.
- Availability of space for plant.
- Capacity and type of plant.
- Size of landholding.
- Area under cultivation and types of crops grown.
- Annual income of beneficiary.
- Amount of loan desired.

It should be seen that gross annual surplus generated by a plant is notional and does not generate any liquid cash for the borrower [11]. The beneficiary can repay loan only from his other income which need be assessed by the bank before loan sanction. Banks normally process application within a period of fifteen days and in the event of inordinate delays or rejection, specific reasons need be conveyed to the applicant. In case of any major difficulty, block development office or local functionary can sort out the matter. Banks normally release loan amounts in stages depending on progress of construction. Following completion, block development office issues completion certificate for release of subsidy. However in some states, release of the subsidy is at the same time as sanction of loan by banks [13].

Reserve Bank of India in 1975 set up a multi-disciplinary group on plant financing by financial institutions [16]. It also made recommendations for refinance by certain institutions, such as Agricultural Refinance and Development Corporation (ARDC) and the National Bank for Agriculture and Rural Development (NABARD) to commercial banks extending financial assistance for biogas projects. Important guidelines laid down are as follows [1, 2, 9].

(i) Biogas projects should be certified for their technical feasibility by agencies acceptable to the ARDC/NABARD.

(ii) Beneficiary should possess adequate cattle for owning a plant of requisite capacity.

(iii) Loan eligibility of the borrower should be worked after deducting the amount of subsidy admissible. Refinance to commercial banks can be provided up to 75 per cent of loan granted.

(iv) The period for loan repayment should not exceed seven years for plants pf 2-4 m^3 capacity range and five years for plants of 6 m^3 capacity. Economics of loan repayment giving details of income, periods for loan repayment and surplus generated; and cost-benefit analysis of different capacity plants as per prevalent rates are given in Table 4.4 and 4.5, respectively [2]. These however should be referred to only as a guideline and used only after inflation accounting.

(v) The rate of interest on ARDC/NABARD's refinance assistance and on loans to the ultimate borrower should be based on ARDC/NABARD notification from time to time.

(vi) Loans should be disbursed in stages depending on the work-in-progress. Payment for gashdolders and appliances can be made direct to the suppliers.

(vii) Loan to the borrower should be secured by the banks after obtaining hypothecation of gasholder and appliances/third-party guarantee and other securities as acceptable to the banks.

(viii) Subsidy provided by the government should be viewed as margin money for the loan. It is the responsibility of an agency of state government to arrange release of subsidy from a bank.

(ix) No down payment needs be recovered from the borrowers so long as the subsidy is available. If the subsidy is abolished, down payment will be at 10 per cent of plant cost for plant capacities up to 3 m^3 and 15 per cent for higher plant capacities. If the subsidy is reduced to less than 10 per cent of the plant cost, supplementary down-payment by borrowers will be so fixed as to bring total subsidy and down payment to 10 per cent of plant cost for plants with capacities up to 3 m^3 and 15 per cent for plants with higher capacities.

(x) Average cost of community, institutional and night-soil based biogas plant corresponding to a retention period of 40 days is given in Table 4.6

Table 4.4 Economics of Loan Repayment Giving Details of Income. Period of Repayment and Net Surplus Generated (at 1981-82 Price Level)

Plant Capacity m^3 (ft^3)	*Number of Animal Required*	*Capital Investment (Rs.)*	*Bank Loan (Rs.)*	*Annual Income (Rs.)*	*Repayment Periods (Years)*	*Annual Instalment (Rs.)*	*Net Surplus (Rs.)*
2 (70)	3	2150	1720	486	7	373	114
3 (105)	4	3000	2400	844	7	519	325
4 (140)	5	3625	2900	1074	7	627	447
5 (210)	7	4650	3750	1788	7	1023	765

Table 4.5 Cost Benefit Analysis of Different Capacity Biogas Plants

Sl. No.	*Measure*	*Capacitywise Income and Cost (Rs.)*			
		$2m^3$ (70 ft^3)	*$3m^3$ (105 ft^3)*	*$4m^3$ (140 ft^3)*	*$6m^3$ (120 ft^3)*
	Annual Expenditure				
1.	Cattle dung required in kg per day.	50 kg	80 kg	100 kg	160 kg
2.	Value of cattle dung in money terms @ 4 paise per kg.	Rs. 730	Rs. 1168	Rs. 1460	Rs. 2336
3.	Repairs &maintanence charges.	Rs. 100	Rs. 100	Rs. 100	Rs. 100
4.	Labour hours needed for miscellaneous purpose including dung handling.	1/2 hr	1/2 hr	1 hr	1 hr
5.	Labour charges in money terms @ Rs. 6 per day.	Rs. 137	Rs. 137	Rs. 274	Rs. 274
6.	Total annual expenditure	Rs. 967	Rs. 1405	Rs. 1834	Rs. 2710
	Annual Income.				
7.	Biogas yield per year in money terms (2).	(365 × 2 × 0.62 × 1.70) Rs. 769 (0.375 × 50 × 365 × 0.1) Rs. 648	(365 ö 3 × 0.62 × 1.70) Rs. 1154 (0.375 × 80 × 365 × 0.1) Rs. 1095	(365 × 4 × 0.62 × 1.70) Rs. 1539 (0.375 × 100 × 365 × 0.1) Rs. 1369	(365 × 6 × 0.62 × 1.70) Rs. 2308 (0.375 × 160 × 365 × 0.1) Rs. 2190
	Total	Rs. 1453	Rs. 2249	Rs. 2908	Rs. 4498
	Surplus	Rs. 486	Rs. 844	Rs. 1074	Rs. 1788

Note: 1. All the costs are worked out at 1981-82 price-level.

2. Biogas yield per year in money terms = (365 days) x (plant capacity) x (kerosene equivalent of 1 m^3 of biogas) x (rate of kerosene per litre).

3. Fertiliser yield per year from plant in money terms = (0.375 kg of fertiliser per kg of dung x (cattle required per day) x (365 days) x (Rs. 0.10 per kg.)

Source: ARDC, Circular on Refinance Facility for Implementation of the Scheme for Setting-up of Biogas Plants, Agricultural Refinance and Development Corporation, Worli, Mumbai; March 20, 1981.

Table 4.6 Average Cost of Community, Institutional and Night-Soil based Biogas Plants

m^3	CBP	IBP	NBP
8	—	—	41500
10	—	—	49400
15	87000	66300	76300
20	112250	86450	96450
25	137500	106600	116660
35	199500	159300	560000
45	301500	191800	625000
60	338800	229900	800000
85	399500	282500	—

Source: MNES, Biogas : A Rural Energy Source, New Delhi, Oct., 1995, p. 7.

D. Sundaresan [19] reviewed strategies and experience of Canara Bank regarding financing of biogas plants in six districts of Karnataka during 1983-84 and 1984-85. Financing was done on the basis that one-third of the credit would come from cooperative sector and remaining two-third from commercial banks such as Canara Bank. In this task, all the semi-urban and rural branches of the bank were pressed into service to finance minimum of five plants by each during 1983-84 and seven plants during 1984-85, with options to finance more if necessary. The bank worked in association with the KVIC which during the initial phase was mainly responsible for the implementation of biogas programme. As part of its keeness of promote biomethanation , it waived off the requirement of landed property as collateral security. Bank however faced difficulties which included lack of adequate publicity and motivation of implementing agencies towards biomethanation, low motivation due to high initial investment and lack of generation of direct income, shortage of skilled manpower for plant construction, inability of plant owners to repay back loans, procedural delays in securing subsidy, and lack of effective infrastructure support [19].

For setting up renewable energy projects such as community biogas plant, availability of funds is necessary. In India funding of development projects over the years is more seen as extension of government schemes of subsidies or grant. In recent years the government has taken some new initiatives for funding renewable energy projects such as the setting-up of Indian Renewable Energy Development Agency (IREDA) and launching of the programme for Acceleration of Commercialisation of Energy Resources. IREDA is a public limited company registered under the Companies Act, 1956 incorporated on March 11, 1987 under the administrative control of MNES for developing, funding and promoting new, renewable and non-conventional sources of energy [31]. IREDA promotes these technologies by providing loans at concessional rates of interest ranging from 9.8-15.5 per cent to both manufacturers and users. Projects financed in past include biogas generation from industrial effluents, biomass gasification, mini and micro-hydel, wind energy, solar photovoltaics and solar thermal energy conversion. IREDA also provides technical expertise in preparing detailed project reports. While extending loan assistance, IREDA accords due priority to the development of economically backward regions and promoting interests of women, scheduled castes/scheduled tribes, ex-servicemen and other disadvantaged categories [31]. The company provided loan assistance to 265 projects in 1995-96 as against 136 projects in 1994-95. It disbursed Rs. 238.39 crores in 1995-96 as against Rs. 130. 29 crores in 1994-95. An additional 176 mw renewable energy capacity was created during 1995-96 through IREDA funded projects [36]. IREDA is likely to obtain 145 million dollars from the World Bank including 30 million dollars to be contributed by the Global Environment Facility and Swiss Development Corporation. Government of the Netherlands, Danish International Development Agency (DANIDA) and the Asian Development Bank are some of the other agencies that make funds available to the IREDA [36]. Wide dispersal of activity centres and high risks of return from renewable energy projects are some distinctive problems of this sector [23]. More details about the financial incentives can be obtained from the MNES, Block No. 14, CGO Complex, Lodhi Road, New Delhi 110 003 and the IREDA, Habitat place, Lodhi Road, New Delhi 110003.

4.7 Methodology for Estimating Cost of Plant Operation

In quite a large number pf cases, biogas plants are set up after securing loans from banks or other financial institutions. Once the loans are obtained, it is necessary to repay them which is generally done in yearly instalments extending over a period of five to ten years depending upon the amount involved and the economic status of plant owners. Loan repayment amount is computed by dividing the capical cost of the plant 'C' by the annuity factor 'a' dependent on rate of interest and period of repayment [20].

Annuity factor can be determined using the following relationship which can be either computed or read from the annuity tables (Table 4.7).

$$a = \frac{1 - (1 + i)^{-n}}{i}$$

where,

i = rate of interest in per cent

n = period of loan repayment in years

For i = 6, n = 7, a = 5.58

Assuming the plant cost (C) as Rs. 10,000, annual loan repayment

amount $\frac{c}{a} = \frac{10,000}{5.58} = 1792$

Total loan amount payable in seven years = Rs. 1792 × 7 = Rs. 12,545

For a 20 year plant life, dispersal of loan liability on yearly basis

$= \frac{\text{Rs.}12545}{20}$ = Rs. 627.25 ≅ Rs. 627

For estimating labour costs, no uniform norms are available. However, according to the UN Guidebook on Biogas Development, labour input ranges anywhere between 7 to 10 minutes daily per m^3 for digester volume. For a 3 m^3 plant capacity, nearly 30 minutes of labour input daily or 183 hours annually

(365 x 0.5= 183) is needed.

Annual labour cost @ Rs. 5 per hour = Rs. 183 × 5 = Rs. 915

Hence, total annual cost of plant operation

= 627 + 200 + 915 = Rs. 1742

Discounted cash flow (DCF) analysis takes into account cash flow and timing of their receipts. It discounts future costs and benefits at a compound rate to estimate their present value. The rate of return is thus based on a time-scale and hence the technique is also known as the time-adjusted return on capital investment. Two common DCF technique are (i) internal rate of return (IRR) and (ii) net present value (NPV). According to first method, rate of interest is applied to future cash flows, which reduces or discounts their monetary size, until they are equal to the initial cost of capital in total. Trial and error approach is generally used to determine the IRR of a project. The difference between cash inflows and outflows would be zero if the IRR is correctly chosen. If the net value of receipt-costs is positive, a higher rate should be tried and if negative a lower rate, which after successive trials and interpolation leads to the identification of corrent rates. In the net present value method, the discounting principle is applied conversely. The future yields or returns are discounted at an admissible rate of return usually the prevailing rate of interest. The difference between the total discounted value of future returns and the cost of investment gives the net present value of the investment which if positive indicates that money cab be invested in the project [13, 15].

DCF in the context of a biogas project can help to evaluate various options. Estimation of annual cash flow includes contributions for gross revenue, operating costs, depreciation and taxes. Gross revenue is the value of gas and other products of the digestion process. Operating costs are expenses incurred directly for the operation of the plant. These expenses are usually calculated on annual basis and include material, manpower, administration, taxes and overhead costs. The primary raw material for the digestion system is the raw feed. Manpower costs are related to the size of the plant. Administration and overhead costs are mostly relevant for community plants. Depreciation is a function of the plant life and total plant investment. Taxes are determined from the taxable income.

Net present value of a project refers to the discounted sum of stream of net income (returns minus cost) during its expected life. Discounting is done by applying the compound interest principle in reverse order. Rate of discount needs to be specified in estimating NPV. A positive value of NPV indicates that there are overall net economic gains from the project [3].

$$\text{Net present value} = \sum_{t=0}^{n} \frac{R_t - C_t}{(1+r)^t} = 0$$

where, R_t = Returns in period t.

C = Operating costs in period t.

C_0= Cost of initial investment.

t = Time period in years.

n = Total economic life of the plant in years.

$\frac{1}{(1+r)^t}$ = Discount factor (d_t) for annual compounding.

Discount factor (d_1) for continuous compounding is given by $d_t = e^{-rt}$ where r is the expected rate of return. Discount factors for continuous compounding are lower than for annual compounding.

The internal rate of return is that discount rate which makes the NPV of the net income stream (benefits minus cost) from a gas plant equal to zero. It represents average rate of earning on the money invested in the plant. A higher internal rate of return than the interest charged on the capital investment is an indication of sound financial gain from the plant [13].

$$\text{Internal rate of return} = \sum_{t=0}^{n} \frac{R_t - C_t}{(1+r)^t} = 0$$

where all the constituent terms are taken as they are included in NPV except r which refers to that rate of return which makes NPV equal to zero.

The benefit-cost ratio indicates the returns per rupee invested during the entire life of plant. This can be estimated by discounting the benefit and cost stream separately and taking the ratio. A ratio greater than unity indicates that plant is economically viable.

$$\text{Benefit-cost ratio} = \sum_{t=0}^{n} \frac{R_t/(1+r)^t}{C_t/(1+r)^t}$$

where all the quantities are same as in NPV

The payback period is the time it will take the investor to recover the initial investment in the plant from the net earning. It is not a discounted measure and does not account for the time value of money and net return after the payback period. It is useful for fixing the terms of loan repayment (maturity period).

A commonly used scaling relationship for estimation of the plant cost is the power factor rule which is as follows

$$(\text{Cost of plant A}) = (\text{Cost of plant B}) \cdot \left[\frac{\text{Capacity of plant A}}{\text{Capacity of Plant B}}\right]^X$$

Normally the value of scaling constant x varies from less than 0.2 to greater than unity depending upon the type of plant scaled. In the absence of precise data for estimation of x, a value of 0.6 is usually found appropriate fit.

ESSO Research and Engineering Company in USA developed a procedure for calculating the Unit Gas Cost (UGC) based on the utility financing method used by the American Gas Association. The average UGC as per this method is given by [22].

$$\text{UGC} = \frac{N + 0.05\,(C - W) + 0.5\,[p' + \frac{t}{1-t}(1-d)\,r](C+W)]}{G_y}$$

Where,

N = Total annual operating cost

C = total capital requirement
W = working capital
P = fractional return on rate base [$p' = di + (1 - d)\ r$]
t = fractional federal income tax rate
d = fraction debt
r = fractional return on equity
i = fractional interest on debt
G_y = annual gas production, MMBTU/Year
UGC = unit gas cost per MMBTU

The first term of the numerator on the right side corresponds to the net operating cost; the second term is due to 5 per cent year straight line depreciation, and the third term accounts for the return on rate base as well as federal income tax. The sum of these terms gives the total average annual revenue requirement. The UGC is obtained by dividing the average annual revenue requirement by the annual gas yield. In this model, no operating cost escalation during the project life is assumed [22].

4.8 Economic Viability of Biomethanation

Controversies continue to surround economics of plant operation. According to S. S. Bhatnagar, *et al.* [5], unless subsidy and other concessions continue, burning dung along with supplementary wood in an improved *chulha* is more cost-effective than biomethanation. However since burning dung-cakes pollutes environment and completely destroys fertiliser potential of animal wastes, biomethanation is beneficial in view of its overall benefits, better health, improved hygienic conditions and enriched fertiliser. A set of following four variants were evaluated.

- Dispose off dung and meet all requirement by supplementary purchases of wood, kerosene and chemical fertiliser (option 1).
- Compost available dung to obtain fertiliser and make supplementary purchases of fuel for cooking, kerosene for lighting and fertiliser for farming (option 2).
- Burn dung for cooking and make supplementary purchases of kerosene for lighting, wood for cooking and fertiliser for farming (option 3).
- Utilise dung for producing biogas (option 4).

Based on the economic analysis [5], if burning of dung is to be avoided considering that fertiliser component is lost and it is injurious to health, biomethanation option is most advantageous even if annualised cost of biogas plant becomes double or more.

According to N. Ashok Kumar [12], direct and indirect benefits of biomethanation do not justify large scale promotion of biogas programme. According to him plant operation is cumbersome and economically non-viable due to several reasons. While analysing plant economics 10 kg dung per cow and 15 kg dung per buffalo is usually taken as generation norm whereas in reality wastes are generated not more than six kg per cow and 10 kg per buffalo.

Thirty kg of dung is needed to, produce one cubic meter of gas per day and since 15 kg of dung from one buffalo is assumed, four buffaloes are needed to produce two cubic meters of gas, which is considered adequate for a household of six members. In reality however a plant of required size may need feed from six to eight buffaloes. Thus the plant may generate only half of gas output. According to N. A. Kumar [11, 12], an expenditure of Rs. 400 per year on fuelwood produces energy equivalent to four cubic meters of biogas per day. At a capital expenditure apportioned infinitely at 12 per cent rate of interest, recurring annual cost for plant of this size is more than Rs. 1500. Even if the entire plant cost is subsidised, cost of running the plant will be more than Rs. 800. Based on estimates made by N. A. Kumar [11], for a plant of any size biogas is always costlier than an energy-equivalent quantity of fuelwood or dung-cakes, a fact which makes the technology economically less attractive to the farmers.

Ludwig Sasse, Editor, *Biogas Forum* (*BORDA, Germany*), champion of biomethanation and guiding spirit behind biogas programmes of several countries, made a powerful rejoinder to the two critical articles on biogas published in two important Indian newspapers (i) 'Biogas Cumbersome Technology' by N. Ashok Kumar, *Times of India*, Jan. 19, 1990 and (ii) 'Biogas: The Foolish Flame,' by N. N. Narayan Rao, *Maharashtra Business Herald*, Aug. 30, 1990. He began his rejoinder by observing that 'both articles were very critical of economic viability of biogas. If such articles are presented to the public, they cannot go without comment'. His main argument for defence of biomethanation was that it represented 'Appropriate Technology' vital for the growth of several developing countries [30]. He conceded that biomethanation might not make economic sense in some situations but suggested that it did make in most situations. L. Sasse argued that one should propagate biogas plants only where they made sense. There were still more senseful situations than we could manage to serve, he argued [30].

For making a more realistic and fact-based defence of biogas, L. Sassee, put forward the findings of a biogas survey carried out by the BORDA in 1988 based on working of 645 biogas plants set up under nine programmes of seven countries. The survey showed that a functioning biogas plant is by itself economically viable for the farmer and more so for the national economy (*Biogas Survey* - 1988 BORDA, GATE, Eschborn 1, Bremen, Germany). If 30 per cent of construction costs are subsidised by the government, there will still be a net profit for the national economy if ecological or foreign exchange savings are taken into account. However, according to the survey, if the failure rates in a programme or non-functioning plants are high it may mean a direct loss for the farmers and the national economy [30]. L. Sasse also emphasised that in any cost-benefit exercise inflation rate should be taken into account along with the interest rates. Any cost of an availed loan or long term investment depends largely on the relation between the inflation rate and the interest rate [30].

According to L. Sasse, amount of water a biogas plant consumes is rightly criticised which needs to be changed. In many places people have realised that it is uneconomical to feed water to the plant and keep at the same time animals

Table 4.7 Values of the Annuity factor for Varying Rate of Interest (i) and Payment Period in Years (n)

Payment Period in Year (n)	*Rate of Interest i (Per cent)*											
	3	*4*	*5*	*6*	*7*	*8*	*9*	*10*	*11*	*12*	*13*	*14*
5	4.58	4.45	4.33	4.21	4.10	3.99	3.89	3.79	3.70	3.60	3.52	3.43
6	5.42	5.24	5.08	4.92	4.77	4.62	4.49	4.36	4.23	4.11	4.00	3.89
7	6.23	6.00	5.79	5.58	5.39	5.21	5.03	4.87	4.71	4.56	4.42	4.29
8	7.02	6.73	6.46	6.21	5.97	5.75	5.53	5.53	5.15	4.97	4.80	4.64
9	7.79	7.44	7.11	6.80	6.52	6.25	6.00	5.76	5.54	5.33	5.13	4.95
10	8.53	8.11	4.72	7.36	7.02	6.71	6.42	6.14	5.89	5.65	5.43	5.22
11	9.25	8.76	8.31	7.89	7.50	7.14	6.81	6.50	6.21	5.94	5.69	5.45
12	9.95	9.39	8.86	8.38	7.94	7.54	7.16	6.81	6.49	6.19	5.92	5.66
13	10.63	9.99	9.39	8.85	8.36	7.90	7.49	7.10	6.75	6.42	6.12	5.84
14	11.30	10.56	9.90	9.29	8.75	8.24	7.79	7.37	6.98	6.63	6.30	6.00
15	11.94	11.12	10.38	9.71	9.11	8.56	8.06	7.61	7.19	8.81	6.46	6.14
16	12.56	11.65	10.84	10.11	9.45	8.85	8.31	7.82	7.38	7.97	6.60	6.27
17	13.17	12.17	11.27	10.48	9.76	9.12	8.54	8.02	7.50	7.12	6.73	6.37
18	13.75	12.66	11.69	10.83	10.06	9.37	8.76	8.20	7.75	7.25	6.84	6.47
19	14.32	13.13	12.09	11.16	10.34	9.60	8.95	8.35	7.84	7.37	6.94	6.55
20	14.88	13.59	12.46	11.47	10.59	9.82	9.13	8.51	7.96	7.47	7.02	6.62
25	17.41	15.62	14.09	12.78	11.65	10.67	9.82	9.08	8.42	7.84	7.33	6.87

Note: Annuity factor 'a' is given by the expression

$$a = \frac{[1-(1=i)^{-n}]}{i}$$

Source : UN, Guidebook on Biogas Development, New York, 1980, p. 97.

thirsty and the cows and the buffaloes with little milk production. The more sensible course is to give animals more water to drink to keep them healthy and produce more urine which can be collected and fed to biogas plants. In such a course no additional water shall be needed.

L. Sasse also differed from the suggestion that LPG should replace dung-cakes to make kitchen smokeless and use it instead of biogas. He argued that biomethanation should be appreciated as a process of wastes-recycling and epitome of self-reliance in villages. The viewpoint that farmers, expecially rich ones use LPG and poor ones who cannot afford to pay the high investment cost of a biogas plant and therefore regard that LPG is cheaper, is also true. This is true with poor people who cannot afford the most economical and sensible solution because they have no capital. This is exactly the reason why the Indian government rightly subsidises the construction cost of biogas plants.

G. K. Criner [7] studied the impact of variation in unit price of oil and electricity on economic viability of biogas plants. For a large sized plant in an agricultural farm utilising animal wastes, it was analysed that plant operation was not cost-effective if the farm had less than 137 cow equivalents even when oil and electricity prices prohibitively expensive.

C. S. Sinha and T. C. Kandpal [32] developed a framework for financial evaluation of household biogas plants based on estimation of incremental benefits. These benefits can be quantified in terms of quantity of fuelwood substituted by biogas for cooking, quantity of kerosene replaced with biogas for lighting, and the diesel-fuel displaced with a biogas-fulled dual-fuel engine for motive power. Biogas produced is utilised for cooking, lighting, and/or as a part substitute for diesel in IC engines. For estimating incremental benefits, it is assumed that prior to the installation of the biogas plant the total dung available was being used either as dung-cakes or as a farmyard manure (FYM). Benefits accruing from biogas should be evaluated in terms of the monetary value of the energy source it replaces for a given end-use [32].

If a fraction, F, of the dung previously used fir preparing FYM is deverted as feed material, it produces both biogas and enriched fertiliser. Finanacial benefits resulting from this course could be mathematically expressed as [32] :

$$= 365\ Vn_p\ Fg\ \frac{Q_b \cdot \eta_{bc}}{Q_f \cdot \eta_{fc}} \cdot p_f \qquad \text{... (1)}$$

Where,

V = capacity of plant, m^3

g = fraction of gas used for cooking

Q_b = calorific value of biogas (kJ/kg)

Q_t = calorific value of fuelwood (kJ/kg)

η_{bc} = efficiency of biogas cookstove

η_{fc} = efficiency of fuelwood cookstove

P_f = price of fuel wood (Rs./kg)

The balance dung is consumed in the form of dung cakes for cooking. The incremental benefits arising out of the use of this fraction would depend on availability of additional useful thermal energy. Thus the annual incremental benefits resulting by using this fraction of dung as a feed material is [32] :

$$= 365\ Vn_p\ (1 - F)_g\ \frac{Q_b \cdot \eta_{bc} - d_u \cdot wQ_d \cdot \eta_{dc}}{Q_f \cdot \eta_f} \cdot p_f \qquad \text{... (2)}$$

Where,

Q_d = calorific value of dry dung (kJ/kg)

d_n = quantity of wet fresh dung in kg to produce 1 m^3 of biogas

w = dry to set dung weight ratio

η_{dc} = efficiency of cookstove when dry dung is used as a fuel.

The total financial benefits from cooking will be sum of Eq. (1) and (2) which on simplification can be written as [32] :

$$= 365\ V n_p \cdot g \cdot p_f \cdot \frac{[\,Q_b \cdot\cdot \eta_{bc} - (1 - F) \cdot d_u \cdot W \cdot Q_d \cdot \eta_{dc}\,]}{Q_f \cdot \eta_{fc}} \qquad \text{...(3)}$$

The incremental benefits from the use of biogas as a lighting fuel is assumed independent of its use as dung cake or manure which the dung was earlier put to. The total annual monetary benefits from biogas as lighting fuel is [32] :

$$= 365\ .\ V \eta_p\, m\ .\ \frac{Q_b \cdot \eta_{bl}}{Q_k \cdot \eta_{ki}} \cdot p_k \qquad \text{...(4)}$$

Where,

m = fraction of gas used for lighting

Q_k = calorific value of kerosene (kJ/litre)

P_k = price of kerosene (Rs./litre)

η_{bl} = efficiency of biogas as lighting fuel

η_{kl} = efficiency of kerosene as lighting fuel

The incremental benefits from the use of biogas as a part substitute for diesel in IC engines in dual-mode is independent of dung use for cooking or FYM prior to its utilisation in a biogas plant. Total annual monetary benefits from biogas as a cooking fuel are [32] :

$$= 365\ V n_p \cdot (1 - g - m) \cdot \frac{sfc_d}{sfc_b} \cdot r_d \cdot df_d \qquad \text{... (5)}$$

Where,

1-g-m = fraction of biogas used as a fuel for running IC engines

sfc_d = specific diesel consumption in independent mode (litre/kwh)

sfc_b = specific biogas consumption in dual-fuel mode (litre / kwh)

r_d = diesel replacement ratio of biogas

$£_d$ = unit price of diesel (Rs./litre)

d = duration of IC engine in the dual-fuel mode.

The total annual monetary benefits from the use of biogas is the sum of Eq. (3), (4) and (5).

C. S. Sinha and T. C. Kandpal [32] also developed a model for evaluating benefits of biogas fertiliser. Financial benefits from biogas fertiliser from the fraction, F, of dung previously used for preparing FYM may be expressed as :

$$= 365\, V\, d_u\, F\, (\, n/46\,)\, (\, h_b - h_c\,)\, p_u \qquad \text{...(6)}$$

N value of urea is taken as 46

where,

d_u = quantity of dung to produce 1 m^3 of biogas

n = elemental nitrogen present in fresh dung indicated by its N value.

P_u = price of urea (Rs./kg)

For dung fraction (1-F) which was previously burnt, incremental annual benefits also include benefits of P and K values apart from N value of the spent slurry. The annual incremental benefits can be computed as :

$$= 365\, V\, d_u\, (\, 1 - F\,)\, [\, (\, n/46\,)\, p_u + (\, p/20\,) p_p + (\, k/60\,)\, p_m\,] \qquad \text{...(7)}$$

Where n, p and k are the N-P-K values of the spent slurry, respectively while urea, superphosphate and muriate of potash are priced at p_u, p_p and p_m and their N – P – K values are taken at 46, 20 and 60, respectively [26, 32]. The total annual monetary benefits from spent slurry can be obtained by summing up Eq. (6) and (7).

4.9 Findings of a Plant Economics Study Carried Out by the IIM, Ahmedabad.

During 1975, T. K. Moulik and U. K. Srivastava [13] carried out socio-economic evaluation of biogas plants in Gujarat. For analysing economic-viability of plants, cost and income streams were compared after discounting at successive intervals. It was done by computing net present value, internal rate of return, benefit-cost ratio and pay-back periods. Economic life of the plant was taken as 30 years. Analysis was carried out for three different discount rates 10, 13 and 15 per cent. Estimates were based on the KVIC data and sample data collected during the study. It was found that for a 60 ft^3 capacity plant with a 15 per cent dicount rate, it was difficult to recover plant cost within the entire life-span of plant. Thus for 60 and 100 ft^3 capacity plants, banks should introduce system of differential maturity periods to circumvent this problem. NPV was found to be positive both in respect of the KVIC data ad the sample data, except for 60 ft^3 capacity plant with 15 per cent discount rate. NPV based on sample data was lower than based on the KVIC data. NPV was found to increase with increase in plant capacity which is equivalent to saying that large capacity plants are more cost-effective. Similar rising trends with increase in plant capacity were observed in respect of IRR and BCR also (Table 4.8 and 4.9). This shows that community plants are more cost-effective, than family units. For improving cost-effectiveness of smaller plants thus more subsidy should be provided. For higher cost-effectiveness, differential rate of interest should be charged for different capacity plants and varying maturity periods. In this study conclusions were derived from the view-point of owners and society at large. From owners angle, subsidy is viewed as reduction in cost but from society's point-of-view subsidy is an economic burden on people [17].

Table 4.8 Internal Rate of Return on Investment for Different Capacity Plants

Plant Capacity (ft^3)	*Economic Analysis in per cent*		*Financial Analysis in per cent*	
	Based on KVIC Data	*Based on Sample Data*	*KVIC Data with 25% Subsidy*	*Sample Data with 25% Subsidy*
60	19.99	13.53	23.23	19.04
100	26.39	21.45	35.57	29.35
150	38.23	33.34	51.15	44.95
200	41.65	37.53	55.63	50.44
250	46.04	40.53	61.43	54.38
300	53.80	50.90	71.67	68.04
350	51.66	45.66	68.85	61.12
500	53.97	52.15	71.93	69.71
1250	63.94	58.97	85.18	78.73

Table 4.9 Benefit Cost Ratios for Different Capacity Plants

Plant Capacity (ft^3)	*Discount Rate (Per cent)*	*Benefit-Cost Ratio*	
		Based on KVIC Data	*Based on Sample Data*
60	10	1.23	1.10
	13	1.12	1.01
	15	1.05	0.96
100	10	1.55	1.34
	13	1.42	1.34
	15	1.32	1.24
150	10	1.87	1.66
	13	1.73	1.54
	15	1.64	1.47
200	10	1.97	1.78
	13	1.82	1.65
	15	1.72	1.57
250	10	2.07	1.84
	13	1.92	1.71
	15	1.82	1.63
300	10	2.23	2.07
	13	2.07	1.94
	15	1.98	1.86
350	10	2.21	1.96
	13	2.05	1.83
	15	1.95	1.75
500	10	2.29	2.16
	13	2.12	2.02
	15	2.02	1.93
1250	10	2.51	2.29
	13	2.34	2.15
	15	2.23	2.05

4.10 Concept of Opportunity Cost in Plant Operation

Analysis of opportunity cost provides important insights for economic evaluation of biogas plants. According to this approach input and output of a system should be valued in terms of loss to the identified goal if these were to become part of the alternative investment. This procedure involves identifying possible alternatives to biomethanation and evaluating financial implications of each. It is just possible that feasibility of a biogas plant in one location may not be economically-viable in another [3]. According to recent study, biogas plants may not operate cost-effectively in several situations [8]. When loans are involved, their repayment schedule becomes an important aspect in economic evaluation. It is possible that although many of the opportunity cost benefits of plant investments may be significant, cash value of these returns may not be sufficient to pay back the required loans. An analysis solely based on cash transactions associated with an investment may thus not be the best strategy. For viable economics, it is important that beneficiaries receive non-recoverable grants and loans at low interest rates with easy terms along with technical and maintenance support. Economic-evaluation can become more realistic if intangible gains such as cleaner environment, reduced deforestation, added employment opportunities and improvement in rural hygiene are also taken into account. Plant operation will be economically viable when inputs have a low opportunity cost, plant efficiency is satisfactory and when alternatives to outputs from biogas plants have a higher opportunity cost. A low opportunity cost of inputs implies that agricultural wastes are adequately available with opportunity cost at least no more than when it is used as a fertiliser or fuel, industrial wastes are plentifully available with opportunity cost less than their alternative usage, night-soil can be conveniently used for biomethanation without any social-stigma, water supply is adequate and can be easily connected to the digester, capital is available with a low opportunity cost so that its alternative usage is not much attractive , and labour for plant operation is available at a low rate. As for plant, higher efficiency is achievable in larger plants, feed materials are consistently available, digester can be maintained at 35°C, technical and maintenance support is available, and plant has needed efficiency level [3].

As for alternative to plant outputs having high opportunity cost, this implies a search for options that can be compared with biogas. Possible options can be as follows [3.13] :

- Biogas is mostly used as a fuel in rural areas, hence it should be compared with other fuels such as fuelwood, vegetable wastes, kerosene, etc.
- Biogas plants generate enriched fertiliser and therefore it is appropriate to compare them with an anaerobic composting process or with other processes that generate chemical fertiliser.
- Biogas substitutes activities such as burning of dung or fuelwood. Whereas dung can be directly used as fertiliser, consumption of fuelwood causes deforestation and also soil-erosion.
- Biomethanation also represents efficient system for waste disposal.

Hence, it is appropriate to compare biomethanation with alternative waste disposal processes including disposal through lagoons and septic tanks.

- Investment in biogas plants also represents effective way of utilising wastes many of which otherwise have no other application.

One of the ongoing criticism of biogas programme lies in unattractive economies of scale it generates [4, 18]. The US experience has been no different considering that wastes are widely scattered for the economics to be attractive although biomethanation makes sense for helping to solve disposal problems of organic wastes. As an instance-in-point, US Department of Energy in past funded a biogas project at Bartow, Florida which though technically sound provided poor economic returns. Another company that sold gas produced by anaerobic fermentation of manure from cattle feedlots operated for a few years and finally closed due to poor economic returns [6]. D. L. Klass [10] in his US based study concluded that biomethanation is not a very attractive proposition. According to R. W. Ward [21], plant effluent which though has considerable nutrient value for supplementing cattle feed but feeding trials showed poor weight gain and somewhat inferior quality of the meat. H. R. Bungay [6] concluded that biomethanation might not play a major role in the USA in the manner it played in several developing countries due to very different socio-economic conditions.

4.11 Concluding Remarks

Economic-viability of biogas plants is important for commercial success of biomethanation in future. A number of studies carried out in past indicated unattractive economies of scale which plants were found to generate. The problem is compounded by the fact that gross annual surplus generated by a plant is merely national income which does not create any liquid cash in the hands of the borrower. Plant owners can repay loans from income from other sources which many may not afford. Given this context government subsidy as a non-recoverable grant is essential to support the programme. More subsidy need be provided to owners of smaller plants for improving their economic-viability. Differential rate of interest should be charged for loan repayment for different capacity plants and varying maturity periods. In wake of globalisation, private entrepreneurs and NGOs should be increasingly involved in the implementation of the biogas programme.

REFERENCES

1. ARDC, *Circular on Refinance Facility for Implementation of Biogas Programme*, Letter No Plan 32517/MS(37), Mumbai, 1976-77.
2. ARDC, *Circular on Refinance Facility for Implementation of the Scheme for Setting up of Biogas Plants* (1980-81), Letter No Plan 1299/MS (37), Mumbai, Mar. 20, 1981.
3. Barnett, A., *et al.*, *Biogas Technology in the Third World : A Multidisciplinary Review*, International Development Research Centre, Ottawa, 1978.

4. Bembalkar, N. L., Energy Crisis : Indian Response, *Science Express (Indian Express)*, Visakhapatnam, Feb. 26, 1991, p. III.
5. Bhatnagar, S. S., *et al.*, Biogas Option of Rural Family, *Urja*, **4**, 63-65 (1989)
6. Bungay, H. R., Commercialisation of Biomass Conversion, *Energy Exploration and Exploitation*, **6(1)**, 61-69 (1988)
7. Criner, G. K., Economic Feasibility of Anaerobic Digesters, *Bio Cycle*, **28**, 51-53 (1987).
8. *DNES, A Communication*, New Delhi, Feb. 16, 1989.
9. Khandelwal, K. C. and Mahdi, S. S., *Biogas Technology : A Practical Handbook*,, Tata Mc Graw Hill Publishing Co. Ltd., New Delhi, 1986, p. 61-67.
10. Klass, D. L., *Science*, **223**, 1021-1028 (1984).
11. Kumar, N. A., *Economics of Biogas Evaluation*, Times Research Foundation, Pune, 1988.
12. Kumar, N. A., Biogas Cumbersome Technology, *Times of India*, New Delhi, Jan. 26, 1990.
13. Moulik, T. K. and U. K. Srivastava, *Biogas Plants at the Village Lvel: Problems and Prospects in Gujarat*, CMA Monograph No 59, Centre for Management in Agriculture, IIM, Ahmedabad, 1975.
14. Ramarao, A., Biogas Programme in India, *Khadi Gramodyog*, June, 1988, p. 405.
15. Rao, K. V., Merits of Discounted Cash Flow Technique, *The Hindu*, Nov. 6, 1986, p. 18.
16. RBI, *Report of the Inter-Institutional Group on Financing Gobar Gas Plants by Banks*, Credit Planning Cell, Reserve Bank of India, Mumbai, 1976, p. 41-48.
17. Reddy, A. K. N., Energy Planning: Time for Tough Decisions, *The Hindu*, Visakhapatnam, Nov. 23, 1990.
18. Srivastava, D. H., Development Strategy: Energy Options, *Economic Times*, Jan. 10, 1989, p. 5.
19. Sundaresan, D., Canara Bank's Strategies and Experiences in Financing Biogas Plants, *Khadi Gramodyog*,, **30(8)**, 33-337 (1984).
20. UN, *Guidebook on Biogas Development*, Energy Resources Development Series No 21, An ESCAP Publication, United Nations, New York, 1980, p. 95-97.
21. Ward, R. W., in 'Wise, D. L., *Fuel Gas Development*, CRC Press, 1983', p. 161-188.
22. West, C. E. and E, Ashare, *Feasibility Study for Anaerobic Digestion of Agricultural Crop Residues*, Report No TR- 98195-1, SERI, Colorado, March, 1981, p. 3-11.
23. Bhakthvatsalam, V., Promoting Renewable Energy: Role of Banking System, *Proc. Seminar on Integrated Management*, Hotel Ashok, New Delhi, Dec. 21-22, 1992.
24. Chottani, A. A., Linking Biogas Programme with Dairy Development Programmes, *Proc. National Workshop on Biogas Development*, Vigyan Bhavan, New Delhi, Feb. 1982.
25. Datta, K., Energy Use in Agriculture, *Economic Times*, New Delhi, July 21, 1986, p. 5.
26. ICAR, *Handbook of Agriculture*, Indian Council of Agricultural Research, New Delhi, 1987.
27. KVIC, Post-Installation Routine Maintenance (PIRM) Scheme, *Biogas Forum*, (1992), KVIC, 3 Irla Road, Vile Parle (West), Mumbai.
28. Mazumdar, A. and Gopalakrishnan, N. K., *Role of Cooperatives in the Diffusion of Biogas Technology*, A Brochure, TERI, Mumbai, 1982.

29. Parikh, M. P., Installation of Plants on Turnkey Basis, *Proc. National Workshop on Biogas Development*, Vigyan Bhavan, New Delhi, Feb., 1982.
30. Sasse., L., Biogas : The Foolish Flame, *Biogas Forum*, **IV** (**43**) 13-15 (1990).
31. Shekhar, R. C., Institutional Financing for Renewable Energy Development in India, *Urja Bharati*, **3**(**3**), 53-54 (1993).
32. Sinha, C. S. and Kandpal, T. C., A Framework for the Financial Evaluation of Household Biogas Plants in India, *Biomass*, **23**, 39-53 (1990).
33. TERI, *Managing the Biogas Programme (Sponsored by the MNES), New Delhi*, 1994.
34. TERI, *Biogas Technology (Sponsored by the MNES)*, New Delhi, 1994.
35. TERI, *Biogas : A Source of Rural Employment (Sponsored by the MNES)*, New Delhi, 1994.
36. TERI, *TEDDY* : *1995-96*, New Delhi, 1996.
37. Venkataramana, P., National Project on Biogas Development, *Urja Bharati*, **3**(**3**). 33-35 (1993).
38. AKRSP, AKRSP's Biogas Programme : Some Impact Studies, background papers, Workshop on '*Policy Issues in Biogas*' conducted by the Aga Khan Rural Support Programme, New Delhi, August 22, 1995.

Chapter 5

SOCIO-ECONOMIC EVALUATION OF BIOGAS SYSTEMS

5.1 Introduction

Biogas programme is very relevant for a developing country like India where it not only augments fuel supplies, keeps environment clean and saves fuelwood consumption but also creates considerable employment opportunities for rural youth. A number of socio-economic studies have been carried out in India in past from time to time which have helped a great deal in analysing social relevance of biogas programme, difficulties in implementation, economics and economic-viability of plant operations, functional status of plants, factors contributing to non-functioning of plants, applications of biogas and its impact on rural economy, and as a measure for *sustainable development.* Some important socio-economic studies carried out in past were those by T. K. Moulik and U. K. Srivastava (1975), K. S. Parikh, *et al.* (1976), S. Bhavani (1976), R. Bhatia (1977), T. K. Moulik, *et al.* (1978), S. P. Jain (1979), Planning Research and Action Division (1980). T. K. Moulik, *et al.* (1983), National Council of Applied Economic Research (1985 and 1987). Operations Research Group (1985), Aga Khan Rural Support Programme (1987), K. M. Mital and Prem Vrat (1991), S. Kumar, *et al.* (1992), AKRSP/Apporva Oza (1993), R. M. Myles and S. Sampat Kumar (1994), and AKRSP (1995). In addition to these studies during 1986, the Comptroller and Auditor General of India (CAG) carried out a detailed performance audit of the NPBD and brought out a comprehensive report. Some of the critical observations in the CAG report were later contested by T. K. Moulik [12], Bharat Bhusan [2] and S. Parikh [13] who maintained that shortcomings were of a very trivial nature and on the whole the programme made a very significant headway during past few years. In this chapter is has been attempted to summarise and report important findings of these studies and a detailed account of socio-economic evaluation of Janata Biogas Plants (JBPs) carried out by K. M. Mital and P. Vrat (1991).

5.2 Socio-Economic Survey Carried Out by IIM, Ahmedabad (1975).

During 1975, T. K. Moulik and U. K. Srivastava, Centre for Management in Agriculture (CMA), IIM, Ahmedabad (IIMA) carried out a study entitled '*Biogas Plants at the Village Level: Problems and Prospects in Gujarat*' in association with Gujarat Agro-Industries Corporation (GAIC), Ahmedabad. The study sought to analyse problems faced by beneficiaries in installing and maintaining plants, determine economic viability of plants under field conditions, evaluate organisational effectiveness of *Khadi Gramodyog Boards* (KGB) in carrying out these tasks and determine research and development needs in alleviating users' difficulties. It included a cost-benefit analysis based on very comprehensive data collected and that derived from a previous KVIC study. It also covered programme financing by KVIC and commercial banks and problems of loan repayment. Although most findings are relevant on all-India basis the study was based on primary and secondary data derived from Gujarat only. Primary data was collected from 174 individual plant-owners and 25 institutional plants. The plants covered were of size form 60 to 1250 cubic feet. The secondary data was culled out from records of KGBs, KVIC and district supervisors. It was found that beneficiaries were generally households in upper income group. Promotion of community plants wherever applicable was suggested for meeting fuel needs of low income group families. Several plants covered in the study were facing technical, socio-economic and organisational problems. It suggested coordinated approach by different agencies involved in financing, installing and follow-up services. It also proposed a single agency in each district for producing plant components, installing plants on turn-key basis and providing follow-up services on regular basis.

For meeting plants targets along with maintenance support, technological service and extension packages were proposed. Plants were grouped according to their location in coastal or non-coastal districts which in turn were classified as high, medium and low performers depending upon plant density in each. For instance, Vadodra, Surat and Broach were categorised as high performers, Rajkot as medium performer and Ahmedabad, Gandhinagar, Junagarh, Mehsana and Jamnagar as poor performers. For study purposes, sample of plants was made at random from a stratified list. At the time of study, 147 plants were found to be functional and 52 non-functional out of a total of 199 plants. It covered 174 household and 25 institutional plants.

5.3 Socio-economic Survey Carried Out by the IIM, Ahmedabad (1978)

During 1978, T. K. Moulik, U. K. Srivastava, P. M. Shingi and R. Agarwal of the CMA at IIMA carried out a study entitled '*Biogas Systems in India: A Socio-Economic Evaluation*' [11]. As part of this study, a total of 173 plant owners were interviewed comprising 44 from Uttar Pradesh, 45 from Haryana, 44 from Andhra Pradesh and 40 from Madhya Pradesh. In each of these states three districts were selected having minimum, medium and maximum plants except in M. P. where only two districts were selected. Etawah, Varanasi and Nainital were selected in U. P.; Nizamabad, Krishna and Gundur in A. P.; Sonipat, Ambala and Karnal in Haryana; and Ratlam and Narsingpur in M. P. Most of the beneficiaries were rich in economic status as evident by their large land-holdings with average as 31 acres per beneficiary. Level of education of beneficiaries was found to be equally high. The number of cattle possessed by beneficiaries ranged from a minimum of three to a maximum of hundred with average being 12 animals. As intuitively expected, number of animals possessed had a direct relationship with plant capacity. Beneficiaries in general had large families with average number of members being seven. Distance between household and plant varied from 2 to 30 metres. In Uttar Pradesh and Haryana plants were located at somewhat greater distance than M. P. and A. P. In 41 per cent of the cases, owners themselves prepared and transported slurry. In some cases women also assisted in this task. External labour was engaged in one-fourth cases for preparing and shifting slurry. Dung was collected either once or twice a day. This frequency was not found to be related to any socio-economic factor or plant capacity. Biogas was mainly used a a cooking fuel, with its use for lighting only as a secondary objective, and its use for running I. C. engines a rare practice.

In this study, it was also attempted to identify factors which led to plant shut-down. It was found that in most cases break-down occurred due to accumulation of moisture in connecting pipelines, gas leakage from gasholder and pipeline joints, corrosion of steel parts, accumulation of water in plant vicinity during rainy season, low gas output during winter and faulty choice of operating parameters. It is suggested that network of service-centres be created and strengthened wherever already existing. Promotional activities for popularising the programme were suggested to be intensified. It was also stressed to create a unified central agency for pursuing the programme, an idea which later was implemented with the creation of the CASE and later the DNES/MNES. Due to high cost of biogas plants, individual ownership of plants by rural households was becoming increasingly difficult. The study pointed out that there was a need to improve facilities for setting-up biogas plants and improve follow-up services. Furthermore, considerable scope for improving technological, promotional and organisational aspects of biogas programme was found. For success of this programme, installation of biogas plants on turn-key basis, adequate training for the concerned personnel and organisational framework for improved follow-up services were sought to be provided.

5.4 Socio-Economic Survey by National Institute of Rural Development (NIRD), Hyderabad (1979)

During 1979, S. P. Jain [6] of the NIRD carried out a study entitled '*Biogas: A Study of the Working of Plants in Andhra Pradesh, Haryana and Karnataka*'. The study covering three states was undertaken with a view to analyse organisational framework for implementing the biogas programme, examine socio-economic background of beneficiaries and find out peoples' views regarding implementation of the programme. A total of 200 plants were studied comprising 75 plants from Krishna district in Andhra Pradesh, 50 plants from Ambala district in Haryana and 75 plants from Dharwar district in Karnataka. The study evaluated the impact of certain socio-economic factors such as age, family size, education, land-ownership, occupation, livestock-ownership, etc. on the biogas programme. According to the study findings, acceptance of biogas scheme was found to be related to socio-economic factors of the region. It confirmed findings of other studies that biogas programme found greater acceptance among economically affluent rural households. In addition the study analysed certain aspects of adoption behaviour of beneficiaries such as period of plant ownership, sources of finance, organisational and operational aspects and social acceptance of biogas technology.

5.5 Socio-Economic Study of Janata Biogas Plants by Planning Research and Action Division, Lucknow (1980)

During 1980, PRAD, Lucknow, carried out a study entitled '*Janata Biogas System: An Evaluation*'. This study was restricted to Janata Biogas Plants (JBPs) installed in U. P. [14]. It was carried out with a view to analyse reasons for initial high acceptance of JBPs; assess fuelwood and kerosene needs of beneficiaries for cooking and lighting prior to the installation of biogas plants; study pattern of biogas generated and assess level of satisfaction achieved; obtain beneficiaries feedback concerning plant operation; and analyse economic returns from plants. A majority of the plant-owners were satisfied with their performance. Available wastes generated by household animals provided feed for these plants. Considering the fact that JBPs were relatively inexpensive, the main reason for their having not too wide acceptance was inadequacy of animal faeces generated within households. It was, therefore, suggested that plant capacity should be estimated realistically after taking into account available wastes and gas demand. In other words, beneficiaries should be discouraged from having large plants which may short of animal or agricultural wastes to feed them [1].

5.6 Socio-Economic Study of Community Plants by the IIMA (1983)

During 1983, T. K. Moulik, *et al.* [5] at the IIMA carried out a study on socio-economic evaluation of community biogas plants in India. According to study findings, experiments with community biogas plants did not make much headway till date and in fact many of them were not economically viable. The study found a number of shortcomings in the implementation process, especially with regard to target-setting and organisational parameters. Agencies in their anxiety to get physical targets completed, sought community participation by making promises which they could not fulfill. For twenty-one community plants which were set up in first instance, Department of Science and Technology bore initial cost of construction and operation for three years. It recommended that plants should be run by a duly-constituted board on commercial lines rather than ad. hoc. basis through voluntary participation.

5.7 Audit Report by the Comptroller and Auditor General of India (1986)

During 1986, CAG brought out a report on biogas programme under the National Project for Biogas Development (NPBD). It highlighted cases of misreporting of number of plants set up with many of them defective and remaining uncommissioned [4]. It indicated that some 13,401 plants out of a total of 3,55,000 were in excess of plants actually installed and 17,388 plants were not supported by completion certificates. It further pointed out that 6,238 plants were not functioning properly owing to variety of defects, 3,383 plants were not commissioned, 412 plants were lying incomplete and 844 plants did not physically exist. In Madhya Pradesh, 35 per cent of 7847 plants did not function on account of dung scarcity alone. In Bhatinda district in Punjab, more than 40 per cent of 653 plants were found defective and another 25 per cent functioned only partially. It also pointed out that levy cement was either allotted in excess of requirements or diverted for other applications. Similarly in some states subsidy was paid well in advance and at higher rates than prescribed rates and even to persons who owned no cattle. Delays in disbursement of subsidy and bank loans were considerable with 15 per cent of the subsidy amount remaining unutilised [4].

T. K. Moulik [12], Bharat Bhusan [2] and S. Parikh [13] in their respective rejoinders defended implementation of biogas programme in a very convincing and forceful manner. Essence of their defence was that most of the criticism was of trivial nature and most of the so-called weaknesses were no worse than normally faced in implementation of any government programme. According to T. K. Moulik [12], GAG's figures on plants reported in excess of those installed should be taken in perspective of normal bureaucratic system of reporting and not very specific to biogas programme alone. In any target oriented programme, requests for plants which are already sanctioned and under the process of construction are tempted to be shown as installed to meet the target. One should thus not be surprised if some of these plants shown to be installed in excess of the actual were already installed at the time of CAG's data collection for which the state headquarters might not be having updated information [12]. It was argued that number of incomplete and non-existent plants as reported by the CAG were strikingly low. Moreover, incomplete ones may be completed in near future depending on follow-up by users. A small figure of 844 non-existent plants as compared to 1.8 lakhs already installed was no scar on big success of this programme, particularly if we compare with other programmes [12]. Bharat Bhusan [2] in his rejoinder observed that setting up biogas plants involved a highly decentralised effort with technology transfer each time on a new site using skills of local masons, techinicians and users themselves. Within this backdrop if several lakh biogas plants have been set up under the NPBD with 85 per cent of them in working order, then it is no insignificant achievement. As for misreporting of figures, it was pointed out that functioning and over 90 per cent of 844 non-existent units were planned by the KVIC between 1974-75 and 1981-82 and under the NPBD at all [2]. S. Parikh [13] in his rejoinder supported and endorsed the defence of T. K. Moulik [12] in the context pf country's agricultural economy. S. Parikh [13] also underscored the need to set up an independent biogas board to oversee implementation of biogas programme.

5.8 Other Studies carried out during Eighties and Nineties

According to a study carried out by K. C. Khanna [7], practise of connecting household laterines to plants is not widely followed. Similarly dung produced by most cattleheads is either lost in countryside in open or used in dung cakes which is not a very efficient way to use it. If all these wastes are available for biomethanation, some 300 million cubic meters of gas and 200 million tonnes of humus-rich fertiliser can be produced annually. With this availability of wastes as feed, there exists scope for setting up some 40 million biogas plants in the country [7].

Studies conducted in 14 states in 1985 by agencies such as the National Council of Applied Economic Research, Operations Research Group and the Kirloskar Consultants indicated that on an average 85 per cent of the plants functioned satisfactorily with prospects of a third of the balance 15 per cent restorable to working order after repair and rectification. As per field study conducted by the KVIC in 14 districts in Maharashtra, Tamil Nadu and Bihar, it was revealed that out of 13,216 plants set up between 1974 and 1982, 72.5 per cent of them were found to be satisfactorily working. The ratio of non-working to working plants varied from 45 per cent in Bihar to 15 per cent in Maharashtra. The reasons were low gas output during winters, corrosion of gasholder and defective central guide-frame in KVIC model, cracks in masonary domes, choked pipelines, water-logging due to sub-soil seepage, and lack of adequate maintenance support. Structural defects were more common in fixed dome type plants. On the other hand the KVIC plants with floating drum were more prone to corrosion. It was pointed out that by revamping infrastructural facilities for repair and maintenance, biogas programme can make bigger success. This, among others, can be conveniently achieved by involving and training local youth in plant construction and maintenance [2].

A socio-economic evaluation of plants in rural areas in Goa set up under the NPBD was carried out by a study group. Important observations and findings of the study are as follows [18].

- Out of 117 plants studied, 93 were found operational, 14 non-operational, 3 not commissioned at all and 7 under construction.
- About 26.5 per cent of the plant owners were not satisfied in respect of technical guidance available and number of field visits paid by technicians for this purpose. About 9.4 per cent of beneficiaries reported that no technician ever visited them during plant construction phase whereas 11.1 per cent reported than no one visited then even at later stage during plant operation.
- Ownership of adequate bovine animals is a prerequisite for setting up a biogas plant. However, it was found that in 56.4 per cent cases, plants were allowed to be installed even without meeting this requirement. This showed that subsidies were released to beneficiaries without first ascertaining the adequacy of animals possessed by them.
- As for selection of right capacity, it was found that 70 per cent of the beneficiaries had set up plants of larger capacity.

- As for applications of biogas, it was mainly used for cooking although some households had plans to use it for lighting and energising irrigation pump-sets as well.
- As for sufficiency of gas generated, 52.3 per cent of the beneficiaries reported that available gas was adequate to meet their requirement, balance 47.7 per cent said that it was not so.
- During field investigations, it was observed that in most cases plants were managed by a single member of the household with very limited support by other members. The plant operation was consequently affected whenever the key family member was not available.
- As size of households and number of animals possessed are major factors for determining plant capacity, it was emphasised that it should be done carefully and subsidy be released only after ascertaining its correctness.

A study was carried out by the *Aga Khan Rural Support Programme* (AKRSP) in Junagadh district in Gujarat to assess the impact of drought on the biogas programme with which it was afflicted in 1987-88. Drought affected gas yield following its impact on animal health, their survival and faeces output. The survey covered 61 of 73 plants located in five villages in the region. It was found that drought had a crippling effect on the programme with animals population having depleted from 498 to 301. Drought made a damaging impact on fodder availability in villages resulting in fall in dung output by over 60 per cent causing decline in gas yield. In several cases even individuals had to reduce their meal intake. However, despite the adverse impact, biogas programme in the region did not lose momentum. As against 73 plants installed between 1986 and 1987, 150 plants were planned in 1988. If biogas programme could gain momentum during a drought year surely it should do very well during normal periods. Considering wide ranging benefits and popularity of the programme, it was recommended that nothing should be done that might decelerate its tempo [17].

Zubedabehn, a Master Extension Volunteer (MEV) of the AKRSP played a notable role in February, 1995 in motivating *Galiben*, a poor woman in forties to construct and own a biogas plant of her own despite lack of initial family support. *Galiben*, her husband *Devabhai* and their four children are landless *Rabari* immigrants to Haripur, a village of 300 households in the semi-arid Junagadh (Gujarat). Before the onset of 1987-88 drought *Galiben* had a cattle population of thirty-five which provided reasonable income from milk sale (Rs 20,000 in one particular year). It even enabled her to buy a gold necklace. However, following drought many of *Galiben's* cattle died. From among those who survived, *Galiben* was forced to sell all but four buffaloes and a cow for feeding her family. As a result milk yield was barely sufficient for meeting her family needs with no scope of any income from it. Undeterred by calamity, *Galiben* tried to make earnings from cattle dung. Galiben could accumulate a truckload of dung each month which she then sold for six hundred rupees for eight months a year. *Galiben's* husband started taking cattle to other villages for grazing which enabled to earn on an average fifteen rupees per animal per month. These efforts helped achieve a combined income of Rs. 8,000 – 10,000 per annum which however was for from adequate in meeting household expenses. Within this backdrop, *Zubedabehn*, an MEV from the AKRSP entered the scene in February 1995 and tried to motivate *Galiben* to construct and own a biogas plant of her own without any family support. AKRSP has been active in the village since 1989 and during this period it motivated over one hundred households to go in for biogas plants. With help from Purushotam, another MEV from the AKRSP, *Zubedabehn* finally convinced *Galiben* to own a plant who built a plant by March 1995. Availability of biogas saved her husband's time in fuelwood collection from forest and reduced her own cooking time by nearly half. A full truckload of plant effluent every fifteen days admixed with local and household wastes for use as enriched fertiliser fetched Rs. 600. *Galiben's* monthly income of Rs. 1200 for eight months a year pushed the family's combined annual earnings above Rs. 15,000, apart from added gains of clean environment, freedom from smoke, and dispensing with the tedious process of cooking with fuelwood. In a short spell she could save Rs. 2000 in postal certificates apart from a monthly saving of Rs. 20 in the *Mahila Vikas Mandal.* The case shows that future beholds a lot from biomethanation for such people [21].

Night-soil based biogas plants (NBPs) have become very popular and socially acceptable in Maharashtra in recent years. S. Kumar, *et al.* [19] carried out a survey of 40 NBPs installed under the aegis of the Maharashtra Energy Development Agency (MEDA). The survey of 8 to 85 m^3 capacity NBPs revealed that 75 per cent of them were working satisfactorily. In total 715 m^3/day biogas was being produced which met cooking fuel needs of 1790 persons daily saving 1100 tonnes of fuelwood per annum. Uninterrupted plant operation was found to largely depend on regular cleaning of slurry pit, maintenance of toilets and carrying out minor repairs locally from time to time. Failure of the plants were due to under-utilisation or poor unkeep of toilets leading to inadequate slurry, and negligence of beneficiaries towards occasional minor repairs, The institutional NBPs were found to be operating more efficiently following availability of improved infrastructural facilities as compared to those run by *gram panchayats* and municipal councils.

In India, biogas is predominantly used as a cooking fuel than as a lighting fuel. Realising this a few non-governmental voluntary organisations (NGVOs) have made efforts to intensify biogas usage as a lighting fuel. *Vikas Bharati*, an NGVO working with the AFPRO as a programme-implementing-partners for NPBD located at Bishunpur, a tribal village in Gumala (Bihar). *Vikas Bharati* has been successful in promoting biogas as a lighting fuel in tribal areas where it operates. More than 200 families are regularly using biogas lamps. *Vikas Bharati* has played a key role in motivating tribals to adopt the biogas technology; arranged subsidies and loans from concerned agencies; provided technical services; and trained users in maintaining biogas plants, lamps and stoves for trouble -free operation. One notable contribution of *Vikas Bharti* is promotion of biogas as a lighting fuel [20].

With this background in view, AFPRO has recently carried out as a survey to study, (i) perception of beneficiaries on biogas lighting, (ii) operation and maintenance of biogas lamps, and (iii) impact of biogas lighting on rural life.

The survey covered 20 tribal houses where biogas was used both for lighting and cooking purposes. Out of 20 households, 9 used gas only for lighting and 11 used both as a lighting and cooking fuel. Prior to changeover to biogas lamps, all the twenty households surveyed were using kerosene for domestic lighting. An average 104 litres of kerosene was consumed per month by 20 families before using biogas lamps which later came down to 40 litres thereby reducing kerosene consumption by 61 per cent. Total utilisation hours of 25 lamps were reported as 164.5 hours per day ranging from minimum of 3 hours to a maximum of 11 hours for each lamp. The average utilisation of biogas lamps was found as 6.5 hours per day. The use of burners is not so common, being generally used for preparation of tea and light food only. The total cooking time with 11 burners was only 14.5 hours per day. Average life of biogas lamp mantle was reported as 40 days ranging from as low as eight days to a maximum of six months [20].

Biogas lamps are used for longer duration during winter than summer. The winter days are shorter and cool, the warmth provided by biogas lamps was liked by the residents and therefore they would prefer to use it for longer duration. However, reduced biogas generation during winter presented resident from using biogas lamps throughout the night for warmth. Maintenance of lamps was simple and all family members could repair minor faults themselves [20]. *Vikas Bharati* has proved that with determined efforts biogas lighting can become as popular as biogas cooking.

Benefits of biogas cannot be assessed merely in monetary terms. *Labhuben,* a woman from village Bamanbore (Surendranagar) felt that biogas plant has added 20 years to her life. It was the economic benefits that led to the beginning of biogas programme by the AKRSP. The value of dung-cakes and wood traditionally used as fuel, value of improved manure from slurry, value of the trees that were saved from felling, etc. were the primary concern. Labhuben's statement is only a pointer to the fact that benefits of a biogas plant are much more than economic. Advent of biogas saves time in cooking, collecting fuelwood, and making dung-cakes; increases hours available for sleep as cooking could be started late in the morning; reduces hardships and drudgery involved in collecting fuelwood from distant forests; achieves freedom from intimidation by forest guards, snakes and scorpion bites, and inhaling of smoke and burning of eyes; minimises space requirement for storing fuelwood and dung cakes; increases time availability for attending household needs; and increases school enrollment by girls who do not require to go out for collection of fuelwood/dung [22].

According to a study carried out by S. Suresh of the AKRSP in Junagadh, biogas plants save 15 per cent of women's daily activity time of 14-16 hours. Women who slept only 4-5 hours daily could rest for an hour or two more by getting up at 6.30 AM instead of 5.00 AM. Women labourers could reach in time from their work and not miss out any part wages. In many families children also accompany the women for fuelwood collection leaving little time for school. Biomethanation programme thus improves the child's especially the girl-child's prospects of attending a school. Without biogas 2 to 3 hours were spent every week in repasting of blackened kitchen walls. Cleaning of blackened vessels was equally time consuming. Following availability of biogas, time and effort on these activities has been substantially reduced [22].

Ms. Bina Agarwal (TERI) and Ms. Meena Bilgi (AKRSP) presented a paper on gender aspects of biomethanation at the National Workshop on '*Policy Issues in the Biogas Programme*' which was organised by the AKRSP and sponsored by *Aga Khan Foundation* and *European Commission* in New Delhi during August 21-22, 1995. They reported that 90 per cent of women's energy are spent on cooking with considerable portion of time spent on fuelwood and dung-cake collection. With growing deforestation they have to travel longer distances each day for fuelwood collection and face health problems on account of smoke. According to a WHO study, 74 per cent of women's time is spent indoor in kitchen related activities. Accordingly they have less time available for other household activities such as earning for family and health care. According to the AKRSP study families owning biogas plants enabled higher enrollment rates for girls in school as their labour was not needed all day to collect wood and cook. Biomethanation involves gender issue as it saves cooking time, saves time in fuelwood collection, provides smoke-free environment, minimises expenditure on traditional fuels, and generates enriched fertiliser for kitchen garden leading to improved health and possibly higher income. The fact that women are the immediate beneficiaries of biomethanation as well as the end users, has important implications for success of biogas programme. Women needs should be taken into consideration while developing new plant designs or evolving programme implementation strategies. Women should also be trained in plant maintenance particularly for carrying out small repairs which will help induce gender equity in biomethanation [21].

AKRSP constructed plants for 82 out of 120 households in Dedakiyal village in Mendarda taluka of Junagadh district (Gujarat). AKRSP later carried out socio-economic evaluation of 15 biogas plants of 2-3 m^3 capacity range. The study showed that biomethanation brought considerable improvement in quality-of-life of beneficiaries and also could save 82 mandays in a year which they were earlier spending on fuelwood collection and preparation of dung cakes. As a result women were able to complete their cooking in 3 hours as against 6 hours prior to the introduction of biogas. Biogas achieved a saving of Rs. 900 per year by replacing kerosene, fuelwood and dung cakes. Farmers needed fuelwood and kerosene only during monsoon season when biogas yield was less. Biomethanation also led to family harmony as there were less quarrels in family over delays in food preparation. Following absence of smoke biomethanation also led to improvement in rural hygiene. It also led to generation of 1.5 to 2.0 cartloads of manure as against 1 cartload of manure before introduction of biogas. Earlier some 300 bullock-carts of fuelwood were collected as against 15-20 bullock-carts following biomethanation [23].

Out of 15 households surveyed, 11 households faced no operational problems, one reported a dysfunctional plant, and 3 reported minor defects in accessories as well as water clogging. Biomethanation achieved 15 per cent reduction in cooking time with total cooking time reduced by 32.5 hours per day - an average of 2.15 hours per family ranging from a minimum of one hour to a maximum of 5 hours [22].

Plant owners were divided into two categories. Category I included affluent farmers owning farmhouses (*wadis*) and large capacity biogas plants. Category II included small farmers or individuals engaged in other trades like carpentary, masonary, etc. Factors studied by the AKRSP included : cattle population, family size, pattern of cooking time before and after plant installation, fuels used, utilisation of plant effluent, and benefits accruing to the women [23].

Fuels previously used were generally either dung or wood. Of five families in category I, four obtained wood from their *vadis* (farmhouse) while the other two purchased it from other sources. Eight families in category II who had no *vadi* of their own had to walk as much as 5 km occasionally for fuelwood collection. Furthermore, entry into a forest required a tip of Rs. 20 to the forest guard per visit which averaged about 1 rupee per kg of wood collected. Biogas plants brought more benefits to category II families. Advent of biogas was not only financially rewarding and time-saving but it also prevented harassment and risk involved in fuelwood collection from distant and isolated locations. With an average usage of 5.5 kg of fuelwood per day, biogas plants for category II owners led to a saving of Rs. 5.5 per day or Rs. 2000 annually, Even two families in category II claimed a saving of Rs. 900 per annum on this account [21, 23].

Families in both categories were benefited from a reduction in labour input. A labour reduction of 50 mandays per year for a family in category I corresponded to a saving of Rs. 1000 annual with an assumed labour rate of Rs. 20 per day. Even a reduction of merely 20 mandays led to a financial gain of Rs. 400 per year [23].

Output of digested slurry (enriched fertiliser) depended on amount of wastes fed. Effluent ranged from one to eight cartloads per month. It was utilised by all six families in category I, as enriched fertiliser and soil-conditioner. Digested slurry was also used for making dung-cakes by 4 out of 6 families for meeting extra fuel needs during winter months [23].

Out of the eight families in category II, only three used slurry for cultivation in their small landholdings. Three families sold digested slurry in exchange for wheat which they needed for subsistence. One family exchanged slurry for fodder whereas remaining households preferred to give away effluent to a relative in exchange for his labour during two seasons of the year [23].

Twenty six women from different households were enquired about main advantages perceived from biogas plant. One women replied that health was the primary benefit, three responded that cleanliness was the greatest advantage, eight replied cleanliness and safety combined as the reason, while fourteen claimed that efficient cooking is the greatest benefit. Women from another village of Junagadh district (Dedadiyal) reported that advent of biogas has reduced tensions in the family earlier caused by quarrels over delays in food preparation. Biogas thus provides harmonious and pollution-free environment. Due to reduction in cooking time, women were able to devote their time to other household needs. Following all-round benefits, biogas programme should be viewed as an important intervention for improving the quality-of-life of rural and urban poor [21, 23].

5.9 A Socio-Economic Evaluation of Janata Biogas Plants (1991)

This study carried out by K. M. Mital and Prem Vrat (1991) related to socio-economic evaluation of JBPs which were originally designed and developed by the Planning Research and Action Division (PRAD) of the State Planning Institute, Lucknow. JBPs have no moving parts and accordingly wear and tear and consequent maintenance problems are relatively limited. Accordingly, these plants have a longer working life. Gas yield of JBPs is considered to be higher (0.04 m^3 per kg of cattle dung at 24-27°C) than those of other plants. Effluent obtained from JBPs has the same carbon and nitrogen content, and it is richer in phosphorus and potassium contents by 2.5 to 6.5 per cent and 7.2 to 18.5 per cent, respectively as compared to those from other plants. JBP dispenses with movable steel drum type gasholder. Its gasholder made of bricks and cement is in the form of immovable and fixed-dome cover of the well-like digester as an integral unit. When biogas is formed inside digester, it rises up within dome and pushes slurry down. Displaced slurry level provides necessary pressure for release of gas for its use.

During 1978, Directorate of Extension in Union Ministry of Agriculture and Government of Uttar Pradesh took several steps to popularise JBPs. In 1979, a scheme for setting up 100 JBPs at different Farmers Training Centres, Extension Training Centres, government dairy farms and demonstration centres was approved with 100 per cent central subsidy. The scheme also has support of several voluntary and government agencies. In view of growing popularity, PRAD received inquiries from different state governments, voluntary organisations and individuals for providing technical guidance for installation of JBPs. For giving a fillip to JBPs in U. P., facilities for imparting training to beneficiaries were created at twenty one Extension Training Centres (ETC). Furthermore, in early eighties, U. P. government announced a subsidy of Rs. 550, Rs. 725, Rs. 800, Rs. 900 and Rs. 1000 for 2, 3, 4, 5 and 6 cubic meter capacity JBPs, respectively.

This study related to JBPs set up in Uttar Pradesh in past under the aegis of twenty one ETCs. The study was carried out using a questionnaire distributed among one hundred and fifty plant owners out of which sixty eight responded. The questionnaire was drafted in Hindi as most of the owners only knew Hindi and not English. The study touched all related aspects covering technical, behavioural, financial, promotional and social issues. Out of the sixty eight plants studied, two were set-up in 1976, five in 1978, eighteen in 1979 and forty three during eighties. On the whole the study concluded that advent of JBPs in India had been very promising considering their low cost, ease in operation and high degree of satisfaction derived by the beneficiaries. A record sixty eight (91 per cent) conveyed that they were satisfied and benefited with the scheme [9]. Following are detailed results of this study.

Promotional and Motivational Aspects : It was analysed that twenty plant owners (29 per cent) picked up the idea from promotional activity of Block Development Officers (BDO), fourteen (21 per cent) from Village Social Workers (VSW), fourteen (21 per cent) from ETCs, six (9 per cent) from Area

Development Officers (ADO), three (4 per cent) from Village Development Officers and eleven (16 per cent) were self-motivated. As for plant capacities, plants of 140, 150, 200 and 210 cubic feet capacity were more in common as is evident from results summarised in Table 5.1.

Table 5.1 Capacitywise Distribution of Plants Covered

Capacity (Cubic feet)	*Number of Plants*	*Per cent*
70	2	2.9
100	11	16.1
105	7	10.2
120	1	1.4
125	1	1.4
140	14	20.5
150	11	16.1
200	7	10.2
210	7	10.2
250	2	2.9
290	1	1.4
300	2	2.9

Time Interval Between Date of Application and Actual Installation : The time interval between making application for owning plant and actual installation varied from less than a week to a year (Table 5.2). There was no delay in plant installation for sixty nine per cent of the sample. For the balance who experienced delays, 13 per cent attributed it to cement shortage, 9 per cent to technicians' non-availability, 3 per cent to funds shortage and 6 per cent were unable to specify any reason.

Table 5.2 Distribution of Plants According to the Lead Time Between the Date of Application and Installation

Lead Time Weeks	*Number of Plants*	*Per cent*
Less than 1	7	10.2
1	19	28.9
2	8	11.7
3	7	10.2
4	11	16.1
5	3	4.4
6	2	2.9
8	4	5.8
10	1	1.4
12	1	1.4
14	2	2.9
21	1	1.4
25	1	1.4
More than 25	1	1.4

Financial Aspects of Biogas Plants : Out of sixty eight, twenty seven (40 per cent) did not seek bank loans at all. Twenty from balance forty one who availed loans looked forward to repay them from plant surplus to be generated whereas remaining expressed doubts in this regard. Reasons for difficulty in loan repayment included inadequacy of wastes, poor economic status of owners, high bank interest rate and low gas yield. Eighty per cent of beneficiaries who availed loans, received it within four weeks of date of application. About two third of those who took loans were required to repay it within a period of three years and remaining beyond it. In short, it implied that receiving and repaying loans did not pose any major difficulty to beneficiaries.

Three-fourth of beneficiaries availed subsidy amount varying from Rs. 800 to 1200 (Table 5.3). Distribution of plants according to the amount of bank loan availed and amount invested by the beneficiaries is given in Table 5.4 and 5.5. It can be seen that most common subsidy amount drawn was Rs. 1000 or 1200, common loan range was between Rs 2000 to 3000 and amount invested by owner ranged from 500 to 5000.

Table 5.3 Distribution of Plants According to the Amount of Subsidy Drawn

Amount of Subsidy (Rs.)	*Number of Plants*	*Per cent*
0	5	7.3
500	4	5.8
600	3	4.4
700	3	4.4
800	35	51.4
900	—	—
1000	14	20.5
1100	—	—
1200	3	4.4
1300	1	1.4

Table 5.4 Distribution of Plants According to the Amount of Bank Loan Availed

Amount of Bank Loan (Rs.)	*Number of Plants*	*Per cent*
0	36	52.9
1000	2	2.9
1500	2	2.9
2000	7	10.2
2500	6	8.8
3000	9	13.2
3500	1	1.4
4000	3	4.4
4500	2	2.9

Maintenance and Service Aspects : Of the sixty eight beneficiaries twenty two (32 per cent) were satisfied with available maintenance facilities whereas forty six owners (68 per cent) had some dissatisfaction on this score. As for preventive maintenance, it was found that about 90 per cent of beneficiaries had no such idea. About one-third of the beneficiaries reported no maintenance problem since commissioning. Of the forty four owners whose plants posed problems, twenty four (35 per cent) reported difficulties in functioning of pipeline, sixteen (24 per cent) in lamp, thirteen (19 per cent) in burner and six (9 per cent) in gasholder, an order which is same as reported by S. P. Jain [6] although his study did not cover plants in UP nor was it limited to JBPs. As for availability of maintenance support, twenty one (31 per cent) never needed maintenance services, five (7 per cent) availed services of local mechanics, twelve (18 per cent) called mechanics from a distance of 1-5 miles, twelve (18 per cent) from a distance of 6-10 miles, two (3 per cent) from a distance of 11-15 miles, one from a distance of 16-20 miles and three from 50 to 200 miles. These results are in partial agreement with those reported by T. K. Moulik, *et al.* [11, p. 156]. They reported that in 22 per cent cases of plant breakdown, mechanics are available within a distance of 5 km (3.1 miles), in 33 per cent cases within a distance of 6 to 10 km (3.7 – 6.2 miles), in 17 per cent cases within a distance of 11 to 15 km (6.8 – 9.3 miles), in 18 per cent cases within a distance of 20 to 25 km (12.4 – 15.5 miles) and in 10 per cent cases beyond a distance of 30 km (18.6 miles).

Table 5.5 Distribution of Plants According to Amount Invested by the Beneficiaries

Amount of Money Invested (Rs.)	*Number of Plants*	*Per cent*
0	6	8.8
500	3	4.4
1000	10	14.7
1500	9	13.2
2000	5	7.3
2500	6	8.8
3000	6	8.8
3500	6	8.8
4000	11	16.1
4500	1	1.4
5000	5	7.3

Aspects Relating to Plant Operation and Utilisation of Biogas : Beneficiaries were also enquired with regard to operational problems faced in preparation of slurry, transportation of organic wastes to plant site, application of improved techniques such as use of hot water in slurry preparation and utilisation of other feed materials. For preparing plant slurry, forty eight (71 per cent) reported that this is done by one of the members of family, fifteen (22 per cent) accomplished it by hiring labour and for five (7 per cent) it was collective endeavour. As for shifting of animal wastes to plant sites, forty three (63 per cent) mentioned that it is carried out by one of the family members, twenty one (31 per cent) reported that it is done by hiring labour and four (6 per cent) said that this work is collectively performed by family members and hired labour. These findings are sufficiently close to those reported by T. K. Moulik, *et al.* [10, p. 124].

As for use of hot water in slurry preparation, fifty nine (87 per cent) beneficiaries reported that they did not use warm water whereas nine (13 per cent) reported in affirmative. Regarding use of other feed materials as additives in slurry preparation, fifty nine (87 per cent) mentioned that they did not use any, whereas nine (13 per cent) reported that they used sugarcane bagasse, green leaves, rice-husk, straw, crop residues, etc. In view of large quantities of agricultural wastes being generated as by-product, it is desirable that some promotional efforts be undertaken for popularising their use.

Distribution of number of animals owned by beneficiaries is given in Table 5.6, a pattern which is similar to that reported in other studies. It was found that 5 to 6 animals were possessed by 29 per cent of owners whereas T. K. Moulik, *et al.* [10, p. 116] reported that 27 per cent owners had 5 to 6 animals and according to the PRAD study 30 per cent owners had 6 to 8 animals [13, p. 9].

Table 5.6 Distribution of Plants According to the Number of Animals Owned by the Owners/Beneficiaries

Number of Animals	*Number of Plants*	*Per cent*
1/2	1	1.4
3/4	8	11.4
5/6	20	29.4
7/7	11	16.1
9/10	7	10.2
11/12	5	7.3
13/14	2	2.9
15/16	4	5.8
17/18	3	4.4
9/20	3	4.4
More than 20	3	4.4

Savings in other fuels consumption following introduction of biogas is summarised in Table 5.7 which indicates that biogas could replace other fuels substantially. This is substantiated by the fact that with biogas, 38, 50, 92, 92 and 41 per cent of beneficiaries completely stopped using kerosene, dung cakes, soft-coke, hard-coke and fuelwood, respectively.

Table 5.7 Savings in Consumption of Other Fuels Following Introduction of Biogas

Fuels Replaced	*Number of Beneficiaries Reporting One-fourth, Half, Three-Fourth and Complete Replacement of Conventional Fuels*			
	One fourth	*Half*	*Three-fourth*	*Complete*
Kerosene	18	10	14	26
Dung-cakes	11	13	10	34
Soft-coke	5	—	—	63
Hard coke	1	1	1	63
Fuelwood	12	8	10	28

5.10 Concluding Remarks

The major findings of practically all the cost-benefit studies and socio-economic surveys carried out in India in past indicate that if all the micro and macro-level long term social benefits and tangible and intangible gains are taken into account, biomethanation makes a strong economic sense as a viable national programme. While assessing cost-effectiveness, it is important to take intangible benefits into account, as only then biogas programme appears economically attractive and viable programme. The strength of argument in favour of the biogas programme demands interest free credit flow or even a complete subsidisation of the plant cost [14]. Biogas can become viable alternative to many conventional fuels provided hidden subsidies for the latter and environmental degradation they cause are also taken into account [3]. Amount of subsidy should be related to the prevalent cost of plant-construction [15]. In order to prevent beneficiaries from neglecting or abandoning malfunctioning plants, they should be accountable to repay the subsidy to prevent misuse of the scheme [16]. Side by side remedies to minimise operational problems need be found. Common problems range from leakage in gas pipeline, carbon deposition on gas burners, water accumulation in pipeline, improper and irregular supply of feed and low gas yield. During winter when yield is low, solar heating or using a part of biogas generated to heat digester contents can be resorted. As for fixed dome plants, be it, *Janata* or *Deenbandhu* models, their future is particular bright. This optimism is based on the fact that majority of beneficiaries reported that they were benefited by the installation of such plants and the generated gas used for lighting and cooking. While the majority did not report delay in installation, some had it and identified responsible factors. Nearly half of the beneficiaries who availed loans were confident of repaying it from surplus generated. Finally concluding, it is to be emphasised that there is nothing which should shake confidence in future of the biogas programme. In fact every thing possible should be done to make it a bigger success in coming years. In these endeavours it is always beneficial to undertake socio-economic surveys at grass-root level, from time to time which provide valuable feedback on programme implementation and help apply necessary course-correction when a snag is found.

REFERENCES

1. Bahadur, S. and K. K. Singh. *Janata Biogas Plants*, A Monograph, State Planning Institute, Lucknow, March, 1980.
2. Bhusan, B., Biogas Plant Under Attack: Devaluing Decentralised Energy, *Times of India*, New Delhi, Jul. 31, 1986, p. 5.
3. *Business Standard*, Long Term Plan to Develop Conventional Energy, Calcutta, Jan. 22, 1989, p. 3.
4. *Economic Times*, Disturbing Trends (Editorial), New Delhi, June 25, 1986, p. 5.
5. *The Hindu*, Biogas: Community Needs to be Met, Hyderabad, Aug. 24, 1984.
6. Jain, S. P., Study of Biogas Plants in Andhra, Haryana and Karnataka, *Rural Development Digest*, **3(4)**, 231-274 (1980).
7. Khanna, K. C., Biogas in India and China, *Times of India*, New Delhi, Jan. 28, 1986.
8. Kumar, N. A., *Economics of Biogas Evaluation*, Times Research Foundation, Pune, 1989.
9. Mital, K. M. and P. Vrat, *A Review of Socio-Economic Evaluation of Janata Biogas Plants in U. P., A Working Paper; New Delhi, 1991.*
10. Moulik, T. K. and U. K. Srivastava, *Biogas Plants at the Village Level : Problems and Prospects in Gujarat*, A Report, Indian Institute of Management, Ahmedabad, 1975.
11. Moulik, T. K., *et al.*, *Biogas Systems in India : A Socio-Economic Evaluation*, A Report, IIM, Ahemedabad, 1978.

12. Moulik, T. K., India's Biogas Programme : Last 10 years Experience, *Economic Times*, New Delhi, Aug. 20, 1986, p. 5.
13. Parikh, S., National Biogas Programme, *Economic Times*, New Delhi, Sept. 17, 1986, p. 5.
14. PRAD, *Janata Biogas Systems* : *An Evaluation*, A Report, Planning Research and Action Division, State Planning Institute, Lucknow, March, 1980.
15. Rao, A. R., Biogas Programme in India-1, *Khadi Gramodyog*, June, 399-410 (1987).
16. Rao, A. R., Biogas Programme in India-II, *Khadi Gramodyog*, Dec., 165-168 (1987).
17. Shah, A. C., Impact of Drought on Biogas Programme, *Economic Times*, Mumbai, June 13, 1988, p. 11.
18. *Times of India*, Biogas Fuel Going Waste in Goa, New Delhi, Oct. 27, 1987, p. 15.
19. Kumar, S., *et al.*, Implementation of Night-soil Based Large Size Biogas Programme in Maharashtra, in 'N. K. Bansal (ed.), *Proc. Nat. Solar Energy Conv.*, Dec. 28-30, 1992', p. 393-396.
20. Myles, R. M. and T. S. Kumar, Domestic Lighting through Biogas in Tribal Areas of India, *Biogas Forum*, **1** (No **56**), 15-18 (1994).
21. AKRSP, AKRSP's Biogas Programme: Some Impact Studies, Proceedings and Background Papers, *Workshop on Policy Issues on Biogas*, Aga Khan Rural Support Programme, New Delhi, August 22, 1995.
22. Oza, A., Biogas : A Friend of the Environmental and Women, *Exchanges*, No **3**, Actionaid, 3 Rest House Road, Banglore, 1993.
23. AKRSP, *A Report on the Janger Biogas Impact Study*, Navarangpura, Ahmedabad, 1994.

Chapter 6

COMMUNITY BIOGAS PLANTS : AN OVERVIEW AND SELECT CASE STUDIES

6.1 Introduction

Biogas plants can be constructed either as individual family units or as community plants. Whereas individual plants are suitable for families, community plants are appropriate for meeting fuel needs of groups, institutions or villages as a whole. In order to alleviate problems of lack of adequate wastes and finances available with individuals, option of community plants is often advantageous. Community plants can be operated by a group of families, cooperatives or industrial townships. Religious, cultural and social environment of a region can favour or disfavour operation of community plants. A community plant may also help in running a small-scale industry which may also crate employment opportunities for rural youth. A community plant also makes it possible to run a community kitchen. Some experts even maintain that individual plants should be constructed only when community plant is not feasible [11].

Community biogas plants are socially more egalitarian. Out of 100-odd million households in India, only about 20 million own three-to-four heads of cattle needed to feed family-size biogas plants. According to the report of the Planning Commission's Working Group on Energy Policy (1979), even if all those households who could set-up family-size plants were to do so, still enough cattle would be available to feed 560,000 community plants-approximately one in each village of the country. During sixth plan (1980-85) when idea of community plants received prominence a target of 100 community plants was fixed up.

In a country like India where large majority of farmers cannot own biogas plants due to economic difficulties, promotion of community plants is a viable solution. A farmer who is reluctant to invest a large sum of money for his own private plant may not find it difficult to pay a small deposit and rentals as his share in the running of a community plant. Community plants provide economies of scale which arise from savings in initial capital cost, land utilised and skilled man-hours needed in operating plants. Its economics can further improve if night-soil from community latrines is also fed to it. Animal wastes often lying on village roads and grazing lands should also be utilised. Community plants achieve higher gas yield per kg of feed due to greater economic viability and creation of facilities such as warming up slurry during winter season which is not cost-effective in family units. Increased biogas yield from community plants can also be used for generating power, fuelling rural industrial units, operating irrigation pumpsets and lighting streets [4]. They also offer prospects of large-scale composing of organic wastes generated in villages by mixing them with plant sludge. Input output analysis of a typical community biogas plant is schematically shown in Fig. 6.1.

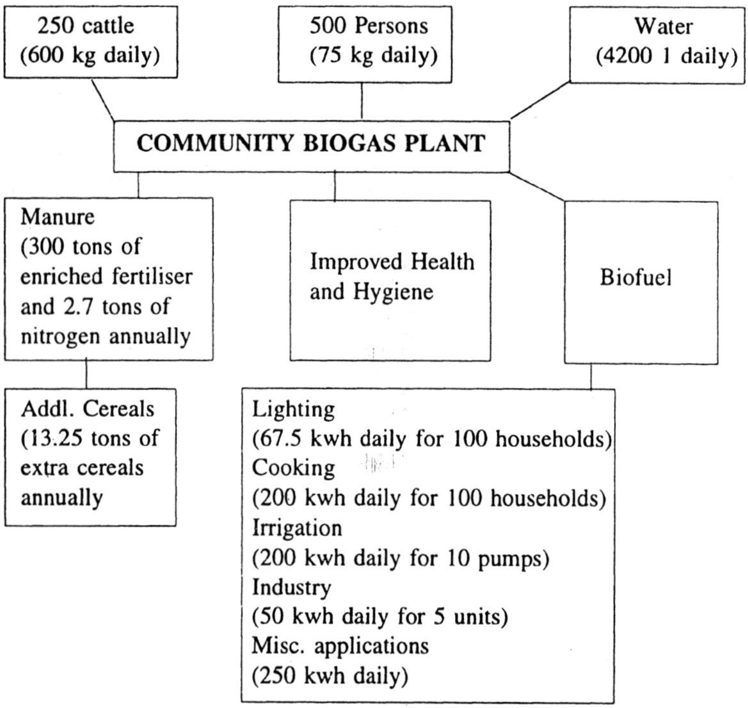

Fig 6.1 Input output energy of a typical community biogas plant.

Source : BORDA/GATE (Germany), Various reports.

6.2 Factors Affecting Operation of Community Plants

There are several social managerial and technical factors which affect operation of community plants. Social aspects deal with issues like relative importance of community plants over other rural projects, society's perception regarding their utility as fuel generating or waste-treatment device, and community's overall concern and participation in its operation. Managerial aspects relate to operation and maintenance of plants, allocation and pricing of gas and digested manure (fertiliser) among users, collection of feed materials for plant operation, supply of water for slurry preparation and funds mobilisation if plant runs into economic difficulties. Technical aspects cover issues such as laying of pipeline for gas supply. For instance, if houses in a colony are widely scattered, distribution of biogas through pipeline may pose economic difficulties. Community plants require higher degree of managerial and technical skills for operation due to wider problems of waste collection, gas and effluent storage, and allocation gas and effluent (fertiliser) among users [13].

6.3 Community Biogas Plant at *Fateh Singh- Ka-Purwa*

Fateh-Singh-ka-Purwa is a tiny village in Bhagynagar block in Etawah (U. P.) and is linked to Bhagyanagar-Dibiapur road by a *kutcha* link road. During 1978 it was assessed that forty seven per cent of the villagers owned less than requisite cattle to own their individual family plants and accordingly it was decided to set up country's first ever community plant in this village. The plant was set up by the Planning Research and Action Division (PRAD), Lucknow, with UNICEF assistance. Site-selection was influenced by the fact that it already had the Integrated Area Development Project in Etawah which provided it considerable infrastructural support. Adequate wastes availability, villagers cooperation and commitment, and fairly uniform economic status of households were some of the other favourable factors for site-selection.

The community plant consisted of two units of 35 m^3 and 45 m^3 capacity with former supplying gas for meeting cooking and lighting needs and latter for running a 5 HP dual-fuel engine to power a 3.5 KVA electric generator. The first unit provided gas during 7-9 AM for preparing breakfast, during 11 AM to 1 PM for lunch and during 7-9 PM for dinner. The second unit enabled to supply electric power to 27 households each having two bulbs of 40 watts each and eleven tubelights for lighting the streets daily for four hours. During initial phase of plant operation, clay-burners were handed over to every household free of charge which were then replaced by steel burners as and when the earlier ones broke down and these were also supplied free. Whereas biogas for meeting cooking need is supplied free, the one used for running the flour mill, chaff cutters and tube-wells is nominally billed. Apart from supply of gas, the resulting plant effluent is also distributed among households in proportion of the cattle-wastes contributed by them [1]. Gas supply was done through underground GI pipeline fitted with adequate water traps and safety valves at appropriate locations.

S. Bahadur and S. C. Agarwal [1] carried out cost-benefit analysis of the plant based on one year of plant operation. In their analysis it was assumed that 45 per cent of available gas was used for cooking, 30 per cent for lighting and balance 25 per cent for other applications.

Animal wastes available = 1524 kg/day
Biogas yield during winter = 38 litres/kg of dung
Biogas yield during summer = 55 litres/kg of dung
Biogas required for obtaining one horsepower per hour for running a dual-fuel engine (80 per cent gas and 20 per cent diesel) = 420 litres
Diesel consumed per horsepower in a diesel engine = 0.25 litres
Heat content of 28 litres (1 ft^3) of biogas = 135 kcal
Efficiency of a standard gas burner = 60 per cent
Effective heat utilisation in burning 28 litres

(1 ft^3) of biogas = 135 × 0.6 = 81 kcal

Heat produced by 1 kg of coal in an open stove with 28 per cent efficiency = 1761 kcal

Amount of kerosene oil equivalent to 28 litres of biogas = 0.018 litres

Based on 1978 price level economic cost of soft cake = Rs. 185/tonne

Biogas Yield During the Year :

Total biogas yield during four winter months = 120 × 38 × 1524
= 69,49,440 litres

Total biogas yield during eight summer months = (365-120) × 55 × 1524
= 205,35,900 litres

Total biogas yield during the year = 6949440 + 20535900
= 2,78,45,40 litres

Utilisation of Biogas for Various Applications

Biogas utilised for cooking @ 35910 litres per day = 35910 × 365
= 1,31,07,150 litres/year

Biogas utilised for lighting electric bulb @ 10080 litres per day = 10080 × 365
= 36,79,200 litres/year

Biogas utilised for lighting gas bulbs

With 10080 litres per day needed to generate 2560 watts and assuming 120 watts equal to 100 candle power and that one gas lamp is sufficient to provide illumination of 100 candle power, number of gas lamps needed:

(2560/100) × (100/21) = 21 gas lamps

Assuming that each gas lamp consumes 126 litres per hour, total gas consumed
= 21 × 4 × 126
= 10,584 litres/day

Hence, total gas consumed = 10584 × 365
= 38,61,700 litres/year

Biogas available for other applications when generator is used = 27485340–(13107150 + 3679200)
= 1,06,98,990 litres

Horsepower available per annum for other uses @ 420 litres of biogas/ H.P = 10698990/420
= 25,474 HP hour

Biogas available for other applications when gas lamps are used = 27485340 – (13107150 ı 3861700)
= 105,16,490 litres

Horsepower available per annum for other purposes @ 420 litres of biogas/HP = 10516490/420
= 25,039 HP hour

Valuation of cooking gas in terms of equivalent heat contained in soft coke

Total gas required in terms of heat utilisation

for cooking @ 60 per cent burner efficiency = 13107150 × 81/28 kcal/year
= 3,79,17,153 kcal/year

Total gas required for cooking in terms of coal equivalent = 37917153/1761 kg of coal/year
= 21,532 kg of coal/year
= 21.532 ≈ 22 tonnes of coal

Total gas required for cooking in money terms= Rs. 185 × 22
= Rs. 4070/year

Valuation of biogas lighting in equivalent illumination by kerosene

Total gas required for lighting with electric bulbs in terms of kerosene equivalent = 3679200 × 0.018/28
= 2365 litres of kerosene/year

Total gas required per annum for lighting in money terms assuming unit price of kerosene @ Rs. 1.70/litre = Rs. 2365 × 1.70 = Rs. 4020

Valuation of benefits as cost of diesel replaced in dual engine with generator.

Horsepower available for other applications = 25,474 HP/hour

Equivalent diesel consumption = 25474 × 0.25
= 6368 litres/year

Equivalent diesel consumption in money terms assuming unit price of diesel @ Rs. 1.85 per litre = Rs. 6368 × 1.85
= Rs. 11,780 per annum

Total benefits with electric generator = Rs. (4070+ 4020 + 11780)
= Rs. 19,870

Valuation of benefits without generator

Total gas required for cooking in money terms = Rs. 185 × 22
= Rs. 4070/year

Total gas required for lighting in money terms = Rs. 2365 × 1.70
= Rs. 4020/year

Horsepower available when gas lamps are used = Rs. 25,039 HP hour/year

Equivalent diesel consumption = 25039 × 0.25
= 6260 litres/year

Equivalent diesel consumption in money terms = Rs. 6260 × 1.85
= Rs. 11,581/year

Total benefit without generator = Rs. (4070 + 4020 + 11581)
= Rs. 19,671

Manurial Value of Effluent

Daily feed at two plants at *Feteh-Singh-Ka-Purwa* is 1229 kg. Feed comprises animal wastes, crop residues and weeds like water-hyacinth. The feed thus on annual basis is about 450 tonnes. Before setting up plant about half of available cattle dung was providing manurial input to field through composing, the other half used as cow dung cakes or wasted. Following setting up community plant, additional 225 tonnes of organic manure is provided to villagers as enriched fertiliser. Out of 225 tonnes of slurry, solid content is about 21 tonnes comprising 4.2 tonnes of nitrogen, 2.1 tonnes of phosphorus and 2.7 tonnes of potassium. The value of enriched fertiliser was estimated as Rs. 15000 per annum..

Total Benefits

Total benefits accruing from Community Plant = Rs. (15000 + 4020 + 4070 + 11780)
= Rs. 34,870 per annum

Total Costs

Total capital cost with generator = Rs. 2,02,000
Operating cost = Rs. 2398
Salary payments to electrician @ Rs. 400 per month= Rs. 4800 per annum
Fuel (diesel) cost = Rs. 613 per annum
Maintenance cost = Rs. 1000 per annum
Total costs = Rs. 8811 per annum

Details of cost benefit analysis intangible are summarised in Table 6.1. It does not take into account several benefits like improved health, sanitation and extra leisure time. If these benefits are also taken into account, benefit-cost ratio can improve further. However, even without taking them into account, the analysis establishes economic viability of the project (It is to be noted that costs relate to 1978 price level and since then these have gone up considerably).

Table 6.1 Cost benefit Analysis of Community Plan at Fateh-Singh-Ka-Purwa

Year	*Capital Cost (Rs.)*	*Operating Cost (Rs.)*	*Benefits (Rs.)*	*Net Benefits (Rs.)*	*Depreciation (10%)*	*Present Value of Capital Cost (Rs.)*	*Present Value of Net Benefits (Rs.)*
1	2,02,000	4405	17435	13080	0.909	183618	11844
2.	—	8811	34870	26059	0.826	—	21525
3.	—	8811	34870	26059	0.751	—	19570
4–20	—	8811	34870	26059	6.028	—	157135
Total						183618	210074

Benefit-cost Ratio = 210074/183618 = 1.14 : 1

6.4 Community Biogas Plant at Kubadthal

This case-study pertains to a community plant at Kubadthal village in Ahmedabad (Gujarat) which was set up by the *Vimla Gram Seva Samaj Trust,* a voluntary organisation, with active support of its founder Mr. Vadilal Lallubhai Mehta. Kubadthal village is located on Ahmedabad-Baroda road at a distance of 35 kilometers from Ahmedabad. Decision to set up the plant was motivated by the need to provide low cost clean fuel to households, reduce pollution and improve hygienic conditions and quality-of-life in the village. Although initially

123 households showed interest in the project, however, following commissioning of plant in 1981, only 51 households comprising 25 affluent farmers, 8 small and marginal farmers and 18 landless labourers came forward to avail the facility after depositing Rs. 125 as security. In a meeting of trust members and participating villagers, it was agreed that whereas villagers shall bear the cost of stoves and gas meters the remaining expenses shall be met by the trust. Following submission of the techno-economic feasibility report by the GAIC, the project began on a 2000 sq. m piece of land acquired at a cost of Rs. 6800 in June 1980 which was completed in Dec. 1980. It was estimated that about 3460 kg of animal wastes could be made available to the plant daily and accordingly 140 m^3. capacity plant was selected. It was decided to purchase animal wastes from villagers at a rate of Rs. 20 per tonne (2 paise per kg) and sell them digested effluent at a rate of Rs. 60 per tonne (6 paise per kg). Gas-pricing was kept different for small and big families with latter required to lay less. According to the revised estimates the project was estimated to cost Rs. 3,94,597 which included a capital investment of Rs. 3,92,620 (Table 6.2) out of which Rs. 1 lakh was to be met by government subsidy and balance amount as bank loan at 11 per cent rate of interest [19].

Table 6.2 Break-up of Capital Investment Involved

S. No.	*Cost-Item*	*Amount (Rs.)*
1.	Land	11,360
2.	Civil Works	1,12,441
3.	Consultancy charges	3,210
4.	Slurry storage pits	800
5.	Pipelines	1,16,066
6.	Gasholders and hot-plates	84,693
7.	Gas engine, air-compressor, etc.	13,433
8.	Gas-meters	46,202
9.	Electric works	2,815
10.	Hand-pump & polythene sheet	1,600
	Total	3,92,620

Cost-benefit analysis details of community plant at Kubadthal are as follows [19] :

Animal wastes generated = 3804 kg/day

Amount spent annually in buying wastes @ 5 paise / kg = 365 × 3804 × 0.05
= Rs. 69,430

Annual expenditure on water for slurry preparation = Rs. 500

Annual wage-bill for engaging 5 labourers @ Rs. 200, Rs. 150, and Rs. 165 per month = Rs. 10,140

Annual expenses on running oil pumps, etc = Rs. 1,200

Total annual operating expenses = Rs. 81,270

Depreciation of civil-works @ 2.5 per cent = Rs. 112441 × 0.025
= Rs. 2811.025 / year

Depreciation on value of gasholder @ 10 per cent	= Rs. 84693 × 0.10
	= Rs. 8469.30
Total cost of other capital items	= Rs. 392620 – 112441 – 84693
	= Rs. 195486.0
Depreciation in value of other capital items @ 5 per cent	= Rs. 195486 × 0.05
	= Rs. 9774.30/year
Total depreciation in value of capital items	= Rs. (2811.025 + 8469.30 + 9774.30)
	= Rs. 21054.625
	or, say, Rs. 21,055/year
Interest on long term loan @ 11 per cent	= Rs. 294587 × 0.11
	= Rs. 32405.57
	or, say , Rs. 32405
Annual repair/maintenance expenses	= Rs. 1000
Annual expenses on stationery items	= Rs. 1000
Total annual overhead expenses	= Rs. (1000 + 1000 + 1000)
	= Rs. 3000
Annual establishment charges	= Rs. 10,140
Monthly establishment charges	= Rs. 10140/12 = Rs. 845
Monthly overhead expenses	= Rs. 3000/12 = Rs. 250
Annual operating expenses	= Rs. 81,270
Monthly operating expenses	= Rs 81270/12 = Rs. 6772.50
Total monthly expenses	= Rs. 845 + Rs. 250 + Rs. 6772.50
Total annual working capital needed based on one month's expense	= Rs. 7867.50 or say, Rs. 7868

Out of Rs. 7868 as working capital, 25 per cent i.e. Rs. 1967 is made available as long term loan and balance 75 per cent i.e. Rs. 5901 as bank loan at 16 per cent interest rate.

Interest payable @ 16 per cent	= Rs. 0.16 × 5901/year
	= Rs. 944.16 or say, Rs. 944
Total annual fixed cost	= Establishment charges + over head expenses + interest on working capital (Rs. 5901 + interest on long term loan (Rs. 294587) + depreciation on capital items
	= Rs. (10140 + 3000 + 944 + 32405 + 21055)
	= Rs. 67,544
Total costs	= Fixed costs+variable operating costs

	= Rs. 67544 + Rs. 81270
	= Rs. 1,48,814
Effluent generated	= 445.3 tonne/year
Money value of effluent @ Rs. 150 per tonne	= Rs. 445.3 × 150/year
	= Rs. 66,795
Biogas yield	= 51,100 m^3/year
Net expenses on biogas production	= Rs. 148814 - Rs. 66795
	= Rs. 82,019
Cost of biogas produced	= Rs. 82019/51100
	= Rs. 1.605/m^3
Cost of kerosene (at 1981 price level)	= Rs. 1.52 per litre
Equivalent cost of biogas based on kerosene price (1 m^3 of biogas is equal to 0.62 litre of kerosene)	= Rs. 1.52 × 0.62
	= Rs. 0.94/m^3

Selling price of biogas was thus fixed at Rs. 0.90 per m^3 i.e. not costlier than kerosene

Earnings from the sale of gas	= Rs. 0.90 × 51100
	= Rs. 45,990/year
Meter charges for 123 meters at @ Rs. 5 per meter	= Rs. 615
Total revenue earnings	= Rs. (45990 + 66795 + 615)
	= Rs. 1,13,400
Total annual expenditure	= Establishment charges + overhead expenses + interest on working capital + operating expenses
	= Rs. (10140 + 3000 + 944 + 81,270)
	= Rs. 95,354.
Net surplus generated by the plant	= Rs. 113,400 - Rs. 95,354
	= Rs. 18,046/year

6.5 Community Biogas Plant at Jainpur

A fixed type community plant consisting of three units of 1000 ft^3 capacity with each unit costing Rs. 6 lakh was set up at village Jainpur in 1982. The plant could meet fuel needs of 60 households out of a total of 126 with each household consuming about 60 ft^3 of gas daily. Gas supply was restricted to six hours daily during summer and four hours during winter. Two labourers were engaged for transporting animal wastes on wheel-barrows from individual households to the plant site. For utilising gas, households were charged at a flat rate of Rs. 25 per month but subject to revision from time to time. Besides meeting cooking needs, gas was also used for domestic and street-lighting and running a 10 KVA generator for operating a chaff-cutter, flour-mill and oil-expeller. A supervisor was engaged to manage plant operation and maintain records of wastes-collection and gas and effluent supply to individual households.

6.6 Community Biogas Plant at Pura

Centre for Application of Science and Technology to Rural Areas (ASTRA), IISc, Bangalore set up a large scale experimental plant at village Pura in Tumkur (Karnataka) in seventies. Decision to set up the plant was based on rural energy survey undertaken to identify energy related activities in the village such as cultivation, animal grazing, cooking, fuelwood collection, drinking water supply, domestic and street lighting which were performed by human-power, bullocks, fuelwood, kerosene and electricity (Table 6.3 and 6.4). Following conversion norms [16] were used for computing the values given in Table 6.4.

Conversion Norms

1 man hour	=	250 kcal
1 woman hour	=	200 kcal
1 child hour	=	120 kcal
1 bullock hour	=	2300 kcal
1 kg fuelwood	=	3800 kcal
1 kwh power	=	860 kcal
1 litre kerosene	=	8980 kcal

Table 6.3 Share of Various Forms of Energy in Different Village Activities at Pura

Energy Source	*Unit*	*Agriculture*	*Domestic*	*Lighting*	*Industry*	*Total*
Manhours	hours	19914	82376	—	16485	118775
Womanhours	hours	14934	113928	—	4245	133107
Childhours	hours	—	59202	—	—	59202
Humanhours (sub total)	hours	34848	255506	—	20730	311084
Bullock hours	hours	5393	—	—	—	5392
Fuelwood	kg	—	207807	—	8930	216737
Kerosene	litres	—	—	1938	156	2094
Electricity	kwh	7264	—	3078	820	11162

Table 6.4 Share of Various Forms of Energy in kcal 'X' 10^6/year in Performing Different Village Activities at Pura

Energy Source	*Agriculture*	*Domestic*	*Lighting*	*Industry*	*Total*
Man	4.98	20.59	—	2.12	29.69
Woman	2.99	22.79	—	0.85	26.63
Child	—	7.40	—	—	7.40
Human (sub-total)	7.97	50.78	—	4.97	63.72
Bullock	12.40	—	—	—	12.40
Fuelwood	—	789.66	—	33.93	823.59
Kerosene	—	—	17.40	1.40	18.80
Electricity	6.25	—	2.65	0.71	9.61
Total	26.62	840.44	20.05	41.01	928.12

Total energy = 928 × 10^4 = 1.079 kwh/year;
2955 kwh/day; = 0.28 kwh/day/capita

Source : Reddy, A. K. N. and Subramanian, D. K., Design of Rural Energy Centre, Proc. Indian Academy of Sciences, Vol C2, Part 3, Sept., 395-416 (1979).

A. K. N. Reddy and D. K. Subramanian [17] while at the ASTRA carried out a detailed analysis to formulate the energy plan ranks (Table 6.5 and 6.6) pertaining to animal and human energy, resepctively. Table 6.5 ranks end uses of inanimate and animal energy according to decreasing energy needs. Table 6.6 summarises end uses of human energy for analysing different options. For example, fuelwood collection becomes redundant with a community biogas plant. With wastes available from 143 cattle, it was estimated to generate 35 m^3 of biogas per day (Fig 6.2). In phase 1 of the energy plan it was planned to meet cooking and lighting needs, and operate a pump to store water in overhead storage tank. In phase II, it was planned to meet total energy needs for low temperature heating, water-lifting by windmills and ethanol production by plating energy crops. In phase III it was planned to produce motive power by utilising biogas, producer gas and ethanol engines. The project envisaged a 5 HP biogas engine with 5.33 hours of engine time used for pumping water for meeting biogas plant needs, 3 hours daily for running a generator for domestic lighting and 2 hours daily to run a ball-mill to grind rice-husk ash, lime and produce saleable cement [16].

Table 6.5 Applications of Inanimate and Animal Energy for Different End Uses at Pura

S.No.	*End-use*	*Input Energy/year (kcal/10^6)*	*Efficiency*	*Output Energy/year (kcal/10^6)*
1.	Heating (95–250^{0}C)	688.9	5	34.4
2.	Heating (55^{0}C)	112.4	5	5.6
3.	Heating (800^{0}C)	23.8	5	1.2
4.	Lighting	20.1	2.5	0.5
4.1	Lighting (electrical)	2.7	10	0.3
4.2	Lighting (kerosene)	17.4	1	0.2
5.	Mobile power	12.4	20	2.5
6.	Stationary power	7.0	80	5.6
6.1	Water lifting	6.3	80	5.6
6.2	Flour milling	0.7	80	0.6
	Total	864.6		49.8

Table 6.6 Applications of Human Energy for Different End Uses at Pura

S.No.	*Activities Involving Human Energy*	*Human Energy Expenditure*		
		Hours/year	*Hours/day/ household*	*kcal × 10^6 / year*
1.	Domestic work	255,506	12.5	50.8
1.1	Livestock grazing	117,534	5.7	23.4
1.2	Cooking	58,706	2.9	11.7
1.3	Fuelwood collection	45,991	2.3	9.1
1.4	Fetching water	33,215	1.6	6.6
2.	Agriculture	34,848	1.7	8.0
3.	Industry	20,730	1.0	5.0
	Total	311,084	15.2	63.8

Source : Reddy, A. K. N. and Subramanian, D. K., Design of Rural Energy Centres, Proc. Indian Academy of Sciences, Vol C2, Part 3, Sept., 395-416 (1979).

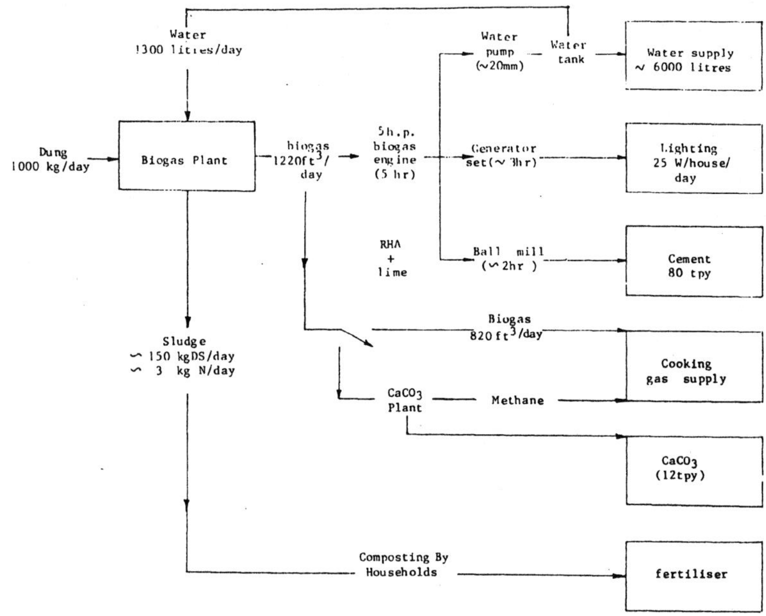

Fig 6.2 Community Plant for Pura Village

6.7 Community Biogas Plant at the KCP Sugar Factory, Vuyyuru

This plant is located in the premises of the KCP Sugar Factory in Krishna district (A.P.) has over 1600 employees. It is fed by wastes from cattle owned by over thirty employees residing in company's township [12]. The factory contributed Rs. 40,000 in setting up the plant which included Rs. 3000 as subsidy from the KVIC. Necessary control in plant operation was provided by the factory management and requisite technical assistance by the *Vidyanam Public Trust.* The trust helped in determining plant capacity in keeping with wastes availability and energy needs. It also provided guidance in obtaining subsidy and constructing the plant. There were over 100 households who could be the possible users of gas. The houses were located in five rows having twenty units each. Initially, fourteen houses were provided with biogas which was later extended to four more households. Distant houses which involved extra pipeline costs were not connected in the first instance. Plant operation was looked after by a full-time paid employee. Gas was supplied three hours from 6 to 7.30 AM and 7 to 8.30 PM. Later when more gas was generated, period was increased. Except for some minor difficulties initially faced, the plant functioned smoothly. There existed feeling of general satisfaction with the plant performance.

In Maharashtra, a biogas plant utilised spent wash generated by the sugar factories as feed. It is managed by the *Shivadasan Griha Nirman Societ Ltd.* A mill discharging 500,000 litres of spent wash daily can generate 15,000 cubic meters of biogas every day which can run its boilers for seven hours, and produce 21,000 tonnes of enriched fertiliser. The plant effluent is dried, supernatant liquid diluted and fed into ponds for cultivating water-hyacinth. The purified water is chlorinated and recycled for industrial applications. Water-hyacinth is chopped and mixed with plant sludge for obtaining rich compost which can be utilised in sugarcane cultivation. The plant is situated near the factory boiler room and the water-hyacinth treatment ponds on a wasteland near sugarcane fields. The entire process was developed as a natural and inexpensive technique of recycling liquid wastes [20].

6.8 Community Biogas Plant at Bidhlam, Sonepat

A community biogas plant of 85 m^3 capacity with technical and financial assistance from the then DNES was set at Bidhlam village in 1987. It is duly linked by roads with Delhi, Sonepat and Rohtak and surrounded by some 40 villages within a radius of 15 km. Decision for setting up the plant was made on the basis of energy survey carried out to assess overall energy demand in the village. For plant maintenance and workers training a workshop-cum-training centre equipped with necessary machines, equipment and tools was set up. A committee comprising representatives of plant beneficiaries was formed for solving day-to-day operational problems such as collection of wastes, schedules for gas supply, disposal of slurry and pricing of gas and plant effluent. From the plant, gas connections through an underground pipelines were given to 36 families to meet their cooking needs. Gas supply was made twice a day for five hours [7, 21].

Biomass gasifier developed by the IIT, Delhi and a five HP engine along with a pump was coupled to a tubewell for supplying drinking water. The engine is run on producer gas from the gasifier that can also be run on biogas. It consumes a dual-fuel of 80 per cent gas and 20 per cent diesel under normal operating conditions. With one charge of biomass such as wood, or other agricultural wastes, the gasifier can supply gas to run the engine for eight hours. When gas supply is cut-off, the engine automatically switches over to diesel without interruption. The engine can be used to pump water for meeting drinking water needs of the village school children. Extra water pumped was proposed to be used in raising community demonstration farm at the project site [7, 21].

8.9 Community Biogas Plant at Methan in North Gujarat

Methan represents a success story of a Rs. 19 lakh community biogas plant run by forming a village cooperative. It demonstrates how the cooperative ventures can successfully run and manage community plants. Methan is a village in Siddhapur *taluk* of Mehsana district with over 600 households and a population of 4000. The main occupation of the villagers is dairy, farming and animal husbandry. The success of earlier two family plants in the village encouraged villagers to go in for a community plant. Earlier successful running of a milk cooperative in the village provided added confidence to run the community plant also as a cooperative. Gujarat Energy Development Agency (GEDA) under the active support of DNES played a lead role in setting up 630 m^3 capacity plant in 1987 which consisted of 8 digesters. These systems were located at three corners of the village for providing easy access to gas at low costs. The funding of the plant was arranged by the MNES [23].

For running the plant a registered society called *'Silver Jubilee Biogas Producers and Distributors Cooperative Society'* was established with all beneficiaries as members. Each member initially paid Rs. 301 for obtaining gas connection and Rs. 100 as membership fee. Animal wastes were supplied by the households and cattleshed owners on daily basis who in turn were given digested slurry (enriched fertiliser) on a pro-rata basis. A tractor with trolly purchased with the money made available by the society is engaged in collecting animal wastes from house to house and carrying it to the plant site. Gas produced by the plant is sufficient to provide 7-8 hours of supply in summer and 5-6 hours in winter. A total of 326 members availed the facility at a flat rate of Rs. 40 per month. The society supplies gas at reduced rates to lower caste and very poor families. The success of the project can be attributed, among others, to its total acceptance by housewives who are the real beneficiaries of the scheme [23].

6.10 Community Biogas Plant at Foshan (P. R. China)

This plant represents a success story of power-generation from night-soil based community plant in China. Originally the plant was set up a waste treatment facility, however, later on around 1974 focus shifted to biogas recovery which was used for generating 30 kw of power. This plant was collectively set-up by Guangzhou Institute of Energy Conversion and the Foshan Environmental Sanitary Management Agency. In view of the fact that this plant grew from a waste-treatment plant, its design is somewhat different from other plants (Fig. 6.3).

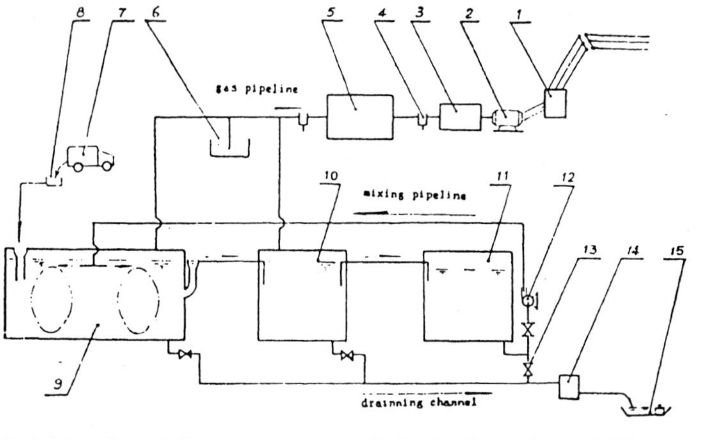

1. Switch board	2. Generator	3. Combustion engine	4. Water seal valve
5. Balloon	6. Safety valve	7. Tank truck	8. Feeding channel
9. Digester	10. Sedimentation tank	11. Storage tank	12. Mixing pump
13. Stop-valve	14. Measuring box	15. Fertiliser boat	

Fig. 6.3 Schematic diagram of community biogas plant at Foshan in P.R. China.

Digesters of this plant consist of three-stage septic tanks which enable easy flow of slurry from one tank to another. Municipal wastes are brought to the plant site in trucks and digested effluents are carried away with boats. Digested slurry is taken away through sedimentation and storage tank. Combined stream after passing through measuring box is finally led to a boat which takes it away. Axial flow pump installed at the storage tank outlet provides effluent-recirculation which leads to higher gas yield. Gas obtained is partly stored in balloons and partly used in running IC engines which are coupled to electric generators. Available power is supplied through rigid. One cubic metre of biogas at 40°C generated 1.5 kwh power. During summer when ambient temperature was 30°C. daily gas yield was 400 m^3 without stirring (0.3 m^3/m^3 of digester tank). With regular stirring yield increased to 720 m^3 a day (0.55 m^3/m^3 of the digester tank). Feed-rate was kept at 60 tonnes of city refuse per day and the resulting gas had 73 per cent methane and 27 per cent carbon dioxide. Attempts were also made for utilising waste-heat for warming up digester contents for obtaining higher gas yield. Burning of 1 m^3 of gas generated 4,600 kcal of waste-heat [18].

6.11 Community Biogas Plant at Fushan Farm in Hangzhou (P. R. China)

Fushan farm is an integrated farm for the planting, raising and processing. It comprises 32.5 hectares of paddy field, 13.7 ha water-shield pools, 7.3 ha fish ponds, 4 ha tea gardens and 1.33 ha grapery in the farm. The farm annually raises over 30,000 hens. 120,000 chickens and 6000 pigs. Animal though a pollution source are also source of renewable energy, fodder and fertiliser (Fig. 6.4). For gainfully utilising chicken wastes, a 200 m^3 capacity community biogas plant was built at Fushan poultry farm in 1988. The project included a

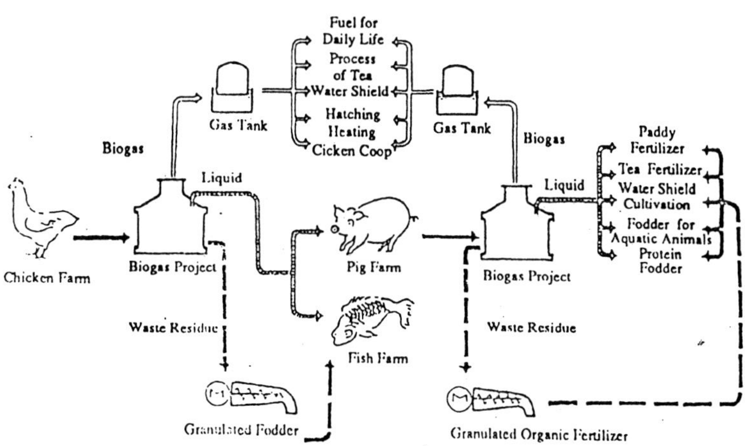

Fig. 6.4 Community biogas project at Fushan Farm

preprocessor, acidification pool, UASB reactor, sediment pool, drying pool, separator, desulphurising tower and gas tank (Fig. 6.5). Dung retention period was kept 8 days at normal temperature in the UASB reactor [22]. In winters digester contents are heated to 20°C. Gas is produced at a rate of 1.2 m^3/m^3.d with COD reduction rate as 91 per cent. The plant produces 240 m^3 gas per day which is supplied to individual households for domestic use and consumed at farm for tea processing, warming of breeding house and chicken house, etc. Digested slurry is used as fertiliser for crops and as fodder for pig and fish-rearing. The community farm at Fushan can supply 85,000 m^3 biogas per year for 120 households and two canteens. For treating wastes from 5000 pigs, a 500 m^3 capacity large-sized plant was set up at Fushan farm which could produce 600 m^3 of gas per day [22].

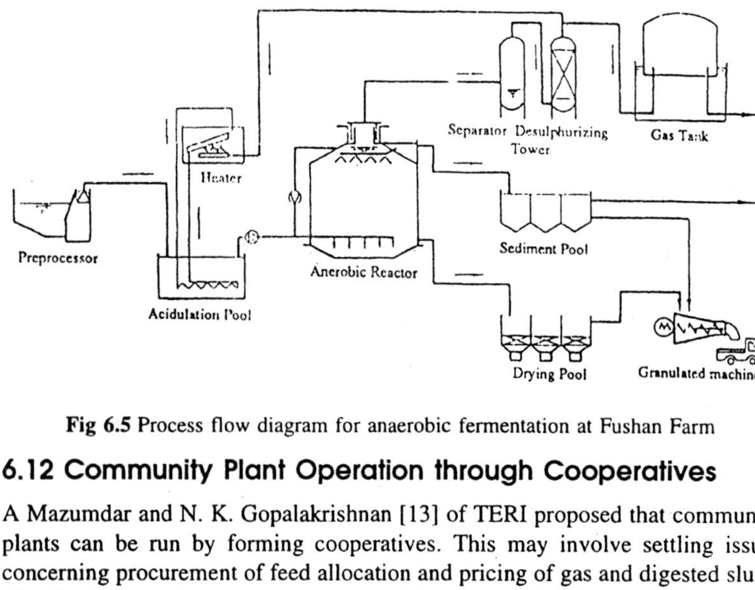

Fig 6.5 Process flow diagram for anaerobic fermentation at Fushan Farm

6.12 Community Plant Operation through Cooperatives

A Mazumdar and N. K. Gopalakrishnan [13] of TERI proposed that community plants can be run by forming cooperatives. This may involve settling issues concerning procurement of feed allocation and pricing of gas and digested slurry (fertiliser), operation and maintenance of the plant and socio-economic aspects.

Collection of Feed Materials : For effective functioning of a community plant, adequate availability of feed materials is important. This may require that burning of animal wastes as dung-cakes be avoided, and other village wastes such as night-soil, crop-residues and water-hyacinth should also be utilised for biomethanation. The cooperative should make suitable arrangements for wastes collection, transportation, and preparation of slurry.

Allocation of gas and digested slurry : Distribution of gas and enriched fertiliser in proportion to their contribution of feed is generally desirable. Cooperatives can adopt following guidelines for supply of gas and fertiliser.

- Households can be charged according to the family size based on standard norms of a gas requirement per person per day.

- Gas may supplied to rural poor free of charge if they are very poor or in exchange for their labour put in for dung-collection or slurry-preparation.
- Gas may be supplied to the beneficiary equivalent to the cost of wastes supplied and can be charged for the gas quantity.
- Enriched fertiliser can be supplied to beneficiaries either in proportion to their contribution of wastes or at some suitable rate worked out.
- It will be the cooperative's discretion to allocate gas for community activities like running a community kitchen, lighting streets or power generation.
- The cooperative should buy wastes at a rate somewhat higher than market rates to attract their use in community plant.
- The unit price of gas should not be higher than equivalent rates of other fuels. In short it should be so priced that it does not become economically disadvantageous to villagers to change over to gas.
- Enriched fertiliser from the plant should not be costlier than the equivalent nitrogenous chemical fertiliser obtainable from the market.

Operational and Maintenance of Plant : Inspite of the fact that biogas is a clean and ideal fuel, well-suited for rural areas, it suffers from problems of storage and of large volume. One effective way of supplying gas to households is through pipelines and installing gas-meters. If use of gas-meters is to be avoided for economic reasons, gas can be supplied during fixed hours in morning and evening. Cooperative should have trained staff for running and maintenance of a community plant.

6.12 Socio-Economic Evaluation of Community Plants

In India, several studies on socio-economic evaluation of community plants have been carried out which included those by C. R. Prasad, *et al.* (1974), S. Bhawani (1976), P. B. Ghate (1979), S. Bahadur and S. C. Agarwal (1980), R. Bhatia and M. Niamir (1979). Salient features of these studies are summarised in Table 6.7 [1, 2, 3, 9, 15]. T. K. Moulik, *et al.* (1983) at IIM, Ahemedabad carried out a study on socio-economic evaluation of community biogas plants in India [14]. According to the study findings, community plants have not made significant headway so far as many of them have been found to be economically non-viable. Several difficulties in setting up these plants from the point-of-view of meeting targets, organisational parameters and socio-economic acceptance have been reported. Agencies in their anxiety to get physical targets completed, tried to seek community participation from people after making unrealistic promises like providing gas free-of-charge. Occasionally, it led to numerous problems including resistance and even lack of cooperation from beneficiaries. For the twenty one community plants which were set up in the first instance, Department of Science and Technology met the initial cost of construction and operating cost for running the plant for three years. The study, among others, recommended that plants should be run by a duly constituted board on commercial lines rather than merely through voluntary participation. It made following large biogas plants [14].

- Floating-drum type plants are more reliable and appropriate than fixed-dome *Janata* design.

- For techno-economic feasibility, active participation of beneficiaries and adequate availability of wastes as feed is crucially important.
- Institutional plants are easier to manage than village community plants.
- Plant management is easier for a socio-economically homogeneous group than a heterogeneous community.
- Intimate knowledge of the community and rapport with it are crucial for successful plant operation.
- Buying feedstock and selling slurry and gas are better options than free voluntary contribution of input for a proportionate free share in output.
- Economic viability of community biogas plants depends on optimal utilisation of both digested slurry and biogas.
- Uncertainities and inordinate delays in plant construction and commissioning tend to shake people's confidence in success of plant operation.
- Community plants offer greater promise of success of biogas programme than family plants.

Bremen Overseas Research and Development Association (BORDA) prepared a manual to help in deciding and setting up community biogas plant [8]. The manual includes guidelines, framed questionnaires, examples of calculations, charts and suggestions for solving managerial, technical and socio-economic problems. Based on information gained at the fourth international symposium on anaerobic digestion held at Guangzhou in China during 1985, D.L. Hawkes [10] provided a detailed review of history, government training policy and construction of full-scale family sized and community biogas plants currently in operation in China.

Decentralised energy systems based on local renewable resources such as those linked to community plants can be viable alternative to many conventional energy systems provided the hidden subsidies for the latter and the environmental degradation they cause are also taken into account [6]. Community plants are ideal for meeting rural energy needs as well as for meeting domestic heating and cooking needs of even urban areas which are not far away from garbage disposal centres. Conventional energy sources can be used in high intensity consumption centres such as industries and commercial establishments and for urban lighting [11]. Cost-competitiveness of community plants can be attractive if real resource costs of supplying grid electricity or diesel/kerosene to rural areas are also taken into account while working out the economics.

6.13 Concluding Remarks

One of the persistent criticism of biogas programme particularly with regard to individual family unit lies in unattractive economies-of-scale it generates, although biomethanation makes sense for helping to solve disposal problems of wastes [5]. Idea of setting up community plants as a remedy to alleviate economic difficulties of individual households in having their independent units is a practical scheme in a country like India. About 1400 community, institutional and night-soil based biogas plants have been installed till 1994-95 in India. Over 200 plants were planned during 1995-96 with an outlay of Rs. 2.48 crores. Community plants provide economies-of-scale which arise from savings in initial capital cost, land utilised and skilled manhours needed in plant-operation. Besides meeting fuel requirement of households either by individual supply or by running a community kitchen, increased biogas yield from a community plant can be used for generating power, fuelling rural industrial units, operating irrigation pump-sets and street-lighting which ultimately may create some employment opportunities for rural youth as well.

Table 6.7 Comparative Features of Socio-Economic Studies of Large Size Biogas Plants Carried Out in India

S. No.	*Author*	*Salient Features*	*Findinngs*
1.	Prasad, C.R., *et al.* (1974)	Aims to evaluate gas in terms of an equivalent amount of electricity used for cooking, lighting, irrigation and industry.	Net annual financial gain valued at Rs. 26,094 resulting from gas and additional manure from 130 m^3 capacity plant.
2.	Bhawani, S. (1976) (for 35 m^3 plant)	Aims to evaluate gas in terms of an equivalent amount of soft-coke used for cooking. Also evaluates benefits of additional manure in terms of equivalent amount of fertiliser.	Financial returns estimated as 10 per cent when distribution cost is assumed to be Rs. 75,492. Financial returns estimated as 20 per cent for a distribution cost of Rs. 33,194.
3.	Ghate, P.B. (1979) (for two plants of 35 m^3 and 45 m^3 capacity at Fateh- Singh-Ka-Purwa)	Aims to evaluate gas in terms of equivalent amount of soft-coke used for cooking, kerosene used for lighting and diesel replaced in a dual-fuel engine. Estimates made assumed that lighting was done both using electricity and gas lamps. It did not consider benefits of additional manure in terms of equivalent amount of fertiliser.	Benefits cost ratios: (a) 0.60 when electricity was used for lighting purposes. (b) 1.54 when biogas was used for lighting purposes. Benefit cost ratio was estimated as 0.60 when electricity was used for lighting purposes.
4.	Bahadur, S. and Agarwal, SC (1980) (for plants at Fateh-Singh-Ka-Purwa)	Aims to evaluate gas in terms of soft coke used for cooking, kerosene used for lighting and diesel replaced in a dual-fuel engine. Estimates made assumed that lighting was done both using electricity and gas lamps. Also evaluates benefits of additional manure in terms of equivalent amount of fertiliser.	
5.	Bhatia, R. and Niamir, M. (1979)	It evaluates gas in terms of (a) equivalent amount of fuelwood, kerosene and diesel displaced and used alternatively, (b) equivalent amount of soft coke needed for cooking under two assumptions (i) lighting done by electricity (ii) lighting done by gas lamps. It did not consider benefits of additional manure in terms of equivalent amount of fertiliser.	Benefit cost ratios under assumption (i) Case a: 0.58 Case b: 0.01 Under assumption (ii) Case a: 0.57 Case b: 0.49

REFERENCES

1. Bahadur, S and S. C. Agarwal, *Community Biogas Plant at Fateh-Singh-Ka-Purwa: An Evaluation,* Planning Research and Action Division, State Planning Institute, Lucknow, 1980.
2. Bhatia, R and M. Niamir, Renewable Energy Sources: Community Biogas Plant, *Soft Energy Nores*, **3(2)**, 1980.
3. Bhavani, S., Biogas for Fuel and Fertilisers in Rural India: A Social Benefit Cost Analysis, *Indian Jour. of Agri. Econ.*, 31, part 3, 1976.
4. BORDA, *Community Biogas Plant Manual*, Bremen Overseas Research and Development Association, Bremen, Germany, 1979.
5. Bungay, H. R., Commercialisation of Biomass Conversion, *Energy Exploration and Exploitation* , **6(1)**, 61-69 (1988).
6. Bembalkar, N. L., Energy Crisis: Indian Response, *Indian Express*, Visakhapatnam, Feb. 26, 1991.
7. *Economic Times*, Bidhlam : First Energy Village in Haryana, New Delhi, Jan. 2, 1988, p. 4.
8. Eggeling, G., *et al.*, Biogas Community Plants Manual, *RERIC News*, **10(1)**, 9-18 (1987).
9. Ghate, P. B., *Biogas: A Pilot Project to Investigate a Decentralised Energy Systems, PRAD*, Lucknow, 1979.
10. Hawkes, D. L., Review of Full Scale Anaerobic Digester in China, *Agricultural Wastes*, **18(3)**, 197-205 (1986).
11. *The Hindu*, Biogas: Community Needs to be Met, Hyderabad, Aug. 24, 1984.
12. Jain, S. P., A Study of the Working of Biogas Plants in Andhra Pradesh, Haryana and Karnataka, *Rural Development Digest*, **3(4)**, 272-274 (1980).
13. Mazumdar, M. A. and N. K. Gopalakrishnan, *Role of Cooperatives in the Diffusion of Biogas Technology*, TERI, Mumbai, 1982.
14. Moulik, T. K., N. Murthy and A. Subramanian, *India's Experiments with Community Biogas Systems*, A Report, IIM, Ahmedabad, 1983.
15. Prasad, C. R., K. K. Prasad and A. K. N. Reddy, Biogas Plants: Prospects, Problems and Tasks, *Economic and Political Weekly*, Aug., 1974.
16. Reddy, A. K. N. and D. K. Subramanian, *Design of Rural Energy Centres*, Indian Academy of Sciences, Vol. **C2**, Part **3**, Sept., 395-416 (1979)
17. Reddy, A. K. N., Energy Planning: Time for Tough Decisions, *The Hindu*, Visakhapatnam, Nov. 23, 1990.
18. Ru-chen, C. H. Cong and X. Zhi-ping, *A Biogas Power Station in Foshan: Energy from Night*-soil, Ganghzhou Institute of Energy Conversion, Chinese Academy of Sciences, P.R. China, 1979.
19. Shah, H. C. Biogas Activities, *Personnel Communication*, June 20, 1981.
20. *Times of India*, Recycle Wastes the Natural Way, New Delhi, Nov. 3, 1987, p. 15.
21. *Times of India*, Brighter Plan for First Energy Village, New Delhi, Nov. 12, 1987, p. 8.
22. REO, Biogas Project of Fushan Farm in Hangzhou, *Biogas Forum*, 23-25 (1993), Rural Energy Office, Hangzhou, Zhejiang, P. R. China.
23. Venkatramana, P., Success Through Cooperation; Community Biogas Plant at Methan, *Urja Bharati*, **3(3)**, 36-37 (1993).

Chapter 7

CONCLUSIONS

7.1 Overview of Energy Needs in Developing Countries

Biofuels like fuelwood/crop residues and animal dung which are also categorised as non-commercial fuels account for 60-70 per cent of energy consumption in rural areas of developing countries and may account for 25 to 30 per cent of the total energy consumption in the developing world. Africa is most dependent on non-commercial fuels followed by Asia and Latin America regions [43]. In view of fossil-fuels not being in abundant supply, dependence on non-commercial fuels is likely to continue undiminished in rural areas and with prospects of conventional energy supplies substituting them also not rated very bright [25, 26]. Given this scenario, households are constrained to burn fuelwood or dung-cakes which apart from causing deforestation or depriving soil of valuable nutrients, also cause eye, throat and lung diseases from the resulting smoke. Biogas provides effective, solution to non-availability of fossil fuels, and environmentally harmful burning of dung-cakes. Biomethanation reduces fuelwood consumption and helps to conserve continually depleting forest resources which if not checked may cause soil-erosion, flash floods, landslides and river-silation in plains. In the context of controlling deforestation, renewable energy sources like biogas assume considerable significance. In few decades the fluid hydrocarbons, oil and gas will no longer be available in unlimited amount as energy sources [1]. This anticipation is based on two global trends: a rapidly expanding global population, especially in areas where economic development is seriously hampered following inadequate availability of commercial energy, and the increasingly dangerous pollution of the atmosphere by human activity [44].

The number of people on earth increase by almost 90 million each year. In 1994 the world population was 5.66 billion which is expected to rise to 6 billion by the year 2000 which may further increase to 8.5 billion in 1025 and 10 billion by the year 2050. India's population growth rate which averaged 2.14 per cent for the period 1981-91 has slowed down to 1.9 per cent in 1992 [2]. However, despite this, India's population which stood at 89.66 crore in 1993-94 is likely to cross 100 crore mark by 2001, 108.2 crores by 2006 and 139.66 crore by 2025. Needless to emphasise that a slower global population growth will relieve pressure on environment, energy and food needed for human subsistence [11]. This will compel, among others, fostering sustainable use of energy, and promoting a diverse and vigourous natural environment. Biomethanation which is a non-conventional energy technology not only augments energy supplies in energy-deficient rural areas but also saves the environment from degradation.

Mrs. Indira Gandhi, the late Indian Prime Minister, in her key-note address at the UN conference on New and Renewable Sources of Energy at Nairobi in 1981 expressed that transfer of renewable energy technologies is likely to emerge as focal-point of international cooperation among nations. She suggested:

> 'Measures must be taken to increase the use of mature technologies of renewable energy such as fuller harnessing of hydropower potential, coal gasification, large scale programmes of reforestation and energy plantation, biogas production, use of solar energy for low and medium temperature requirements like water pumping. Let a powerful call go out from this conference to make energy a focal point of contemporary international cooperation and collaboration'.

The Nairobi Conference resulted in a global consensus on action plan for promoting renewable energy at national, regional and global levels. Consensus emerged in favour of following action points [7].

- Importance of recognising the need for smooth energy transition from conventional energy era to an era of less dependence on hydrocarbons and greater reliance on renewable sources of energy.
- Emphasis on global cooperation and maximum consensus on energy issues; identification of a concrete plan of action to provide a framework for national, regional and international cooperation and evolve guidelines for activities in priority areas.
- Identification of R & D areas pertaining to new and renewable sources of energy.
- Wider application and utilisation of mature renewable energy technologies.
- Concerted efforts towards more efficient utilisation and conservation of energy.
- Reinforcement of infrastructural support for wider adoption of renewable energy.
- Strengthening regional cooperation as well as economic and technical cooperation among developing countries.
- Formation of an inter-governmental body to supervise and monitor implementation of the Nairobi recommendations.
- Establishment of a flexible consortium and consultative mechanism for mobilisation of financial resources to support renewable energy programmes in developing countries.
- According priorities to afforestation programmes coupled with accelerated development and utilisation of renewable energy such as solar, wind, wave and biomass.

India as a country also presented an official paper at the Nairobi Conference covering its energy scene and country's efforts in developing new and renewable sources of energy. In the paper developed countries were called upon to help developing nations by [7]:

- Supporting the efforts of developing countries to strengthen their R & D base.

- Providing financial, material and other support to developing countries at their request for pilot projects, production facilities, etc.
- Participating in bilateral and multilateral arrangements for transfer of technology on equitable terms to developing countries.
- Promoting free-flow of scientific and technological information to developing nations.
- Carrying out such R & D activities which are more relevant for the needs of developing countries.
- Removing tariff and non-tariff barriers against products relating to new and renewable sources of energy exported by developing countries.

In his message to a special sponsoned feature 'Renewable Energy for Sustainable Development' in India Today, August 31, 1994, Mr P. V. Narasimha Rao, the former Prime Minister of India looked forward for private sector participation in augmenting non-conventional energy supplies in the country. He noted :

> 'Energy from the sun, water, wind, biomass and the ocean is renewable, clear and environment friendly. It may be considered non-conventional energy today but it is the energy that we must learn to use increasingly if our planet is to survive. It is particularly suited to the requirements of our large country, adundantly endowed as we are with these resources. Non-conventional energy adds a new dimension to the concept of decentralisation.
>
> It is heartening that these new technologies are becoming more reliable and competitive. With the entry of the private sector in power generation and some good incentives offered by the central and state governments, there is a spurt of interest in non-conventional energy projects. The industry is poised for rapid growth in the years ahead. Entrepreneurs should seize this opportunity to contribute to the environment and the economic growth of the country.'

The Government of India has placed growing emphasis on development of non-conventional sources of energy as is evident by increasing financial outlays earmarked for sixth (Rs. 50 crores), seventh (Rs. 415 crores) and eighth plan (Rs. 860 crores). During the eighth plan in the budgeted amount Rs. 557 crores have been allocated for central sector and Rs. 303 crores for the state sector. As per revised plan estimates, more than Rs. 1000 crores are proposed to be spent on non-conventional energy during eighth plan [16]. In comparison during the same period power sector received Rs. 81,200 crores, petroleum Rs. 33,500 crores and coal Rs. 15,850 crores. Non-conventional/renewable technologies account for less than one per cent of funds allocated for energy sector (Rs 1,16,230 crores) during the eighth plan. During this plan period Rs. 320 crores have been allocated for biogas programme which represents 37 per cent of the MNES budget. This implies that only Rs. 64 crores are available per annum to MNES to be spent on biogas programme which is not a big sum considering the vast untapped potential of biomethanation. Renewable energy technologies should be given much higher fund allocation then made so far. According to Mr. S. krishna Kumar, the then Minister for State for NCES, the country adopted incremental approach in past rather than looking at NCES development as a high priority activity and attempting a leap frogging approach.

NCES are widely available on renewable basis, eco-friendly and add no pollutant or heat to the environment. According to Mr. S. Krishna Kumar, cost-effectiveness of NCES should be evaluated not only in terms of the initial cost but also other factors such as short gestation, favourable environment impact and no appreciable cost-escalation. Recent experience has shown that certain NCES such as wind-power, biomass-based power systems and small hydro projects are economical and competitive even at today's cost. Other technologies are fast becoming cost-effective as technology advances and volume production take place [16]. NCES become cost-effective on account of zero fuel cost, low operation and maintenance costs, and small gestation period. NCES can be put up as stand-alone systems in rural areas where the cost of power supply from grid may be quite high. NCES involve somewhat higher costs in the beginning but since operational costs are virtually zero, they pay-off later. If hidden costs and subsidies available to NCES are taken into account, several NCES can easily compete with the conventional systems.

Mr. L. M. Menezes, the then Secretary, MNES and then Chairman, IREDA, in a write-up, 'Renewable Energy: Time for Initiative', *India Today,* August 31, 1994 expressed that private investment in renewable has been spurred by tax incentives, concessional tariffs, and other facilities extended by state governments. Projects involving biomethanation from industrial wastes, solar thermal systems, photovoltaics, cogeneration and wind energy are looking increasingly profitable. The economics can only improve with better technology and the rising costs of conventional energy generation.

While addressing the national convention of Solar Energy Society of India at Jadavpur on Dec. 26, 1990, Dr. R. K. Pachauri, Director, TERI, expressed that historical changes in consumption pattern of fuels in the world have made the transition from conventional energy to renewables like biofuels absolutely necessary. For certain categories and situations economic viability of renewable energy options is no longer a matter of doubt which include options such as community biogas plants at all locations, wind-farms in coastal areas and photovoltaic power at remote locations. Decentralised energy systems based on local renewable resources can be viable alternative to many conventional energy systems particularly when the hidden subsidies for the latter and the environmental degradation they cause are also taken into account. For instance, for a village 40 km from the grid with a minimum load of 10 kw, photovaltaic systems are more economical than laying tie-line from the grid [22, 23].

Energy is too serious a subject to be left to energy planners alone. Energy policy should be *flexible* and based on dynamic and fruitful interaction among representatives of industry, government, and research and academic institutions. Energy policy should not be static but constantly reviewed and modified from time to time reflecting changing priorities. Taskforces and committees set up by the government from time to time should act a think-tanks to keep the content and direction of existing policies under constant review. There is a need to convert the government policies and programmes into peoples movement by involving citizen groups with broad-based membership and professional interest groups like CII (Confederation of Indian Industry), FICCI (Federation of Indian Chambers of Commerce and Industry) and ASSOCHAM (Associated Chambers of Commerce and Industry). It is vital for any new energy policy to recognise that the future of mankind lies in development of renewable energy sources as coal, oil and natural gas may not last for ever [27].

7.2 Application of Computers in Biomethanation

Computer-applications in diverse fields of biomethanation are fast growing. These can be broadly divided into two areas covering (i) planning, scheduling and implementation of biogas programme by different agencies and (ii) fast expanding technical field of biomethanation. In India, planning and scheduling of the NPBD is carried out by the MNES, state nodal departments/agencies, NGOs, autonomous organisations like the KVIC, implementing agencies at the state and district levels, and turnkey agencies operating at the district and block levels. Computer applications in technical field of biomethanation broadly relate to selection of design parameters for maximising biogas yield under varying operating and environmental conditions, data collection and data management for optimisation of design parameters, modelling of microbiology and process biochemistry, project management for setting up community and institutional plants, economics of plant operation, studies relating to the functional status of biogas plants and socio-economic evaluation of biogas programme. In developed countries, CAD/CAM (computer aided design/computer aided manufacturing) are being used for monitoring the day-to-day operations of municipal solid wastes-based plants and landfill gas-based projects. Computer softwares have been developed for simulating fermentation process for achieving process control, evaluating economics of plant operation, and for analysing impact of operating and environmental parameters on biogas yield [17]. Digesters in Denmark in the capacity range of 2500-3000 m^3 are controlled and monitored by the CAD/CAM system. A computer-aided plant comprising three municipal refuse-based large-sized digesters was set up at Shanwayao, Chengdu (China). In Second International Exhibition on New and Renewable Energy held in May 1992, Prof. A. Hamzech and Prof. M. Horanl presented their new CAD package developed for medium and large capacity digesters of range 50-500 m^3. The system covers parameters like daily electricity load, performance data of digesters, gas yield, heating requirement, environmental parameters, feed data, digester volume [17]. In India, A. Chand, *et al.* [6] developed a computer-based management information system for monitoring the implementation of the NPBD at the district level on trial basis for a cluster sample of plants in the Marathawada region in Maharashtra.

7.3 Scope for Research in Biomethanation

There is considerable scope to carry out studies in several areas to improve the working of biogas plants and make them more cost-effective. This may involve studies concerning use of different feed materials; development of more efficient and low-cost plant designs and biogas appliances; and varied applications of biogas and plant effluent in diverse fields. Some possible areas where studies can be carried out are as follows:

- Numerous studies have been carried out in past concerning impact of different feed materials on biogas yield but very few studies are carried out in relation to specific sites as a pre-installation activity where large capacity community plants have to be installed. Studies of this nature can enable more rational pre-evaluation of plants to be set up.
- Microbiology of anaerobic (able to live without oxygen), aerobic (able to live under oxygen), and facultative bacteria (able to live under different conditions, needs to be further studied to analyse interaction among them and study impact of environmental and operating factors on their growth-rates, metabolism activities and nutrient requirements [33].
- For improving the performance of city refuse digesters, there is a need to develop more efficient techniques for separating inert and non-digestible components before feeding to the digester which clog and create scumming problems in them [20].
- In view of large scale dissemination of biogas technology and for optimising the performance of individual plants there is need to develop low-cost accurate and sensitive instruments for measuring parameters like feed rate, effluent outflow rate, pH of digester contents, digester temperature and gas flow rate [15].
- In view of seasonal variation in biogas output and for more uniform round-the-year utilisation, adequate gas storage facility should be developed particularly for large community plants. Storage of biogas unlike LPG is troublesome in view of difficulties in liquefaction and higher proportion of carbon dioxide content compared to other gaseous fuels [20].
- Appropriate technologies for obtaining animal feed from biogas plant effluent for some animals need to be developed. Technology for producing unconventional products like vitamin B_{12} from plant effluent need to be made commercially viable [9].
- Biogas apart from being used as cooking, lighting and engine fuel can also be used for making products like ammonia and methanol with processes involving successive decomposition and re-cycling. There is a need to develop such technologies for possible use in future.
- Application of biogas as an engine fuel needs to be studied further from the view-point of its long term impact on engine optimising life, range of engine parameters for efficient performance, design changes in engine carburettor and ignition system, and changes in material specifications to prevent corrosion occurring due to high proportions of carbon dioxide and impurities like hydrogen sulfide in biogas [10,18].
- Most agricultural farms need electric power for operating farm machinery, energy requirement of which in many cases can be met by power supply from biogas-based decentralised energy systems. It is seen that power generation by decentralised systems in many cases is not cost-effective.

Studies need be carried out to improve the economics of power generation by these systems. After improving the cost-effectiveness, possibility of power supply from biogas-based decentralised power systems to the power grid can be explored [42].

- Prospects of integrated farming systems covering cultivation for food and energy crops, algal cultivation, pisciculture and biogas generation wherever applicable need be studied and promoted [36, 37].
- For periodic evaluation of biogas programme in different countries, socio-economic survey need be undertaken regularly by organisations like the IDRC (Canada), BORDA (Germany), ESCAP (Bangkok), FAO, UNDP, UNDO, etc. for which these agencies can develop an appropriate schedule identifying the region or nations each will cover along with the periodicity of surveys. Some of the important aspects which these surveys should highlight may include: number and type of plants set up during the period, number of operational and non-operational plants; plant category whether family, institutional or community type; pattern of plant ownership and owners economic background, annual biogas yield and effluent produced; pattern of biogas usage for different applications like cooking, lighting or operating farm machinery, and analysis of socio-economic and motivational factors contributing to the success of biogas programme [30, 31].

7.4 Integrated Rural Energy Programme (IREP)

IREP was launched by the government during the sixth plan (1980-85) for implementing rural energy plans and projects. Coordination and implementation of this scheme has some time back been transferred from the Rural Energy Cell of the Planning Commission to the MNES. It started as a pilot scheme under which projects were launched in few select blocks on trial basis. Following successful experience gained, IREP was later introduced as a regular plan scheme from the seventh plan onwards. Till 1993-94 end, 450 blocks were covered under the IREP. The programme covers host of activities for selected blocks such as socio-economic surveys, project report preparation, arranging funds, training, R & D and project monitoring. Realising the importance of training and R & D in success of the IREP, the government has set up training-cum-R & D centres at national and regional levels. This included the Centre for Integrated Rural Energy Planning at village Bakoli near Delhi at national level and four regional training centres at Lucknow, Bangalore, Shillong and Kheda in Gujarat. In several states centres have been set up at district level also. In some states the IREP is being implemented by linking with other development programmes like IRDP (Integrated Rural Development Programme). Goals of rural energy planning under the IREP are as follows [40].

- To supplement/repalce the commercial energy supply by renewable energy such as biogas.
- To augment energy supply for productive activities.
- To improve quality-of-life by upgrading/substituting the existing energy systems.
- To reduce fuelwood consumption for controlling environmental degradation.

- To improve biomass availability by energy plantation to meet rural energy needs.

During the eighth plan (1992-97), the IREP is being implemented with following objectives in view [40]:

- Extension of the programme to cover at least 100 blocks per year.
- Providing minimum energy needs for cooking heating and lighting in each IREP block and cent-per-cent coverage for the weaker sections in the identified blocks.
- Supply of right energy mix such as biogas, solar and wind energy which can be planned and implemented on sustainable basis.
- Implementing the programme with direct involvement of *panchayats*, voluntary agencies and NGOs including the beneficiaries and people at large.
- Dovetailing the programme with other ongoing national and state level programmes.

A separate provision of Rs. 500 crores has been made for meeting the energy needs of the economically weaker sections of people in the IREP blocks. A provision of Rs. 250 crores has been made for programme implementation in states. This has been done to build institutional mechanism for setting-up IREP cells at the state, district and block levels; national and regional training-cum-R & D centres, and undertaking training, demonstration and extension programmes. In short, IREP activities basically aim at introducing appropriate technologies which can at least meet the rudiments of energy needs in rural areas. In the context of overall rural energy scene IREP objectives converge to (i) estimate demand of various forms of energy, (ii) determine the application and distribution of these resources among different consuming sectors and industries, (iii) determine the possibilities for inter-fuel substitutions and energy conservation, and (iv) establish minimum levels of energy supply that could support basic activities [19]. The government proposes to broadbase the IREP to meet the rural energy needs through the optimum mix of all types of locally available energy sources. There is a need to develop proper linkages among the IREP, rural electrification programme and rural development programme for which the MNES, Ministry of Power and Ministry of Rural Development should work with proper coordination among them. Mr. L. M. Menzes, the then Secretary, MNES [12] while addressing a meeting of concerned officials of states and union territories in New Delhi on September 7, 1994 called upon the integration of various ongoing renewable energy programmes including biogas, improved *chulha*, biomass and solar energy with the IREP which has already brought a significant change in life-styles of people in blocks.

Renewable energy systems which are decentralised in operation fit in very well with the country's pattern of rural and agriculture economy. Renewable energy promotes decentralised rural economy which favours rural development rather than urbanisation. These technologies are appropriate to promote welfare of those who are least able to afford rising costs of energy from conventional sources. *Urjagram*, a programme for self-sufficient energy villages aims at an integrated approach for decentralised energy supply in villages by harnessing locally available renewable energy sources such as solar, wind and biomass. Selection of technologies is made on the basis of surveys carried out to identify the specific energy needs. By 1992 end a total of 170 *Urjagrams* had been established in 13 states and another 225 were under different stages of progress. Urjagram projects are implemented by the state nodal agencies in association with educational institutions, research centres, industry and NGOs [32]. Normally unelectrified villages within the population range of 500-1000 with high concentration of SC/ST and other backward communities are selected in the first instance.

In past, MNES has sponsored a number of projects in different part of the country under the Urjagram scheme. By 1992, a total of 1626 rural energy surveys practically coverings all the state and union territories were completed. As on December 31, 1992, survey work in some 400 more villages was in progress. Apart from these urjagrams three experimental reference urjagram projects have been taken up for implementation. These include projects at Kalyanpura in Gujarat, Ramachanoi Khalkapat in Orissa and Idayamvillai in Tamil Nadu [32]. Experience of the previous projects is taken into account while implementing subsequent projects. Under the urjagram scheme till March 31. 1993, a total of 209 projects were already completed and another 203 were in progress.

The total duration of an *Urjagram* project is three years with equipment installation completed within first six months. Funding of renewable energy projects is done in accordance with the prevalent MNES norms. System and equipment installed in *urjagram* projects are adjusted against overall allocation/ targets of various ongoing programmes in different states. Funds for controlled operation and managements of projects are provided for a period of three years under urjagram budget head. During first year of project execution, MNES arranges requisite funds whereas for the next two years it is jointly done by the MNES and the state nodal agency on equal sharing basis. Ten per cent of the project cost is earmarked for miscellaneous expenditure on construction, travel, commissioning, repair and maintenance. The cost of a complete *urjagram* project ranges between Rs. 10-12 lakhs. Progress of *urjagram* projects is monitored through regular review meetings and periodic inspection visits [32].

A joint bioenergy project with India contributing Rs. 14.2 crores and the UNDP $ 5.5 million, has recently been formulated. The project aims at developing an institutional framework at the national level to generate the necessary awareness and capabilities to provide impetus to the bioenergy development programme by utilising a high rate biomethanation process. The government is also setting up a National Bioenergy Board (NBB) under the MNES to act as a nodal agency for development of bioenergy. The NBB will also oversee the activities connected with the UNDP assistance and other projects for utilising wastes for energy recovery. The NBB shall promote biomethanation as part of well-coordinated national strategy for bioenergy development [45].

7.5 Biomethanation and Sustainable Development

Renewable energy technologies form the basis of sustainable development [34]. The term sustainable development was for the first time used in the report (1987) '*Our Common Future*' presented by the World Commission on Environment and Development (the Brundtland Commission) which was established in December 1983. The report argues that it we continue to use up natural resources as we do at present, if we ignore the plight of the poor, then we are surely heading for a decline in the quality-of-life [3]. The report suggests that the concept of sustainable development should become an acceptable principle of both national and international development and policies, and that national governments should begin a systematic integration of economic and environmental considerations. Sustainable development is described as development that seeks to meet the needs and aspirations of the present generation without compromising the ability to meet those of future generations. Several major conclusions of the Brundtland Report were discussed at UN in 1989 the recommendations of which ultimately led to the convening of the UN Conference of Environment and Development at Rio de Janeiro, June 1992, which has come to be known as *Earth Summit*-92. At this conference a comprehensive and far reaching programme for sustainable development was agreed upon in the form of Agenda 21 comprising 38 chapters [24]. At the *Earth Summit* over 150 nations agreed that it was essential to reduce greenhouse gas emissions. This has, among others, given urgency and fillip for extensive use of renewable energy for sustainable development.

In essence sustainable development aims to ensure preservation of earth's life-support systems. Development should not lead to environmental pollution. Sustainable development alone can keep a check on harmful effects of environmental degradation such as greenhouse effect, global warming, acid rain and ozone layer depletion. The greenhouse effect is the build-up of pollutant gases in the atmosphere which prevent heat to escape into space and instead bounce it back to the earth's surface. Greenhouse gases commonly include carbon dioxide, chloroflurocarbons (CFC), nitrous oxide and methane. UNDP has assigned index numbers to various fuels as an indicator for greenhouse gas emission level. With 100 as the base for emission from petrol, methanol from coal is placed at 198, electricity from coal (126), diesel (81), natural gas (81), hydrogen from non-fossil fuel sources (o) and electricity from non-fossil fuel sources zero [5]. According to the World Bank Report for 1992-93, trends in nitrous oxide emissions show increase in nearly every city monitored. They cause damage to lung tissues, stinging eyes, coughing and chest discomfort, increased asthama attacks and greater susceptibility to infection [8]. Global warming and ozone layer depletion pose hazards to all which cannot discriminate between the rich and the poor. With so much beneficial to mankind from time immemorial, it is hard to believe that following global warming and ozone layer depletion, even the sun rays may turn quite destructive and cause skin cancer.

Acid rain has been defined as rain that is unnaturally acid as a result of the pollution of the atmosphere with oxides of sulphur and nitrogen from the burning of coal and oil. In recent years in several developed countries it has caused massive damage to trees, crops, and plants, polluted rivers, lakes and springs, damaged human health as well as wild-life and defaced some of the most beautiful and cherished monuments [8].

The numerous health hazards and ailments caused by air and water pollution are well-known. What is not fully realised is that in near future probably within our life span the levels of pollution can cause irreversible damage to earth which will make life disaster prone and at its very extreme even threaten the existence of mankind. In recent decades increase in pollution has taken place at frightening levels. The entire industrial progress till 1960 had led to only 20 per cent addition to greenhouse gases since pre-industrial levels whereas in the next 30 years (1960-90) it reached 40 per cent mark in relation to pre-industrial levels [8]. Unless corrective measures are taken it may reach 100 per cent mark above this level by the year 2030. If greenhouse gas emissions are not controlled, the global mean temperature would go up by three degree Celsius before 2100 AD creating catastrophic changes including extinction of several species and plants with chain effects, rising of sea-levels, shortage of drinking water, expansion of deserts and climatic extremes [8].

Dr. S. D. Sharma, President of India, while inaugurating a three day National Conservation Congress in New Delhi on Nov. 21, 1994 organised by the World Wide Fund for Nature-India observed that sustainable development is very important in alleviating environmental problems. He said that from *1972 Stockholm Conference* to the *1992 Rio Summit*, the world took two decades to conceptually visualise that sustainable development was an effective solution to the environmental problems. Recalling Mahatma Gandhi's strong attachment to nature, Dr. Sharma said that the Father of Nation had cautioned against harmful effects of industrialisation such as pollution. Dr. Sharma underscored the importance of renewable sources of energy such as biogas in energy planning. He noted:

> 'Minimising pollution, avoiding wastage and conserving non-renewable resources must become both personal and national goals. Until environmental consciousness becomes intrinsic to our thinking on development the world will continue to grope for an answer even as its resources get depleted'.

Dr. S. D. Sharma, while inaugurating the 23rd World Conference of the International Federation of Training and Development Organisations held in New Delhi on Nov. 8, 1994 stressed the importance of sustainable development in national planning. He said :

> 'Few issues encapsulate the relationship between change and stability as does the preservation of the environment. The depletion of natural resources which has accompanied the expansion of industry has brought into question the very stability of our planet. That poverty and pollution strengthen one another is evident. Our common approach for a solution to this global problem has resulted in support for sustainable development. Carrying this process forward is of utmost importance.'

Mr. P. V. Narasimha Rao, the former Prime Minister of India, in his valedictory address to the State Power Ministers Conference held in New Delhi on Jan. 9, 1993, observed that new and renewable sources shall become a compulsive necessity in years to come. Renewable sources of energy such as biogas, solar, wind and wave energy can alone meet the energy needs of future on sustainable basis He said:

> 'Whether we depend on one source or another, all conventional sources are finite. At some point of time we will run out of them and the entire globe will run out of them. We better not run out of them if we really want the human species to survive and not end within the next five or ten generations. You have to think of a source which is infinite. That is the natural power, the natural energy that is available from the sun, from the wind, from the waves and so on. At least it can co-exist with the earth. As long as the earth survives, the power that is available to the earth from the sun would also survive'.

Mr. S. Krishna Kumar, the then Minister of State for NCES in his message to the special issue of *Urja Bharti* on rural energy, MNES Journal (Vol 3, No 3 March 1993, stressed the need for developing sustainable energy systems for efficient harnessing of locally available renewable energy sources. Energy availability in rural areas of developing countries cannot be isolated from basic need for sustenance like food, shelter, drinking water and clean environment. Given this context biogas programme is very relevant for meeting rural energy needs on sustainable basis.

In the context of agricultural production whereas continued rapid growth of agricultural and aquatic productivity is vital to avoid starvation, improved farm practices involving chemical fertilisers and irrigation practices adversely affect the environment and ecology. The solution lies in pursuing sustainable development by ensuring a sustainable resource base for future. Prof. U. R. Rao, former Chairman, Space Commission, in the 29th Founder Memorial Lecture (1993) organised by the Shriram Institute for Industrial Research, New Delhi, highlighted the importance of sustainable development particularly in the context of ever growing demand for energy. He said [30]:

> 'Sustainable development of natural resources has to be based on maintaining the fragile balance between productivity function and conservation practices through constant monitoring and identification of problem areas which require application of improved farm practices, crop rotation, use of bio-fertilisers, integrated pest management techniques (IPM), energy efficient farming methods and reclamation of under-utilised or degraded lands. It calls for the integration of various renewable and non-renewable energy resources, characterisation of coherent zones of agricultural practices and identification of constraints/ecological problems at the micro-level instead of the present practice of beneficiary oriented, arbitrary sectoral approach.'

According to Dr. R. S. Venkataraman, C & MD, Engineers India Ltd, New Delhi, a country's leading design engineering organisation in petroleum sector, energy and environment are inseparably linked to one another and thus development should be carried out in a manner without degrading the environment. In this context energy conservation and increasing use of non-conventional forms of energy assume special significance as these measures not only help to conserve scarce hydrocarbon resources but also save the environment. Global warming, acid rain, ozone layer depletion and greenhouse gas emissions are all different manifestations of environmental pollution caused by increased use of commercial energy in prolific and unplanned manner. As the development cannot be stopped, solution lies in pursuing sustainable development [41]. Biomethanation is eco-friendly and conserves forests by reducing dependence on fuelwood.

According to Dr. A. K. N. Reddy, a former distinguished faculty from the I.I.Sc., Bangalore, energy for sustainable development is more worthwhile objective than sustainable energy development. This is because sustainable energy development can be pursued even without any development of society which can never be the complete goal of any national programme [31]. India's per capita energy consumption has nearly doubled from 110 kgoe (kilogram of oil equivalent) in 1973 to 218 kgoe in 1990 indicating a significant growth in economy [35]. According to Dr. Reddy, emphasis should shift from per capita energy consumption as indicator of economic growth to the supply and dispersal of energy among the poor i.e. emphasis should shift from the consumption of energy to the provision of energy services. In this context it is crucial to ensure that supplies are not restricted to the conventional centralised source alone but renewable energy sources which are decentralised, are also covered. Level of service enjoyed by the population should be true indicator of sustainable development.

According to the UNDP's Human Development Report (1991) energy consumption per capita measured in gigajoules is 30 in Latin America, 21 in Asia and 12 in Africa whereas the world average is 56 gigajoules. The world average is exceeded by several developed countries such as USA (28)), the Netherlands (213), Commonwealth of Independent States (194), Germany (165), Britain (150), Japan (110) and France (109). Countries whose energy consumption levels are much below the world average are Brazil (22), China (22), India (8), Nigeria (5), and Tanzania, Ethiopia and Mali (1 each). According to the UNDP report the solution of the problem cannot be to restrict the south's development to conserve the world's energy [14]. Rich countries will have to consider cutting back consumption to enable developing countries will have to consider cutting back consumption to enable developing countries to achieve their full potential [24]. The challenge is much bigger which calls for change and moderation in life-styles of rich. For sustainable development it is necessary that gap in people's lifestyles in developed and developing countries is minimised [21].

Sangita Agarwal (TERI) in her paper on biogas and forestry, presented at the '*National Workshop on Policy Issues in the Biogas Programme*' in new

Delhi during August 21-22, 1995 highlighted the importance of biomethanation in forestry programmes. The paper was based on an impact assessment study carried out by TERI for plants set up by the AKRSP at village Pingot (Bharuch). In this study, TERI focused on three village areas covering a total of 304 hectares of forest land under Joint Forest Management Scheme (JFMC). The village had a high cattle population with adequate water and space available for setting up biogas plants. Several AKRSP built plants were found non-functioning due to missing or damaged parts or due to underfeeding. Application of plant effluent as fertiliser was not found extensive due to problems of transportation despite difficulties in use of chemical fertilisers also as it tended to wash away on sloping fields. Many women however used plant effluent in their rocky-soiled kitchen garden which increased garden productivity by an average 20 per cent. TERI estimated that the average family size plant saved 1-3 tons of fuelwood per year. Biomethanation helped minimise conflict between villagers and forest department. TERI concluded that households in forest areas with substantial cattle holdings are ideally-suited to have their own plants. Biomethanation was however not part of the JFMS perhaps due to ignorance and lack of awareness about its promise for minimising deforestation and developing wastelands [46].

7.6 Concluding Remarks

Biogas represents appropriate technology eminently-suited for socio-economic development of rural areas of developing countries [28]. It meets basic need of rural households as cooking and lighting fuel, utilises local skills and resources in plant construction and operation, reduces workload and drudgery of women, improves environment and quality-of-life [43]. Biogas technology is eco-friendly and forms the basis for long term sustainable development of rural areas. Biomethanation is a source of energy, fodder and fertiliser and promotes decentralised and self-reliant rural economy [13]. Decentralised energy systems based on bioenergy provide viable alternative to electric power for a substantial number of India's 567,000 villages particularly those which are difficult to be connected through transmission lines [11,29]. Dissemination and implementation of biogas programme, and biogas utilisation is one of the areas in which India can proudly claim to be far ahead of the western world. The pace of progress of biogas programme has been very impressive in India and numbers have grown phenomenally from 80,000 in 1980 to 23.6 lakh plants in 1995-96. This is entirely due to the success of the NPBD which has made rapid strides with turnkey approach in recent years. As per the turnkey approach, plan implementation is done by TKWs, private entrepreneurs and NGOs. Taking note of the success achieved by TKWs and NGOs in meeting the plan targets, MNES from 1994-95 onwards proposes to meet 75 per cent of plan targets with their involvement only [38]. In consonance with the new economic policy, MNES in all fields of non-conventional energy including NPBD is encouraging market-driven strategy rather than subsidy-driven approach. Over the years subsidies available to NPBD have been progressively reduced and are likely to be completely withdrawn in near future. However, possibility of retaining the subsidy for a rural-based appropriate technology programme like NPBD needs to be re-examined. There are many rural families who have adequate animals to generate feed-materials but do not have enough funds to set up biogas plant [29]. In the absence of requisite capital to own a plant, such families are constrained to burn either dung-cakes or fuelwood as domestic fuel. This is the reason why the government has been providing subsidy all along to alleviate the financial hardships of poor in owning a biogas plant. It is unclear as to low under the new economic policy capital needs of poor farmers in owning the biogas plants suddenly disappear. It is thus suggested that subsidy available to NPBD should continue unabated in spite of the new economic policy of the central government whereby subsidies in different economic sector are being gradually phased out. In the absence of subsidy success of NPBD in future shall largely depend on NGO's, private entrepreneurs and TKWs, who in any case whether the subsidy remains or goes need be given all possible incentive or support by the government. Biogas programme despite initial difficulties has achieved phenomenal success and is poised for further expansion. Reliance on biogas as a fuel involves neither dependence on valuable foreign exchange reserves nor diminishing supply of hydrocarbon resources. Whereas fuelwood (unless replanted), charcoal, kerosene, petroleum and coal resources may not last for ever, biogas shall be available to mankind as long cattle plays a role in agriculture and milk production.

REFERENCES

1. *ABE, Towards a Perspective on Energy Demand and Supply in India in 2004/5*, Advisory Board on Energy, New Delhi, May, 1985.
2. Agarwal, R., Population Growth Rate Slowing, *Economic Times*, New Delhi, Aug. 24, 1994.
3. Brundtland Commission. *Our Common Future*, World Commission on Environment and Development, Oxford University Press, Delhi, 1987.
4. Buren, A. V., *Transferring Biogas Technology to the Rural Areas of the Third World*, Inter. Inst. of Env. and Dev., London, 1980.
5. *Business Today*, Alternative to Greenhouse Gases, New Delhi, May 22 - June 6. 1993, p. 16.
6. Chand, A., *et al.*, District Level Management Information Systems for Biogas Programme : An Analysis, *Eco. & Pol. Weekly*, **23**, May, M 80-84 (1988).
7. Dayal, M., Development of New and Renewable Sources of Energy, *Financial Express*, New Delhi, Dec. 20, 1982.
8. Dogra, B., Wanted Fresh Air and Clean Water, *Hindustan Times*, New Delhi, Aug. 25, 1994, p. 14.
9. FAO, *China: Azolla Propagation and Small Scale Biogas Technology*, FAO Soil Bull. No. 41, Food and Agriculture Organisation, Rome, 1978, p. 69.
10. Gupta, R. K. and N. V. Iyer, Use of Biogas in Diesel Engines, *Agri. Engg. Today*, **1 (9)**, 7-10 (1977).
11. *Hindustan Times*, Population Touches 89.66 Crores Mark, New Delhi. Aug. 18, 1994.
12. *Hindustan Times*, Rural Energy Plans to be Broad-based, New Delhi, Sept. 12, 1994, p. 12.

13. Joshi, V., Rural Energy Planning: Issues and Dilemmas, *Urja Bharti* **3(3)** 9 (1993).
14. Kasbekar, S.R., Plummeted by Profligacy and Poverty, *Economic Times*, July 30, 1992.
15. Kishore, V.V.N., V.V. Rangaro and P. Raman, *Some Problems of Implementation of Biogas Technology in Rural Areas*, A Report, TERI, New Delhi, Oct., 1987.
16. Krishnakumar, S,. Non-Conventional Energy Sources Must become Conventional Sources, F. E. Symp., *Financial Express*, New Delhi, May 18, 1994, p. 5.
17. Libin, W. and P., Maegaard, Computerisation in Anaerobic Digestion Technology, *Biogas Forum (Jour. BORDA)*, **4 (55)**, 13-17 (1993)
18. Mehta, M. M., *et al.*, Biogas Utilisation in Tractor Diesel Engines for Stationary Applications, *Agri Engg. Today*, **3**, 13-16 (1979).
19. Murraym B. and D. R., Fowler, *Development and Application of Decentralised Energy System Utilising Non-Conventional Energy Sources,* Jet Propulsion Laboratory, Calif. Inst of Tech., Pasadena, Aug. 29, 1980.
20. NAS, *Methane Generation from Human, Animal and Agricultural Wastes*, Report of an Ad. Hoc. Panel of the Advisory Committee on Technology Innovation, National Academy of Sciences, Washington, DC, 1977.
21. *Observer*, India Calls for Equitable Sharing of Earth Resources, New Delhi, May 27, 1994.
22. Pachauri, R. K. and L. Srivatava (eds.), *Energy Policy Issues*, Wiley Eastern Ltd., New Delhi, 1986.
23. Pachauri, R. K., L. Srivastava, and S. Gupta, 'Fuelish' Trends and Wise Choice Options for the Future, *Jwala Dhwani* (House Magazine of Cochin Refineries Ltd), **28(1)**, 13 (1993).
24. Powar, K. B., Environmental Issues and Universities , *Univ. News (Jour. of Asso. of Indian Universities)*, **31 (47)**, 1-2 (1993)
25. Ramana, G. V., Energy Policy Issues-I, *Economic Times*, New Delhi, May 2, 1988, p. 5.
26. Ramana, G. V., Energy Policy Issues-II, *Economic Times*, New Delhi, May 3, 1988, p. 5.
27. Raghavan, B. S., Non-Conventional Energy Need for Policy Shifts, *The Hindu*, Jan. 1, 1988.
28. Ramana, P. V., Biogas Programme in India, *TIDE*, **1 (3),** 1-18 (1991).
29. Rao, G., *National Project for Biogas Development: Role of ARDC and Bank*, A Report, Agri. Refinance Dev. Corp., Mumbai, 1982.
30. Rao, U. R., Space Technology for Achieving Socio-Economic Revolution, *29th Founder Memorial Lecture*, Shriram Institute for Industrial Research, New Delhi, April 26, 1993, p. 7-9.
31. Reddy, A. K. N., Energy Choices for the Future, *The Hindu Survey of Environment*, Chennai, 1992, p. 6-11.
32. Singh, G. R. and Singh, N. P., Urjagram: A Programme for Self-sufficient Energy Villages, *Urja Bharti*, **3(3)**, 18-19 (1993).
33. SPOPD, *Proc. of the Chengdu Seminar on Biogas Technology and Utilisation*, Sichauan Provincial Office of Biogas Development, Sichuan, P.R. China, 1979.
34. TERI, *Annual Report*: 1992-93, New Delhi, p. 29.
35. Tewari, G. C., V.K. Chawla, and U. Maheshwari, Energy Audit Experience in Indian Crude Oil Refineries, *paper presented at the Environmax Asia Watermax Asia Conf.*, World Trade Centre, Singapore, Oct., 1993.

36. Thukral, K., *et al.*, *State of Renewable Energy Technologies in India*, A Report, Washington, DC, Sept., 1986.
37. Thukral, K., *et al.*, *Problems and Prospects for Expanded Commercialisation of Mature Non-Conventional Energy Technologies*, A Report, World Bank, Washington, DC, Dec., 1986.
38. *TIDE*, Sustainable Energy Development, **1 (2)**, 16-20 (1991).
39. UN, *World Development Report*, NY, 1992.
40. *Urja Bharti*, Integrated Rural Energy Planning Programme (Based on excerpts from the eighth plan document of the Planning Commission, New Delhi), **3 (3)**, 13-16 (1993).
41. Venkataraman, R. S., Foreword to 'Energy and Environment', *A Compendium of Articles Brought Out during Oil Conservation Week, Jan. 4-11, 1993*, EIL, New Delhi.
42. Vinodkumar, T. M. and D. R. Ahuja (eds.), *Rural Energy Planning*, Wiley Eastern Ltd., New Delhi, 1986.
43. WB, *Renewable Energy Sources in the Developing Countries*, 1818 H Street, N. W., Washington, DC, 1980.
44. Winter, C. J., Solar Cities, *Renewable Energy*, **4 (1)**, 15-26 (1994).
45. Pathak, B., Methane as Energy, *Hindustan Times*, New Delhi, Sept. 22, 19 (1994). p. 16.
46. AKRSP, *AKRSP's Biogas Programme : Some Impact Studies,* Proceedings and Background Papers, Workshop on 'Policy Issues in Biogas' conducted by the Aga Khan rural Support Programme, New Delhi, August 22, 1995.

Appendix I

BIBLIOGRAPHY

Abdel Nasser, M., Egypt Studies Refuse Composting and Biogas Recovery, *Biocycle*, **24(3)**, 26-28 (1983).

Acharya, C. N., Your Home Needs a Gas Plant, *Indian Farming*, **8(2)**, 27-30 (1956).

Acharya, C. N., Preparation of Fuel and Manure by Anaerobic Fermentation of Organic materials, *ICAR Res, Ser.*, 15, New Delhi, 1957.

Ader, C. and B. I. Wheatley, *Conversion of Biomass to Fuels by Anaerobic Digestion–Review and Preliminary Assessment*, A Res. Report, Energy Tech. Support Unit, UK, 1979.

Adrichen, P., Biogas Applications in Rural Thailand, *Biogas Nepal*, No. **13**, Summer, 1981.

AERE, *Renewable Energy Sources for Developing Countries: An Initial Appraisal*, prepared for the Ministry of Overseas Development, Energy Tech. Support Unit, AERE Harwell, Oxford Shire, Sept. 1978.

AFPRO, *Proc. of IRMA-AFPRO-CHF National Workshop on Implementation of Janata Biogas Programme Through Voluntary Organisations*, Anand, Nov. 25-26, 1986.

AFPRO, *Women Speak Out on Biogas*, 25/1A Institutional Area, D - Block, Pankha Road, Janakpuri, New Delhi 110058, 1995.

Agromica Foundation, *Biogas, the Brown Revolution ? Social Aspects and Introducing Biogas Technology in Developing Countries with Special Emphasis on the Potentials and or Consequences for the Rural Poor,* Holland China Friendship Association, Stadium General, The Netherlands, 1984.

Ahmed, S., Methodology of Biogas Programme Implementation Adopted Uttar Pradesh, *Proc. Nat. Workshop on Biogas Development*, Vigyan Bhavan, New Delhi, Feb. 26-27, 1982.

Ainea, E. K., Status of Biogas Technology at CAMARTEC- Arusha, *Biogas Forum* (BORDA), **I (52)**, 14-17 (1993).

AKRSP, AKRSP's Biogas Programme : Some Impact Studies, Background Papers, Workshop on '*Policy Issues in Biogas*' conducted by the Aga Khan Rural Support programme, New Delhi, August 22, 1995.

Amato, T., *et al.*, A Dynamic Linear programming Approach to Market Allocation of Renewable Resources in the Italian Energy System : The Case of Solar Thermal and Biogas Technologies, *Energy Research*, **9**, 33-51 (1985)

Anonymous, Cow Dung Gas Plant is in the News again, *Indian Farming*, No. **9**, 1959.

Anonymous, Why the Biogas Plant has not Caught On With Farmers, *Indian Farming*, No. **3**, 1964.

Anonymous, Case for the Cow Dung Gas Plant, *Productivity News*, **12(7)**, 7 (1974).

Anonymous, Brighter Prospects for Gas Plants, *Jagriti*, Aug. 15, 1974.

Anonymous, Biogas Plants : Performance and Prospects in India, *Science Tech. Bulletin*. **1(4)**, 25-26 (1975)

Anonymous, Biogas May be Possible Answer to Energy Requirements, *Jour. of Industry and Trade*, **25(9)**, 44-45 (1975).

Anonymous, Integrated Programme for Biogas System Proposed, *World Science News*, **12(19)**, 403 (1975).

Anonymous, Gobar Gas for Rural Development : A Feature, *Commerce*, **131 (3358)**, 30-31 (1975).

Anonymous, Gobar Gas Catching Up, *Tribune*, Dec. 24, 1975.

Anonymous, Gobar Gas (Methane) Industry, *Commece*, **132 (3380)**, 23-24 (1976).

Anonymous, Biogas Plants, *Financial Express*, Dec. 29, 1977.

Anonymous, Can Biogas Provide Energy for India's Rural Poor?, Nature, **281**, 5726, Sept. 9, 1979.

Anonymous, Biogas Utilisation in Malaysia : A Country Report, *Khadi Gramodyyog*, **27(3)**, 193-194 (1980).

Anonymous, Biogas from Kitchen Refuse, *The Hindu*, Dec. 24, 1980.

Anonymous, Biogas for Agro-Engines, *The Hindu*, Apr. 30, 1980.

Anonymous, Biogas as Fuel for Generator Sets, *The Hindu*, Oct. 15, 1980.

Anonymous, Biogas for Poorer people, *Commerce*, **141 (3613),** 565-570 (1980).

Anonymous, Biogas : A Potential Energy Source for the Rural Poor, *Bhagirath*, **28**, 149-151 (1981).

Anonymous, Biogas, *Gobar Gas Newsletter*, Aug., 1981.

Anonymous, Research in Biogas Technology, *The Hindu*, Sept. 9, 1981.

Anonymous, Combustible Gas from Cotton Waste Developed, *Business Standard*, Oct. 16, 1981.

Anonymous, Biogas from Textile Wastes, *Economic Times*, Apr. 27, 1982.

Anonymous, Biogas Lights at Railway Station, *Hindustan Times*, June 2, 1982.

Anonymous, More than a Lakh of Gobar Gas Plants *The Hindu*, May 20, 1981.

Anonymous, Biogas Plants for Varied Conditions, *The Hindu*, May 20, 1981.

Anonymous, BHEL Runs Engine on Biogas, *Hindustan Times*, Nov. 28, 1981.

Anonymous, *Biogas Information Sources*, paper presented at the Nat. Workshop on Biogas Dev., New Delhi, Feb. 26-27, 1982.

Anonymous, Biogas Potential in India: An Overview, *Science for Villages*, Jan, 23-24 (1983).

Anonymous, India, China and Africa : Learning from Asia for Biogas in Africa, *Env. and Energy*, **1**, Sept. 4, 1986.

Anonymous, Development and Utilisation of Biogas Technology in China, *RAPA Bull.*, **1**, 10-17 (1987).

Appana, C. H., March of the Biogas Programme, *Khadi Gramodyog*, Dec., 165-168 (1987).

APRBRTC, *Biogas Technology : UNDP-ESCAP-FAO China Biogas Training Course*, Asian Pacific Regional Biogas Res. Training Centre, Chengdu, China, 1983.

Apte, D. P., Biogas : An Alternative Source for the Use of Electricity in Rural Maharashtra, *Proc. Seminar on Rural Energy : Problem of Crisis*, Bangalore, Jan. 23-24, 1985.

Arbuthnot, J. B., *A Scheme for Cow Dung Gas Plant*, A Report, PRAD, Lucknow, 1958.

Arceo, T. L., Biogas Digesters in the Philippines, *NSTA Technology Jour.*, **8(4)**. 26-32 (1983).

Arokjaswamy, N. S. S., Management of Energy for Rural Development, *Energy Management*, **2(3)**, 175-184 (1978).

Arya, O. P., Gearing Up for Gobar Gas Plants, *Financial Express*, Monday, June 16, 1980, p. 2.

Awasthi, R. K., Biogas Plants in Perspective, *Financing Agriculture*, **7(3)**, 1-5 (1975).

Babu, C. K., and K. C. Krishnamurthy, Rural Refuse in Northern Karnataka as an Economic Resource : Composting Biogas and Fish Culture, *Proc. Nat. Symp. on Recycling of Residues of Agri. and Industry*, PAU, Ludhiana, Mar., 1989.

Bahadur, S., Role of Biogas in Rural Development, *Proc. Research Planning Workshop on Energy for Rural Dev.* Ahmedabad, Dec., 1979, p. 21-79.

Bahadur, S. and S. C. Agarwal, *Community Biogas Plant at Fateh Singh-Ka-Purwa: An Evaluation,* PRAD, State Planning Institute, Lucknow, 1980.

Bahadur, S. and K. K. Singh, *Janata Biogas Plants*, PRAD, State Planning Institute, Lucknow, March, 1980.

Bahadur, S., Technology and Extension Support for Biogas programme, *Proc. Nat. Workshop on Biogas,* Vigyana Bhawan, New Delhi, Feb. 26-27, 1982.

Bahadur, S., Impact of Biogas Technology on Weaker Sections, *Financing Agriculture*, **14(2-3)**, 62-63 (1982).

Bala, B. K. and M. N. Hossain, Economics of Biogas Digestion in Bangladesh, *Energy*, **17(10)**, 939-944 (1992).

Bala, B. K. and M. A. Shattes, System Dynamics Modelling and Simulation of Biogas Production System, *Renewable Energy - An Inter. Jour.* **1 (5 & 6)**, 723-725 (1991).

Banerjee, M. K. and L. Jayaprakash, Community Factor in Community Biogas Plant. *Urja*, **6(1)**, 53-55 (1979).

Bansal, N. K., A Techno-Economic Assessment of Solar-Assisted Biogas Systems, *Energy Sources*, **10(4)**, 213-229 (1988).

Baoquan, L., Daily Operation and Management of Large or Medium Sized

Plants for Central Biogas Supply, *Biogas Forum (Jour. BORDA)., 2nd* Special Issue, 8-10 (1992).

Bapna, S., Economics of a Biogas Plant, *Jagriti*, **26(4)**, 11-13 (1982).

Barnett, A., *The Social & Economic Assessment of Biogas Technology*, Inst. of Dev. Studies, Univ. of Sussex, UK, 1976.

Barnett, A., L. Pyle, and S. K. Subramaniam, *Biogas Technology in the Third World*, Inter. Dev. Res. Center, Box 8500, Ottawa KIG 3 HP, Head Office: 60 Queen Street, Ottawa, Canada, 1978.

Batty, J. C., Agricultural Wastes and Energy : Some Problems and Potentials, *Utah Science*, June, 1974.

BEI, *Energy Guide-Book*, Biomass Energy Inst. Winnipeg, Manitoba, Canada.

BEI, *Biomass Energy Success Stories: A Portfolio Illustrating Current Economic Uses of Renewable Biomass Energy*, Biomass Energy Inst., Winnipeg, Manitoba, 1978.

Bell, C., *et al.*, *Methane : Fuel of the Future*, Prism Press, Chalmington, Dorchester, Dorset DT2 OHB, UK, 1973.

Benson, P. H., *et al.*, Regional and Community Approaches to Methane from Biomass and Waste : An Industrial Perspective, in '*Energy from Biomass and Wastes*, X, Elsevier Applied Science Pub. Ltd., Barking, Essex, 1987', 987-1008.

Bente, P. F., *Biogas Energy Directory*, Biogas Energy Council, Washington, DC, 1979.

Berger, L. *Economics of Gobar Gas*, Report prepared for the United Mission Eco. Dev. Agency, United Mission to Nepal, Butwal, Nepal, 1976.

Bharde, T. S., Gobar Gas to Energise Village Economy, *Khadi Gramodyog*, **20(7)**, April, 335-339 (1974).

Bhat, S. K., S.M. Laharia and S..S. Patel, Extent and Nature of Adoptions of Biogas Plants in Haryana, *Agri. Situation in India*, **41(2)**, 75-79 (1986).

Bhatia, R.. Economic Appraisal of Biogas Units in India : Framework for Social Benefit Cost Analysis, *Economic and Political Weekly*, **12(32 & 33)**, Special Number, 1503-1518 (1977).

Bhatia, R. and M. Niamir, *Renewable Energy Source. The Community Biogas Plant*, Proc. Semmar at the Dept. of Applied Sci., Harward University Press, Cambridge, 1979, Also in 'Hayes, P. and Drucker. C., *Community Biogas in India*, Soft Energy Notes, **3(2)**, April, 1980.'

Bhatnagar, R. C., Role of Banks in Biogas Programme, *Proc. National Workshop on Biogas Development*, Vigyan Bhavan, New Delhi, Feb. 26-27, 1982.

Bhatnagar, S. S., *et al.*, Biogas Option for Rural Family, Urja, **4**, 63-66 (1989).

Bhatnagar, S. S., A. N. Mathur and S. Jain., Biogas Option for Rural Family : Economics Against Other Alternatives in Indian Context, *Urja*, **27(30)**, 30-32 (1990).

Bhatt, M. S., Biomass Energy Technology for Rural Areas, *Proc. Seminar on Rural Energy Futures, Perspective for Policy Alternatives*, Bangalore, Nov. 24-26, 1983.

Bhatt, M. S., Alternate Biofuels for Rural Transportation, *Changing Villages,* **6(5)**, 384-399 (1984).

Bhattacharya, B. K., Biogas Development Programme in Karnataka, *Financing Agriculture,* **14(2-3)**, 30-31 (1982).

Bhavani, S., Biogas for Fuel and Fertilisers in Rural India: A Social Benefit Cost Analysis, *Indian Jour. of Agri. Eco.*, **31(3)**, 219-231 (1976)

Bhave, C. B., Review of Biogas Programme in Maharashtra, *paper presented at the Nat. Workshop on Biogas Dev.*, Vigyan Bhavan, New Delhi, Feb. 26-27, 1982.

Bhushan, B., Biogas Plan Under Attack : Devaluing Decentralised Energy, *Times of India*, New Delhi, Jul. 31, 1986, p. 8.

Biogas of Colorado, *Methane on the Move*, 5620 Kendall Ct., Arvada Co., 1977.

Biogas of Colorado, *Report to the Four Corners Regional Commission : Energy Potential Through Bioconversion of Agricultural Wastes*, 5620 Kendall Ct., Arvada Co., 1977.

Biswas, T. D., Cow Dung Gas Plant for Energy and Manure, *Fertiliser News,* **19(9)**, 3-7 (1974).

Biswas, T. D. and O. P. Chawla, Biogas as an Alternative Source of Energy, *Workshop on Biogas Tech. and Utilisation*, New Delhi, Sept., 1975.

Biswas, T. D., Biogas Plants : Prospects and Limitations, *Invention Intelligence,* **12(1)**, 71-78 (1977).

Biswas, T. D., Biogas in India : Retrospects and Prospects, *Proc. Symp. on Biogas Tech. and Usage*, New Delhi, Nov. 29-30, 1977.

Biswas, T. D., Utilisation of Animal Excreta and other Agricultural Wastes for Manures and Fuels, *Proc. 64th Indian Sci. Cong.*, Bhubaneshwar, pt 3, Sec X, 116 (1977).

Biswas, T. D., Biogas Technology and Rural Development, *Talk Delivered at the Univ. College of Agri.*, Calcutta, Nov. 8, 1979.

Blobaum, R., Biogas Production in China, *Proc. of a Seminar on Biomass Energy for City, Farm and Industry*, Emmaus, J. G. Press, 212-216 (1980).

Bogardup, E. R., New Strategies for Biogas Recovery, *Biocycle,* **22(5)**, Sept.-Oct., 30-34 (1981).

Bohn, H. L., A Clean New Gas, *Env.*, **13(10)**, Dec. 4-9, 1971.

BORDA, *Biogas : Guide for Housewives*, Bremen Overseas Res. and Dev. Assoc., Bremen, Germany.

BORDA, *Report on the Project Biogas Survey* 1988, Bremen, Germany, April, 1989.

BORDA, *Biomethanation : A Developing Technology in Latin America*, Bremen, Germany, 1993.

Bowonder, B., *et al.*, *Community Biogas Plants : An Evaluation,* A report, Centre for Energy, Env. and Tech., ASCI, Hyderabad and the Institute of Public Enterprises, Hyderabad, 1984, 48 p.

Brower, M., *Biomass Energy*, Union of Concerned Scientists, Cambridge, Mass., 1990.

BRI, *The Potential of Renewable Energies in Planning the Development of Rural Areas*, Report I. 126, Brace Research Institute, Mc. Gill Univ., Quebec, Canada, 1978.

BRI, *Renewable Energy Applications in Agricultural Processes*, Report R. 149, Brace Research Institute, Mc. Gill Univ., Quebec, Canada, 1981.

BRI, *A Worldwide Survey of Village project using Renewable Energy Sources,* Report R. 152, Brace Research Institute, Mc Gill Univ. Quebec, Canada, 1982.

BRI, *Utilisation of Renewable Energy in Post-Harvest Technology*, Report R. 161, Brace Research Institute, Mc. Gill Univ. Quebec, Canada, 1982.

BRI, *Energy Requirements for Cooking in the Developing Areas of the World,* Report F. 38, Brace Research Institute, Mc. Gill Univ. Quebec, Canada, 1986.

BRI, *Renewable Energy Applications and Implications in Developing Areas,* Report R. 189 Brace Research Institute, Mc Gill Univ, Quebec, Canada, 1987.

Bulmen, A., *et al.*, *Gobar Gas Survey in Nepal*, DCS, Nepal, Jan., 1979.

Bulmen, A., *et al.*, *Biogas Challenges and Experiences from Nepal*, Vol. I & II, United Mission to Nepal, P.O. Box 126, Butwal, 1985.

Buren, A. V., Chinese Development of Biogas Programmes and its Applicability to East Africa, *Proc. Conf. on Energy and Env. in East Africa*, Nairobi, May 7-11, 1979.

Buren, A. V., Transferring Biogas Technology to the Rural Areas of the Third World : Achievements, Obstacles and the Next Steps for an Underdeveloped Country, *paper presented at the Seminar on Reutilisation of Excreta for Biogas and Innocuous Biofertiliser Production*, Panajactel, Aug. 25-29, 1980.

Buren, A. V., Biogas Beyond China, *Ambio*, **9(1)**, 10-15 (1980).

Caceres, R. and Chiliquinga, B., *Experience with Rural Biodigester in Latin America*, Latin Amer. Energy Org., Quito, Ecuador, 1985.

Calzada, J. F., Small Scale Biogas Applications in Central America, *Proc. Bio-Energy World Congress and Exposition*, Atlanta, Apr 21-24, 1980.

Calzada, J. F., *et al.*, Biogas Research Projects in Central America, in 'D. L. Wise, *Fuel Gas Production from Biomass*, CRC Press, Florida, 1918,' p 48-49.

CBSI, *Summary of the Economic Benefits of Production and Utilisation of Biogas at de-Yand Country Horticultural Farm,* Chengdu Biogas Scientific Institute, Beijing, China, 1980.

Chakraborty, R. K. and D. Guha, Biogas : An Overview, *Research and Industry*, **36(2)**, 114-123 (1991).

Chan, C., Integrated Biogas Development in Fiji, in 'L. J. goodman and R. N. Love (eds.), *Biomass Energy Projects : Planning and Management*, Pergamon Press, New York, 1981, p. 120-139,'

Chan, G. L., Socio-Economic Aspects of Various Biogas Systems, *paper presented at the ESCAP Workshop on Biogas Technology and Utilisation*,

Manila, Oct. 13-18, 1975.

Chan, G. L., The Role of Integrated Farming Systems in Tropical Rural Development, *Proc. Inter. Symp. on Biogas, Microalgae and Livestock Wastes*, Taipei, Taiwan, Republic of China, Sept. 15-17, 1980.

Chand, A. D. and N. Murthy, District Level Management System for Biogas Programme : An Analysis, *Eco. and Pol. Weekly*, **22, 23, & 28**, May, M 80-84 (1988).

Chandra, M., Biogas and the ASEAN, *Urja*, **24(4)**, 331-335 (1988).

Chatterjee, S. N., P. Bhattacharya, and R. P. Das., Gobar Gas as an Alternative Source of Energy, *Khadi Gramodyog*, **37(12)**, 477-483 (1991).

Chaturvedi, P., Biogas Programme in India : Obstacles and Solutions, *paper presented at the Regional Workshop on Biomass Energy Management*, Hyderabad, Dec. 27-29, 1983.

Chaudhary, K. C. R., Food, fuel, Fertiliser and Cow Dung Plant, *Farm Jour.*, **16(1)**, 21-26 (1974).

Chawla, O. P., Gobar Gas : The Foresaken Innovation, *Yojana*, **18(1)**, 71 (1974).

Chawla, O. P., Cow Dung Gas Plants, *Yojana*, Annual No, Jan. 26, 1974.

Chawla, O. P., Gobar Gas Plant, *Summer Institute on Waste Mgt. by Recycling, Mumbai, June 28-34, 1976.*

Chen, G., *et al.*, Survey of Rural Biogas Fermention Technology in China, *Proc. 2nd EC Conf. on Energy from Biomass*, A. S. Strub (ed.), Sept., 1982, Applied Science, London, 1982.

Chegappa, R., National Biogas Project : Raising a Stink, *India Today*, July 31, 1986, p. 138.

Chong, C. N., An Assessment of Scope for Biogas Programme in Malaysia, *paper presented at the Workshop on Biogas Technology and Utilisation*, New Delhi, Jul. 28- Aug. 2, 16 (1975).

Chong, C. N., Scope for Bioenergy Programme in Malaysia, *paper presented at the US-ASEAN Seminar on Energy Tech : Biomass, Coal, Solar, Wind Energy Planning*, Bandung, Indonesia, June 7-18, 1982, p. 182.

CMIE, *Current Energy Scene in India : Verbatim Extracts Mainly from Recent Official Documents*, Centre for Monitoring Indian Eco., Mumbai, May, 1984.

Coe, W. B. and M., Turk Economic Evaluation of Asset Recovery from Animal Waste by Anaerobic Fermentation, *Animal Wastes Management Programme*, Hamilton Standard Div., United Tech., Windsor Locks, CN, 1973.

Commonwealth Science Council, Biogas Systems for Rural Dev., *Report of the Training Workshop on Monitoring and Evaluation of Biogas Systems*, Lusaka, Zambia, Aug. 29-Sept. 1, 1983, Malborough House, London, 1983.

Coombs, J., Biogas and Power Alcohol, *Chemistry and Industry*, **4**, April, 239-229 (1981).

Costigane, W. D., J. D. Scharer, and P. O. Silveston, Economics of Digestion of Manure for Methane Production on Small Farms, *Proc. of the Symp. on Uses of Agri. Wastes*, Canadian Plain Research Center, Univ. of Regina, Regina, Saskatchewan, 1974.

Couvnoyer, M. S., Evaluation of an Anaerobic Digester, *Proc. 5th Canadian Bioenergy R & D Seminar*, Canada, March 26-28, 1984, Elsevier, London, 341-344 (1984).

Cui, X. and Xie, Z. H., *An Outline of Biogas Development in China*, China State Biogas Assoc., P.R. China, 1985.

Criden, J. and Marchaim, U., A Comprehensive Programme for the Utilisation of Agricultural Waste by Anaerobic Digestion, *Proc. Inter. Symp. on Biogas, Microalgae and Livestock Wastes*, Taipei, Taiwan, Rep. of China, Sept. 15-17.

CXDCAST, *Biogas Technology in China*, China Agricultural Publishing House Chengdu Xingguang Development Corp. for Appropriate Sci. and Tech., No 17 Yulin East St., P. R. China, Oct., 1995.

Dagnall, S., UK Strategy for Centralised Anaerobic Digestion, *Bioresource Tech.*, **52(3)**, 275-280 (1995).

Daize, H., *et al*, Biodigestion : The Pivot of China's Economic and Agricultural Construction, *Biogas Forum* (*BORDA*), **III (46)**, 11-13 (1991).

Dalal, H. J., Economics of Gobar Gas Plant, *Khadi Gramodyog*, July, 1963.

Dandekar, H., Gobar Gas Plants : How Appropriate are They ?, *Eco. and Pol. Weekly*, May 17, 887-893 (1980).

Das, R., Cow Dung Gas Plant Gets Popular in Uttar Pradesh, *Indian Farming*, **12(9)**, 8-9 (1962).

Das, R., Gobar Gas and Potential for its Utilisation, *Allahabad Farmer*, **36(1)**, 17-21 (1962).

Dasilva, E. J., *et al*, United Nations System and Biogas Activities, *Proc. 1st EC Conf. on Energy from Biomass*, Brighton, Nov. 4-17, 1980, W. Plaz. (ed.), Applied Science, London, 796-809 (1980).

Dasilva, E. J., Biogas : Fuel of the Future, *Ambio*, **9(1)**, 2-9 (1980).

Datt, D., Gobar Gas, *Yojana*, **1**, Feb., 1974.

Dayal, M., Community Biogas System, *paper presented at the Nat. Workshop on Biogas Dev.*, Vigyan Bhavan, New Delhi, Feb. 26-27, 1982.

Deemark, R., Status of Biogas Production Programme in Thailand, *paper presented at the ESCAP Workshop on Biogas, Tech and Utilisation*, New Delhi, Jul. 28- Aug. 2, 1975.

Demuynck, M., Nyns, E. J. and Palz, W., *Biogas Plants in Europe*, D. Reidel Pub. Co., Dordrecht, 1984.

Demunyck, M. and Nyns, E. J., Biogas Plants in Europe, *Inter Jour. of Solar Energy*, **2,** 477-485 (1984).

Dena Bank, *The Gobar Gas Plant Scheme*, A Report, Mumbai, April, 1975.

Desai, D. K., *Biogas Plants : A Challenge to Rural Entrepreneurs*, A Report, IIM, Ahmedabad, 1982.

Deshpande, D., Biogas : Its potential and Economics in India, *Proc. of the 1980 Bio-Energy World Congress and Exposition*, Atlanta, April, 443-446 (1980).

De Silva, E. J., Biogas : Fuel for Developing Countries : An Overview, *ASSET*, **1(11)**, 10-16 (1979).

De Silva, E. J., Biogas : Fuel for the Future, *Ambio*, **9(1)**, 2-9 (1980).

De Silva/UNESCO (1979), *Transferring Biogas Technology to the Rural Areas of the Third World*, A. V. Buren (ed.), Inst. of Eng. and Dev., London, 1980.

De Silva, N. A., *Biogas in South America*, SAIM, Brazil, 1985.

Devkota, G. P., Biogas Research and Development in Nepal, *Proc. Inter. Symp. Workshop on Renewable Energy Sources*, Lahore, Mar. 18-23, 1983, T. N. Veiroglo (ed.), *Elsevier Science. Amsterdam*, 65-68 (1984).

Devkota, G. P., Privatisation in Gobar Gas Installation, *Biogas Forum (BORDA)*, **II (53),** 4-7 (1993).

Dhawan, A. S., *et al.* Constraints in the Functioning of Biogas Programme in Maharashtra State, *Energy Management*, **14(3)**, 34-37 (1990).

Daiz, L. F., Biogasification of Organic Wastes in the Republic of the Philippines, *Proc. of the Seminar on Biomass Energy for City, Farm and Industry*, J. G. Press, Emmaus, 217-219 (1980).

Diaz, L. F. and J. C. Glaub, Overview of Selected Biogasification Installations in the USA., *Proc. of the Seminar on Biomass Energy for City, Farm and Industry*, J. G. Press, Emmaus, 15-26 (1980).

Disney, R., Economics of Gobar Gas Versus Fertilisers : Critique of Intermediate Technology, *Development and Change*, **8**, 77-102 (1977).

Dixit, A. P., Development of Biogas : Experience in Area Planning at Micro Level, *Financing Agriculture*, **14(2-3)**, 45-50 (1982).

DNES, Biogas Development, Ch. 2, *Annual Report* 1983-84, Ministry of Energy, p. 6-10, p. 44-45.

Drozario, R., Gobar : It is Time for a Second Look, *Science Today*, **9(11)**, 36-46 (1975).

DST, *Biogas Technology and Utilisation : A Status Report*, Dept. of Sci. and Tech., New Delhi, 1981.

Dunkerly, J., *et al.*, Consumption of Fuelwood and other Household Cooking Fuels in Indian Cities, *Energy Policy*, **18(1)**, 92-99 (1990).

Dutta, S., N. Kukrety and V. Ramana, Implementation Strategies for the Biogas Programme in Gujarat, in 'N. K. Bansal (ed.), *Proc. Nat. Solar Energy Conv.*, Dec. 28-30, 1992, Tata McGraw Hill Pub. Co. Ltd., New Delhi, 1993', 387-392.

Economic Times, Gobar Gas Scheme: Punjab Catching Up, Mumbai, Dec. 22, 1975.

Economic Times, Energy : Long Term Outlook, A Special Supplement on Energy, New Delhi, June 9, 1988, p. 11.

Eggeling, G. and B. Stephen (ed.), *Biogas in the People's Republic of China Summary of the Project Report*, BORDA, Bremen, 1981.

Eggeling, G., *et al.*, Biogas : Community Plants Manual, *RERIC News*, **10(1)**, 9-18 (1987).

Elgohany, F., The Bio-connection : Biomass, Biogas, Bio-Productivity, *Mazingra*, **5(1)**, 81-84 (1981).

El-Halwagi, M. M., *et al*, *Biogas Technology for Rural Areas of Egypt*, A Report, Pilot Plant Laboratory, National Research Centre, Cairo, 1979, Also presented at the *1980 Bioenergy World Congress and Exposition*, Atlanta, 446-480 (1980).

El-Halwagi, M. M., Village Demonstration of Biogas Technology : An Egyptian Case Study, *Natural Resources Forum*, **6**, 329-343 (1982).

El-Halwagi, M. M., *Biogas Technology : Transfer and Diffusion*, Elsevier Science Pub. Co. Inc., New York, 1986.

Ellegard, A., *et al.*, *Experiences from the Operation of a Full-Scale Anaerobic Digester on a Swedish Dairy Cow Farm.* A Report, Metangruppen, Gothenburg, 1980.

Ellegard, A., *et al.*, *Biogas - Not Just a Technology*, SIDA, Gothenburg, Sweden, 1983.

Ellsworth, D. and T. Abeles, Integrated Farm Energy and Management Systems for Anaerobic Digester Feasibility, *Proc. of the 1980 Bioenergy World Congress and Exposition*, Atlanta, 135-137 (1980).

Emmaus, J. G., *Biogas and Alcohol Fuels production*, J. G. Press, PA, 1980.

Engshuber, M., Biogas Technology : Insights and Problems, *Energietechnik* Leipzig), **37(120)**, 449-453 (1987) (In German).

EPA, *Energy and Economic Assessment of Anaerobic Digestion and Biofuels for Rural Waste Management*, A Report, US Env. Protection Agency, Cincinnati, Ohio, 1978.

ESCAP, *Report of the Workshop on Biogas Technology and Utilisation*, UNIDO and Nat. Committee on Sci. and Tech. New Delhi, July, 1973.

ESCAP, *Report of the Preparatory Mission on Biogas Technology and Utilisation*, Manila/Bangkok, 1975.

ESCAP, *Biogas : Technology and Utilisation*, A Collection of Papers, New Delhi, 1975.

ESCAP, *Renewable Source of Energy - II : Biogas*, A Report on Economic and Technical Cooperation Among Developing Countries, Economic and Social Commission for Asia and Pacific, Bangkok, 1981.

ESCAP, *Review of Rural Applications of Biogas Technology in China, India, Nepal and Thailand*, A Report, Bangkok.

ESCAP, *Guidebook on Biogas Development*, United Nations, New York, 1980.

Eusebio, J. A. and B. I. Rabino, Research on Biogas in Developing Countries, *Compost Science*, Mar.-Apr., 24-27 (1980).

Eusuf, M., Prospects of Biogas Industry in Bangladesh, *World Science News*, **12(19)**, 412-414 (1975).

Eusuf, M., Alternative Energy Resources and their Potential in Rural Development, *paper presented at the 9th Commonwealth Sci. Council Meeting*, Colombo, Nov. 29 to Dec. 10, 1976.

Eusuf, M., *et al.*, Case Study of Biogas Plant, *paper presented at the Workshop on Energy, Rural Dev., Res. and Implementation*, Chiang Mai, Thailand, Feb. 5-14, 1980.

Eusuf, M., Development of Biogas in Bangladesh : Technological Aspects, *Biogas Workshop organised by the Commonwealth Sci. Council and KVIC*, Mumbai, Sept. 8-10, 1980. Also published in *Khadi Gramodyog*, **27(30)**, 182-184 (1980).

FAO, *Proc. of the Workshop on Rural Energy Planning in the Developing*

Countries of Asia, Beijing, China, Food and Agri. Organisation, Rome, 1983.

Farvar, M. T., and D. Bajrachanya, Prospects and Problems of Biogas Technology as an Appropriate Technology for Development in the Rural Areas - A Critical Appraisal, *paper presented at the Second Inter. Workshop on Biogas Tech. and Utilisation*, Manila, Oct. 13-18, 1975.

Financial Express, Priority to Gobar Gas Plants in Sixth Plan, New Delhi, Jul. 19, 1980.

Finlay, J. H., *Report on Second Inspection visits to 95 Gobar Gas Plants Constructed in Nepal*, Dev. and Consulting Services, Butwal, Nepal, 1996.

Fraser, M. D., The Economics of SNG Production by Anaerobic Digestion of Specially Grown Plant Matter, *Symp. on Clean Fuels from Biomass*, Inst. of Gas Tech., Orlando, Florida, Jan. 24-28, 1977.

Fulford, D. J., Report on Chinese Style Gobar Gas Plant, *Biogas Newsletter*, No. **5**, Summer, 5-6 (1979).

Fulford, D. J., A Commercial Approach to Biogas Extension in Nepal, *Appropriate Technology*, **8(2)**, Sept., 14-16 (1981).

Fulford, D. J. and Peters, N., *Survey of Present Gobar Gas Work in India*, DCS, Nepal, 1978.

Fulford, D. J., Gobar Gas in Nepal, Biogas Newsletter (*Nepal*), No. **13**, 6-7 (1981).

Fulford, D. J., Biogas Development in Nepal, *Appropriate Technology*, No **2**, 10-17 (1983).

Fulford, D. J., *Running a Biogas Programme*, Intermediate Tech. Pub., UK, 1988.

Gandhi, I., Key-Note Address at the U. N. Conference on New and Renewable Sources of Energy, Nairobi, Kenya, Aug. 10-20, 1981, *Economic Times*, New Delhi, Aug. 11, 1981.

Gandhi, I., *Renewable Energy in Action*, A Report, Commission for Additional Sources of Energy, New Delhi, 1981.

Gandhi, I., Inaugural Address at the World Energy Conference, New Delhi, Sept. 18-26, 1983, *Economic Times*, Sept. 19, 1983.

Ganesh, S. K. and S. K. Subramaniam, Biogas Plant : Performance and Prospects in India, *Economic Times*, May 7, 1975.

Ganesh, S. K. and S. K. Subramaniam, Biogas Promotional Programme, *Economic Times*, May 6, 1975.

Ganesh, S. K., Biogas Technology : The Next Phase, *Urja* , **1(4),** April, 19-23 (1977).

Ganguly, R. K., Biogas Technology for Rural Development : Performance and Prospects, *Khadi Gramodyog*, **27(3)**, Dec., 153-155 (1980).

Ganguly, R. K., Role of the KVIC in Biogas programme, *paper presented at the Nat. Workshop on Biogas Dev.* Vigyan Bhavan, New Delhi, Feb. 26-27, 1982.

Ganguly, R. K., Biogas Programme : Pioneering Work of KVIC, *Financing Agriculture*, **14(2-3)**, 27-29 (1982).

Gautam, S., Energy Crisis : What can We do in India?, *Khadi Gramodyog*, **20(8)**, 397-401 (1974).

GCC, *Biogas Support Programme : Implementation Document*, Agri. Dev. Bank of Nepal, SNN-Nepal and Gobar Gas Company, Kathmandu, Sept., 1992.

Ghate, P. B., Action Research in Community Biogas : A Task for Voluntary Agencies, *Voluntary Action*, **23(2)**, Sept., 51-155 (1980).

Ghate, P. B., Scope of Biogas in India and Related Issues, *paper presented at the Seminar on Cooking Energy*, New Delhi, Aug. 12-14 1983.

Ghouse C. and R. R. Rayarikar, Role of AFC in Financing Biogas Plants, *Financing Agriculture*, **14(2-3)**, 18-21 (1982).

Gilbert, K. D., How to Generate Power from Garbage, *Mother Earth News*, No. **3**, 1970.

Goel, B. S., *et al.*, Availability and Disposal of Dung in India, *Indian Jour. of Animal Sci.*, **43(8)**, 671-676 (1973).

Goodman, G. T., Biomass Energy in Developing Countries : Problem and Challenges, *Ambio*, **16(2)**, 111-119 (1987).

Gopalakrishnan, N. K. and A. Mazumdar, Biogas Programmes, TERI, Mumbai, 1980, p. 236.

Goplalakrishnan, N. K. and A. Mazumdar, Information and Documentation Support for the National Project on Biogas Dev., *Financing Agriculture*, **14(2-3)**, 83-86 (1982).

Gapalrao, P. R., National Project for Biogas Development : Role of ARDC and Banks, *Financing Agriculture*, **14(2-3)**, 22-26 (1982).

Gosling, D., Biogas for Thailand's Rural Development : Transferring the Technology, *Biomass*, **2(4)**, 309-316 (1982).

Goswami, K. P., Fuel Gas From Agricultural and Industrial Wastes, *FRI News*, **15(2)**, 1965.

Goswami. K. P. and S. Chowdhary, Farm Wastes Can Yield Fuel Gas and Manure, *Indian Farming*, Apr. 18-19, 1967.

Goswami, K. P., Appropriate Technology for Rural India Produce Biogas from Vegetable Wastes, *Energy Sources*, **11(1)**, 59-57 (1989).

Govilkar, R., Role of Extension Agencies in Developing the Programme of Energy Utilisation, *Proc. of the Seminar-cum-Workshop*, Karnal, edited by S. Neelkantan, Oct. 3-10, 1980, 9-12 (1981).

Gowda, K. N., Biogas : A Decentralised Rural Energy Source and its Future, *Proc. of the Seminar on Rural Energy Futures : Prospective for Policy Alternatives*, Bangalore, Nov. 24-26, 1982.

Gowda, M. B., *et al.*, Perception of Non-Adopters about the Existing and New Biogas Plants, *Khadi Gramodyog*, 418-420 (1980).

Griffin. R. S., A Communication Approach to Determining Biogas Needs and Application at Farm level, *paper presented at the ESCAP Workshop on Biogas Tech. and Utilisation*, Manila, 1975.

Griffin. R. S., Recommendations for the Workshop on Biogas Tech. and Utilisation, *World Science News*, **12(9)**, 410-411 (1975).

Gunnerson, C. G. and Stuckey, D. A., *Principles and Practice for Biogas*

System, UNDP Integ. Resource Recovery Report No. 5, World Bank Tech. Paper 49, NY, 1986.

Gupta, R., Red Tape Holds Up Biogas Project, *Proc. of the Panel Meeting on Biogas Technology*, Bardoli, Nov. 5-6, 1981.

Gutterer, B. and L. Sasse. *Criteria for Dissemination of Biogas Plants for Agricultural Farm and Household Systems*, BORDA, Germany, 1993, p. 25.

Gutterer, B. and L. Sasse, *Biogas Survey 1992: Cross Section Analyses of Biogas Dissemination Programmes*, Brief Report, BORDA, Germany 1993.

Hallingdale, A. C. and Coward, L. D. G., *Biogas Tech. in Cold Climatic Conditions in Korea,* Tropical Product Inst., ODA, London WCIX SLU, UK, 1979.

Hamad, M. A. and M. M. El-Halwagi, The Expected Role of Biogas Technology in Rural Areas of Egypt : Energy Considerations, *Proc. of the 4th Miami Inter Conf. on Alternative Energy Sources,* Miami Beach, Dec. 14-16, 1981, *Ann Arbor Science*, 319-33 (1982).

Hamad. H. A., *et al.*, Planning Egyptian Villages in New Reclaimed Areas with Regard to Accommodation of Biogas Technology, *Jour. of Inst. of Engrs.,* **68(3)**, June, 48-52 (1988).

Hamzeh, A., Processing of Human, Animal and Plant Wastes and Reducing its Influence on Public Health by Biomass Digestion, *Proc. First Conf. on Env. Sci*, Sabha, Libya, 1990.

Hansen, S., Danish Biogas Programme, *Proc. of the 1980 Bioenergy World Congress and Exhibition,* Atlanta, Apr., 453-455 (1980).

Harinathan, K., *et al.*, Gobar Gas Plants, *paper presented at the 5th Conf. of Orissa Association for the Advancement of Science*, Bhubaneshwar, Dec., 1975.

Hawkes, D. L. and R. Horton, Potential of Anaerobic Digestion, *Energy World,* June, No. **71**, 10-13 (1980).

Hawkes, D. L., Review of Full-scale Anaerobic Digestion in China, *Agricultural Wastes*, **18(3)**, 197-205 (1986).

Hayes, P, and C. Drucker, Community Biogas in India, *Soft Energy Notes*, Apr. 10-13, 1980.

Hees, N., Diffusion of Biogas Plants in the Valle del Cauca, Columbia, *Biogas Forum (BORDA)*, **1(40)**, 17-18 (1990).

Henderson, P. D., *India : The Energy Sector*, World Bank, New York, 1975.

Heruela. C. S., Biogas in Philippines, *Biogas Forum (BORDA)*, **IV (51)** 7-8 (1992).

Hoffman, R., Anaerobic Night-Soil Treatment Plants in Burundi, *Biogas Forum (BORDA)*, **IV (51)**, 15-19 (1992).

Hohlfeld, J, *et al.*, *Production and Utilisation of Biogas in Rural Areas of Industrialised and Developing Countries*. Ch. 8.2 and 8.3, GTZ, BORDA Eschborn, Germany, 1986.

Hollaender, A., *et al.*, An Inquiry into Biological Energy Conversion, *Report on Workshop held at Univ. of Tennessee*, Knoxville, Oct. 12-14, 1972.

Hollingdale, A. C. and L. D. G. Coward, Systems Study of Biogas Production for Rural Energy Provision in the Rep. of Korea, *Proc. of the ISES Congress*, Brighton, UK, Aug. 23-28, 1981, Pergamon, Oxford, 1338 (1982).

Hong. C. M., M. T. Kuo and T. Y. Chow, Production and Utilisation of Biogas in Taiwan, *Proc. Inter. Symp. on Biogas, Microalogae and Livestock Wastes*, Taipei, Taiwan, Rep. of China, Sept. 15-17, 1980.

Horton, R., *et al.*, An Evaluation of Energy Availability from Organic Wastes, *The Heating and Ventilating Engineer*, **49**, 14 (1975).

House, P., *Biogas Handbook*, Peace Press, California, 1978.

Hruska, R. L. and A. G. Hashimote, *Commercialisation of Anaerobic Digestion in USA*, Agri. Res. Centre. Nebraska (USA), 1993.

Huang, Z. J., *Analysis of Economic Feasibility for Biogas Construction in China*, Energy Research Inst., China State Biogas Assoc., P. R. China, 1985.

IAEU, *Present Status of Methane Gas Utilisation as a Rural Fuel in S. Korea, Inst.* of Agri. Eng. and Utilisation, Suweon, South Korea.

ICAR, *Economics of Cow Dung Gas Plant,* New Delhi, 1976.

ICAR, *All India Coordinated Project on Biogas Technology*, Div. of Soil Sci. and Agri. Chemistry, IARI, New Delhi, 1982.

ICAR, *All India Coordinated ICAR Scheme on Renewable Sources of Energy for Agriculture and Agro-based Industries*, Annual Report, viii, New Delhi, 1983, p. 91.

IDB, *Biogas : Why and How ?*, Ind. Dev. Board, Sri Lanka, July, 1976.

Idnani, M. A. and O. P. Chawla, *Developments in Uses of Gobar Gas*, A Report, IARI, New Delhi, 1962.

Idnani, M. A. and N. P. Dutta, Gobar Gas Plants, *Khadi Gramodyog*, 256-264 (1963).

Idnani, M. A., IARI Study Shows Why the Biogas Plant has not Caught On with Farmers, *Indian Farming*, **14(3)**, 25-44 (1964).

IIMA, *Pilot Project for a Study of Social and Technical Aspects of Community Type Biogas Plants*, A Report, IIM, Ahmedabad, 1983.

Islam, M. N., *A Report on Biogas Programme of Bangladesh/China*, Dept. of Chemical Engg., Bangladesh Univ. of Engg. and Tech., Dacca, Bangladesh, 1979.

Islam, M. N., *Study of the Problems and Prospects of Biogas Technology as a Mechanism for Rural Development : Study in a Pilot Area of Bangladesh*, Dept. of Chem. Engg., Bangladesh Univ. of Eng. and Tech., Dacca, Bangladesh, 1980.

Islam, M. N., Policy Issues in Relation to Biogas Technology as a Tool for Rural Development, *paper presented at the Biogas Planning Meeting Organised by the Commonwealth Regional Rural Tech. Programme*, KVIC, Mumbai, 1980. Also published in *Khadi Gramodyog*, **27(3)**, 167-175 (1980).

Jagrithi, Banks to Install 3500 Plants in Andhra, April 1, 1984, p. 6.

Jain, M. K. and M. M. Mishra. Biogas Plant : Alternative Technology for Energy and Manure, *Food Farming and Agriculture*, **9(10)**, April, 292-296 (1978)

Jain. S. P., Gobar Gas : A Study of the Working Plants in Andhra Pradesh, Haryana and Karnataka, *Rural Dev. Digest*, 231-274 (1980).

Jaiswal, S. C., *A Study of Socio-Economic Psychological Characteristics of Adoption of Biogas Plants in Rural Areas (Karnal district)*, M. Sc. thesis, NDRI, Karnal.

Jasper, A. W. and J. D. Patel, Methane Digester : A Source of Bioenergy, *Proc. All India Seminar on Tech. for Rural Dev.*, Allahabad Polytechnic, Allahabad, Jul. 1977.

Jayashankar, S. R. and K. C. Shastri, Production of Energy from Farm Wastes, *Proc. Seminar on Utilisation of Farm Wastes for Rural Industrial Growth*, Bangalore, Dec. 1975.

Jewell, W. J., (ed.), *Energy, Agriculture and Waste Management*, Ann Arbor Science, Ann Arbor, Michigan, 1975.

Jewell, W. J., Biomass Fuels : Past, Present and Future, *Proc. of the Annual Meeting of the Amer. Section of the Inter. Solar Energy Soc.*, Inc., Denver, Colorado, Aug. 28-31, 1978.

Jewell, W. J., *et al.*, *Anaerobic Fermentation of Agricultural Residue : Potential for Improvement and Implementation, Vol. I and II, Final Reports*. New York State College of Agri. and Life Sciences, Cornell Univ., Ithaca, NY, Sep. 1980.

Jewell, W. J., *et al.*, Economics of Plug Flow Methane Reactors, *Proc. of the Conf. on Methane Technology for Agri*, Cornell Univ., Ithaca, NY, Mar. 18, 1981.

Jha, P. S., Fertiliser from Dung : Key to Rural Development, *Times of India*, July 19, 1974.

Jha, P. S., Energy from Garbage- I & II, *Times of India*, Dec. 13 and 17, 1982.

Ji-Quin, N. and Nyns, E. J., *Biomethanation : A Developing Technology in Latin America*, A Report, BORDA, Bremen, Germany, 1994.

Joglekar, V. R., Biogas Plant : A Source of Manures, Cooking Fuel and Cleanliness of Villages, *Financing Agriculture*, **14(2-3)**, 64-66 (1982).

Johal, H. S. and S. P. S. Saini, Installing Gobar Gas Plants : Stimulants, Barriers and Operational Difficulties, *Khadi Gramodyog*, **26(9)**, June, 417-419 (1980).

Johal, H. S. and A. N. Shukla, Adoption of Gobar Gas Plants in Rural Areas, *Khadi Gramodyog*, **26(2)**, Sept., 542-547 (1980).

Joo, Y. H., Biogas Generation from Agricultural Wastes and Residues, *Proc. Nat. Symp. on Renewable Energy Technology*, Manila, Nov. 5-7, 1985, p. 12.

Joppitch, W. G., *Cow Dung Gas Plants in Germany*, A Report, ICAR, New Delhi, 1953. Also published in *Indian Farming*, **2**, 1953.

Joppitch, W. G., German Farms Too Use Fuel Gas Plants, *Indian Farming*, **6(11)**, 35-40 (1957).

Jose, V. R., Technical and Economic Feasibility Study of Integrated Biogas Plants for the Rural Areas, *Proc. of the Workshop on Biogas Tech and Utilisation*, Manila, 1975.

Joshi, V. K., Cow Dung Plant : Farmers Point of View, *Proc. of Symp. on Radioisotopes, Fertiliser and Cow Dung Gas Plant*, New Delhi, ICAR, 1961, p. 405.

Joshi, V. R., *et al.*, *Techno-economic Viability of the Mobile Unit for Repair and Maintenance of Biogas Plants*, A Report, Council for Advancement of Peoples Action and Rural Technology, 1993.

Joshi, M. V., Dual Use of Cow Dung : Fuel and Manure, *Proc. 32nd Indian Science Congress*, pt **2**, 212-221 (1945).

Joshy, D., Biogas Technology in China, *Biogas Newsletter, Nepal*, No. **2**, Autumn, 6-8 (1978).

Joshy, D., Biogas Plant in Nepal : Past and Present, *Biogas Newsletter, Nepal*, No. **1**, Summer, 1-3 (1978).

Kahlon, A. S. and H. Singh, Economic and Financial Evaluation of Biogas Plants in Ludhiana District, *Agricultural Situation in India*, **34(5)**, 283-285 (1979).

Kala, J. C., *et al.*, Strategy to Meet Energy Demands in Arid Areas of Western Rajasthan, *Urja*, **3(3)**, 73-76 (1978).

Kalia, A. K. and S. S. Kanwar, Biogas Technology in Himachal Pradesh : An Evaluation Study, *SESI Jour.*, **1(1)**, 1-11 (1991).

Kalia, A. K. and Singh, S. P., 1 m^3 Deenbandhu Biogas Plant for Hilly Region, *Biogas Forum*, **I** No **60**, 18-20 (1995).

Kamraj, S., Implementation of Biogas Technology in Village : A Case Study, *Proc. of Seminar on Rural Energy Future Perspective for Policy Alternatives*, Bangalore, Nov. 24-26, 1983.

Kandpal, T. C., B. Joshi and C. S. Sinha, Economics of Family Sized Biogas Plants in India, *Energy Conservation and Management*, **33(2)**, 101-113 (1991).

Karanjekar and T. Kate, Survey of Biogas Plants Operating in Some of the Villages, *Proc. of Seminar on Rural Energy*, Bangalore, Jan. 23-24 1985.

Kariuki, P. N., Biogas : Kenyan Case, Proc. Inter Conf. on Biomass, Brighton, 1980, W. Palz, *et al.*, (eds.), *Applied Science*, Essex, 769-775 (1981).

Karki, A. B., Biogas Programme in Nepal : A Review, *Biogas Newsletter, Nepal*, No. **4**, Spring, 5-7 (1979).

Karki, M. M. S., Biomass : An Affordable Energy Sources, *Khadi Gramodyog*, **28(5)**, 261-264 (1982).

Kashviswanathan, T. K., Role of the Agriculture Refinance Development Corporation in Biogas Programme, *paper presented at the Nat. Workshop on Biogas Dev.*, Vigyan Bhavan, New Delhi, Feb. 26-27, 1982.

Kashkari, C., Potential of Biogas Plants, *Proc. Inter. Conf. on Energy Use Management*, Tucson, Vol. **1**, Oct., 975-980 (1977).

Kashkari, C., Biogas Plants in India, in '*Energy in the Developing World : The Real Energy Crisis*, Oxford, 1980', p. 208-214.

Kaufman, H., Biogas in China : Small Scale Decentralised Anaerobic Digestion, *Alternative Sources of Energy*, No. **51**, 24-27 (1981).

Kellner, C, Biogas Dissemination : The Tanzanian Experience, *Biogas Forum (BORDA)*, **IV (47)**, 4-9 (1991).

Kesarwani, S. K. and S. G. Bhati, *Work Done on Biogas in India : Bibliography*, NEERI, Nagpur, 1978.

Kesarwani, S. K., *Biogas in India : Bibliographical Review of Recent Literature* 1979-82, A Report, NEERI, Nagpur, 1982.

Khandelwal, K. C., Free-fuel and Fertility Gobar Gas Plant, *Haryana Farming*, **3(12)**, 1-3 (1974).

Khandelwal, K. C., Making the Cow Dung Gobar Gas Plants Work Round the Year, *Intensive Agriculture*, **19(10)**, 12-13 (1961).

Khandelwal, K. C., Implementation of Biogas Programme at Village level : A Case Study, *Proc. of the Nat. Symp. of Biogas Tech. and Usage*, IARI, New Delhi, Nov. 29-30, 1977.

Khandelwal, K. C., *Some Information About Biogas Technology and Development in India*, A Report, Ministry of Agri. and Irrig. Krishi Bhavan, New Delhi, 1979.

Khandelwal, K. C., *Community Biogas Plant*, A Report, Ministry of Agri. and Irrig. New Delhi, 1979.

Khandelwal, K. C., *Directory of Organisations Involved in R & D on Biogas*, Directorate of Extension, Ministry of Agri. and Irrig., New Delhi, 1979.

Khandelwal, K. C., Ministry of Agriculture's Biogas Scheme, in '*Renewable Energy Research in India*, TERI, Mumbai, 1980'.

Khandelwal, K. C., Promotion of Biogas Systems : Problems and Prospects, *Proc of the Seminar-cum-Workshop*, Karnal, S. Neelakantan (ed.), Oct. 3-10, 1980, 1981, p. 33-38.

Khandelwal, K. C., Biogas for Better Life, *Changing Villages*, **6(4)**, 274-283 (1984).

Kharbanda, V. B. and Qureshi, M. A., Biogas Development in India and the Peoples Republic of China, *Energy Jour.*, **6(3)**, 51-65 (1985).

Kinje, E., *Biogas in Asia : Introductory Field Study on the State of Development of Biogas Digestion for Household Use in Tropical Rural Communities*, A Report, Consultants for Management of Development Programmes by Hollad, Utrecht, The Netherlands, xiii, Jan., 1984, p. 151.

Kingan, S. G., Integrated Energy Plan for a Small Tropical Staff Village, *paper presented at the Workshop on Biogas Tech. and Utilisation*, Manila, Oct., 1975.

Kirtikara, K., Framework for Alternative Energy Research and Development Programme in Thailand (1982-86), *paper presented at the US-ASEAN Seminar on Energy Tech : Biomass, Coal, Solar, Wind Energy Planning*, Bandung, Indonesia, June 7-18, 1982, p. 182.

Kishore, V.V.N., V.V. Ramgarao, and P. Raman, *Some Problems of Implementation of Biogas Tech. in Rural Areas*, A Report, TERI, New Delhi, Oct, 1987.

Kishore, V.V.N. and N.K. Bansal (ed.), *Renewable Energy for Rural Development*, Tata McGraw Hill Pub. Co., New Delhi, 1989.

Klass, D. L. *Energy from Biomass X*, Elesevier Applied Science Pub., Crown House, Linton Road, Barking, Essex, 1987, p. 1707.

Kozlowki, J., Methane Energy Source of the Future, *Appropriate Technology*, **1(1)**, 23-24.

Kreiter, B. C., Energy Recovery from Municipal and Industrial Wastes, *Conservation and Recycling*, **1**, 71-81 (1976).

Krishna, V. V. and S. S. Solanki, Biogas Technology and its Relevance : A Case Study of Indian Villages, *paper presented at the All India Seminar Organised by the Society of Young Scientists*, AIIMS, New Delhi, Feb. 6-8, 1980.

Krishna, V. V., *et al.*, Biogas in the Context of Appropriate Technology : A Case Study of Haryana Village, *Proc. of the Nat. Seminar on Biogas Tech.*, Ludhiana, Jul. 9-11, 53-58 (1981).

Kristoferson, L. A. and V. Bokalders, *Renewable Energy Technology : Their Application in Developing Countries,* Inter. Tech. Pub. Ltd., London, 1991.

Kulmule, M. G., Institutional Finance for Gobar Gas Scheme, *Khadi Gramodyog*, **25**, 395-400 (1980).

Kumar, N. A., *Economics of Biogas Evaluation*, Times Res. Foundation, Pune, 1988.

Kumar, N. A., Biogas Cubersome Technology, *Times of India*, New Delhi, Jan. 26, 1990.

Kumar, R. and M. Madam, Biogas Technology : A Survey (in Hindi), *Gram Praudyogiki*, **1(1)**, 4-18 (1981).

Kumar, S., *et al.*, Implementation of Night-Soil Large Size Biogas Plant in Maharashtra in 'N. K. Bansal (ed.), *Proc. Nat. Solar Energy Conv.*, Dec. 28-30, 1992, Tata McGraw Hill Pub. Co., New Delhi, 1993, p. 393-396.

Kumar, T. S., S. Singh, and N. K. Bansal, Social Response After Installation of Greenhouse over Community Biogas Plant at Masoodpur, *Biogas News*, Jan., 1985.

Kumar, V. and R. V. Misra, Role of Fertiliser Manufacturers in Promotion of Biogas Plants, *Fetiliser News*, **27(4)**, 57-59 (1982).

Kweiyan, A., Biogas Development Scheme in Maharashtra, *Financing Agriculture*, **14 (2-3)**, 32 (1982).

KVIC, Gobar Gas Industry : A Commerce Feature, *Commerce*, Jan. 13, 1973.

KVIC, *Scheme for Intensive Popularisation of Biogas*, Mumbai, 1973.

KVIC, *Gobar Gas : Retrospect and Prospects*, A Report, Directorate of Gobar Gas Scheme, Mumbai, 1979.

KVIC, *Gobar Gas on the March*, A Report, Mumbai, March, 1979.

KVIC, *Gobar Gas : Why and How ?*, Mumbai, 1979.

Lakshmanan, A. R., *et al.*, Sugarcane Pressmud Charged Community Biogas Plant, *Changing Villages*, **11(1)**, 1-5 (1992).

Lakshmi, S. R., *Biogas Plants for Rural India*, A Report, Summer Institute of Waste Management by Recycling, NDRI, Mumbai, 112-116 (1976).

Lal, V. S., Biogas Technology : Its Promises and Limitations, *paper presented at the Panel Meeting on Biogas Technology Organised by the AFPRO*, Agricultural Tools Res. Centre, Bardoli, Gujarat, Nov. 5-6, 1981.

Lalli, B. S., Planning, Installation and Monitoring of Biogas Programme in a District, *Financing Agriculture*, **14(2-3)**, 87-90 (1982).

Langley, K. F., Renewable Energy Through Anaerobic Digestion, Proc. *1st Inter Symp. on Anaerobic Digestion*, AD 79, Cardiff, Sept., 1979.

Lanox, J. N. and C. G. Wagner, *Excreta Disposal in Rural Areas*, WHO, Geneva.

Lee, K. C., Biogas Programme in Latin American Region and in Jamaica, *Khadi Gramodyog*, **27(3)**, Dec., 185-187 (1990).

Leermakers, M., *Extension of Biogas in Nepal : Strategies for Implementation*, Kathmandu, SNN-Nepal, Dec., 1992.

Leermakers, M., *Extension of Biogas in Nepal: Possible Participation in Implementation*, Kathmandu, SNN-Nepal, Jan., 1993.

Leermakers, M. and van-Nes, W. J., Extension of Biogas in Nepal, *Biogas Forum*, **1** No **56**, 9-15 (1994).

Leiu, S. S., Production of Methane from Farm Wastes, *Proc ESCAP Workshop on Biogas Tech. for Asia and the Pacific*, Manila, Philippines, Oct. 13-18, 1975.

Lewis, C., Bioenergy Systems-1 : The Potential Within the UK, *Energy World*, Jan. 2-4, 1980.

Lewis, C., Bioenergy System - II : The Potential Within Western Europe, *Energy World*, Jan. 5-7, 1980.

Lichtman, R. J., *Biogas Systems in India, VITA/COSTED*, Arlington, Virginia, 1983.

Lockeretz, W. (ed.), *Agriculture and Energy*, Academic Press, NY, 1977.

Loehr, R.C., Anaerobic Treatment of Wastes, *Dev. Indus. Microbiology*, **9**, 160 (1968).

Loehr, R.C., *Agricultural Waste Management, Academic Press, NY, 1974.*

Lovejoy, D., Biogas in China, *Proc. of the 1980 Bioenergy World Congress and Exposition*, Atlanta, Apr., 433-434 (1980).

Lusk, P. D., Comparative Economic Analysis : Anaerobic Digester Case Study, *Bioresource Technology*, **36(3)**, 223-228 (1991).

Mac Garry, B., *Biogas Pumping in Botswana*, Africa Energy Programme, Commonwealth Science Council, London, 1980.

Madiath, J., Community Biogas Programmes of *Gram Vikas, Changing Villages*, **6(4)**, 302-309 (184).

Madiath, J., Strategy in Extension and Social Acceptability of Biogas, *Moving Technology*, Oct., 1989.

Magasiner, N., Agricultural Wastes as an Energy Resource, *Jour. of Energy in Southern Africa*, **4(1)**, 16-21 (1993).

Mahajan, S. S., Biogas from Agricultural and Kitchen Garden Wastes, *Invention Intelligence*, **22(5)**, 188-192 (1987).

Mahdi, S. S. and R. V. Mishra, *Energy Substitution in Rural Domestic Sector: Use of Cattle Dung as a Source of Fuel* , TERI, Mumbai, 1979, p. 17.

Mahdi, S. S., Biogas Programme in India, *Financing Agriculture*, **14(2-3)**, 5-17 (1982).

Mahdi, S. S. and K. C. Khandelwal, Technical Support to National Project on

Biogas Development, *paper presented at the Nat. Workshop on Biogas Dev.,* Vigyan Bhavan, New Delhi, Feb. 26-27, 1982.

Maheshwari, A., Fuel Crisis : Gobar Gas is the Solution, *Chemical Weekly,* **20(11)**, 64-66 (1974).

Mahin, D. B., Biogas Power System in Remote Locations in Developing Countries, *Proc. of the Sixth SOLERAS Workshop,* Lac Cruces, New Mexico (USA), May, 243-249 (1985).

Mai, C., A Bigger Role for Biogas, *Biogas Forum (BORDA),* **III** No **54**, 7-9 (1993).

Mallan, G. M., Energy and Resource Recovery from Solid Wastes, *Resources, Recovery and Conservation,* **1(3)**, 207-216 (1976).

Makhan, D. S., Biogas Plants : Problems and Prospects, *Fertiliser News,* **24(4)**, 25-29 (1979).

Makhijani, A., Energy Policy in Rural India, *Economic and Political Weekly,* Special No., Aug., 1977.

Makhijani, A. and A. Poole, *Energy and Agriculture in Third World,* Ballinger Pub. Corp., Cambridge, Mass., 1975.

Makhijani, A. and A. Poole, Energy and Development Policy, Useful Energy and Biogasification, in '*Energy and Agriculture in the Third World,* Ford Foundation Energy Policy Project Report, 1975', p. 133-159.

Mapako, M. C., *Biogas for Zimbabwe,* Africa Energy Programme, Commonwealth Science Council, London, 1985.

Marawanyika, G., *Country Paper on Biogas Project in Zimbabwe,* Africa Energy Programme, Commonwealth Science Council, London, 1983.

Marawanyika, G., Mapako, M. C. and Mac Garry, B., *Biogas Technology for Rural Development in Zimbabwe,* Tech. Pub. Series No. 242, Africa Energy Programme, Commonwealth Science Council, London, 1987.

Marchaim, U. and Criden, J., Research and Development in the Utilisation of Agricultural Wastes in Israel for Energy, Feedstock, Fodder and Industrial Products, in 'D. L. Wise (ed.), *Fuel Gas Production from Biomass,* Vol. 1, CRC Press, Florida, 1989', p. 95-120.

Mardon, C. J., Production of Methane from Organic Wastes in Australia, *Proc. 2nd Inter. Workshop on Biogas Tech. and Utilisation,* Manila, Oct., 1975.

Martin, J. H., *et al.,* Comparing and Contrasting Anaerobic Digesters in the Republic of China and the USA, *Proc. Inter. Symp. on Biogas, Microalgae, and Livestock Wastes,* Taipei, Taiwan, Rep. of China, Sept. 15-17, 1980.

Massey, V., Biogas : An Economic and Alternative Fuel Study, *Workshop Proc. on Biogas Utilisation in Engines, IIP,* Dehradun, Apr. 19-20, 1978, p. 68-71.

Mathur, T. H., Fuel from Wastes : A Minor Energy Source, *Science,* Nov. **10**, 599-602 (1972).

Mayur, R., Human Biomass in India : Year 2000, *paper presented at the Conf. on Bioresources for Dev.,* Univ. of Houston, Houston, Nov. 5-10, 1978.

Mayur, R., Bioenergy for Development in the Commonwealth Countries, *Khadi Gramodyog,* **27(3)**, Dec., 164-166 (1980).

Mazumdar, A. and N. K. Goplakrishnan, Role of Cooperatives in the Promotion of Biogas Technology, *paper presented at the Workshop on Cooperatives and Technology for 2000 AD*, Pune, Feb. 4-6, 1982.

Mazumdar, A. and N. K. Goplakrishnan, Diffusion of Biogas Technology, *Tamil Nadu Jour. of Cooperation*, **73**, 529-541 (1982).

Mc Carty, P. L., Methane Fermentation : Future Promise or Relic of the Past, *Proc. of the Conf. on Bioconversion Energy Res.*, Univ. of Massachussettes, Boston, 1973.

Mc Queen, H. J., Energy Recovery from Wastes in Canada, *Proc. Inter. Conf. on Conservation of Refuse to Energy*, Switzerland, Nov. 3-5, 149-156 (1975).

Menon, M. G. K., Call to Involve Cooperative Bodies for Popularising Biogas, *Economic Times*, New Delhi, Oct. 25, 1985.

Middle-East Centre for the Transfer of Appropriate Technology, MECTAT's Experience in the Transfer of Biogas Technology, *Proc. Inter. Conf. on Biogas Tech.*, Elsevier, 617-622 (1986).

Mishra, S. P., Economic Feasibility of Gobar Plant : A Case Study, *Khadi Gramodyog*, **21**, 286-289 (1975).

Mital, K. M., Social Responsibilities of Business Organisations, *Vikalpa*, **4(1)**, 39-50 (1979).

Mital, K. M., Measurement of Social Benefits and Costs of a Community Biogas Plants, Ch. 5 (Sec. 5.4), in 'Mital, K. M., *Social Responsibilities of Business: Concepts, Areas and Progress*, Chanakya Publications, Delhi, 1988'.

Mital, K. M. and P. Vrat, *A Socio-Economic Evaluation of Janata Biogas Plants in Uttat Pradesh and a Detailed Review of Various Socio-Economic Studies Carried Out in India*, A Report, New Delhi, 1991.

Mital, K. M., Scope of Industrial Engineering in Promoting CNG as a Vehicle Fuel, *Udyog Pragati* (NITE Jour.), **15 (3 & 4)**, 1991.

Mital, K. M., *Biogas Systems : Principles and Applications*, New Age Inter. (P) Ltd., New Delhi, 1996.

Mital, K. M., Biogas Systems, Ch. 8, in 'Mital, K. M., *Non-Conventional Energy Systems : Principles, Progress and Prospects*, Wheeler Publishing, New Delhi, 1997'.

Mitra, T. and B. C. Bhattacharya, Biogas as an Alternative Source of Energy, *Chem. Eng. World*, **13(2)**, 49-52 (1978).

Khan, M.N.G.A., Rural Energy Problems and Biogas, *Khadi Gramodyog*, **27(3)**, Dec., 151-152 (1980).

Mother Earth News, Interview with Ram Bux Singh of India, Issue No. **18**, 1972.

Moulik, T. K. and U. K. Srivastava, *Biogas Plants at the Village Level : Problems and Prospects in Gujarat*, CMA Monograph No. 59, IIM, Ahemadabad, 1975.

Moulik, T. K., Promoting Biogas Technology in the Villages : The Experience in Gujarat, *Proc. Nat. Symp. on Biogas Tech. and Usage*, IARI, New Delhi, Nov. 29-30, 1977.

Moulik, T. K., Biogas Systems : Alternative Technology for Meeting Rural Energy Needs in India, *paper presented at the Seminar on Mangement Res.*, Ahmedabad, Jan. 19-20, 1978.

Moulik, T. K., *et al.*, *Biogas Systems in India : A Socio-Economic Evaluation,* Center for Management in Agri., IIM, Ahemedabad, 1978.

Moulik, T. K., Biogas and Rural Energy Supply, *paper presented at the National Seminar on Biogas Technology*, Punjab Agri, Univ., Ludhiana, Jul. 9-11, 1-8 (1981).

Moulik, T. K., *Biogas Energy in India*, Academic Book Centre, Ahemedabad, 1982.

Moulik, T. K., Introduction of Community Biogas Plant in a Gujarat Village : A Case Study of Technological and Organisational Inventions, *Changing Villages*, **4(4)**, 1-14 (1982).

Moulik, T. K., Biogas Programme, *paper presented at Nat. Workshop on Biogas Dev.*, Vigyan Bhavan, New Delhi, Feb. 26-27, 1982.

Moulik, T. K., N. Murthy and P. Subramanian, *India's Experiments with Community Biogas Systems*, A Report, IIM, Ahmedabad, 1983, p. 65.

Moulik, T. K., *Biogas Programmes in India and China : A Comparative Analysis of Experience*, IIM, Ahmedabad, 1985.

Moulik, T. K. and P. R. Shukla, *Comparative Evaluation of Operating Performance of KVIC and Janata Models of Biogas Plants in India*, IIM, Ahmedabad, 1986.

Moulik, T. K., *Diffusion of Biogas Technology : Strategies and Policies*, IIM, Ahmedabad.

Moulik, T. K., From a Service Facility to a Profitable Enterprise, *Biogas Forum*, **III (54)**, 20-22 (1993).

Mowli, P. P. and S. V. K. Gyaneswar, Biogas Plants for Rural India : Technology and Economics, *paper presented at the Inter. Conf. on Rural Dev. Tech., An Integrated Approach*, Bangkok, Thailand, June 21-24, 1977.

Mukherjee, J. K., *Gobar Gas Plant : A Study*, Gandhi Peace Foundation, New Delhi, 1974.

Mukherjee, S. K. and A. Arya, Comparative Analysis of Social Cost-Benefit Studies of Biogas Plants, *paper presented at the Research Planning Workshop on Energy for Rural Dev.*, Ahmedabad, Dec., 1979.

Mukhopadhyay, B. K., Biogas Plants on the Indian Soil, *Management Accountant*, **14(3)**, 269-271 (1980).

Mulmule, M. G., Institutional Finance for Gobar Gas Scheme, *Khadi Gramodyog*, **26(9)**, June, 395-404 (1980).

Mulmule, M. G., Socio-Economic Impact of Gobar Gas Plants in Betul, *Khadi Gramodyog*, **32(9)**, 441-452 (1986).

Munde, A. V., Energy Through Gobar Gas for Small Farmers, *Khadi Gramodyog*, **23(12)**, 546-548 (1977).

Myles, R. M. and J. B. Singh, Promotion of Biogas Plants for Small and Marginal Farmers on Individual and Community Basis, *paper presented at the AFPRO Seminar on Integrated Approach for Agricultural Dev.*, Oct. 3-5, 1979.

Myles, R. M., *Biogas Technology : A Reveiw of the Approach Adopted by AFPRO in Systematic Promotion of the Janata Biogas Plant*, AFPRO, New Delhi, 1980.

Myles, R. M and I. Hussain, Community Night-Soil Janata Biogas Plant Systems for Harijan Colony in Midnapore, *Changing Villages*, **3(4)**, 7-20 (1981).

Myles, R. M., *Socio-Economic and Feasibility Report of the Community Biogas Plant System of Village Dadu*, A Report, AFPRO, 1982.

Myles, R. M., Systematic Transfer and Extension of Janata Biogas Plant by Voluntary Organisations, *Changing Villages*, **6(4)**, 29-30 (1984).

Myles, R. M., *A Practical guide to Janata Biogas Plant Technology*, A Report, AFPRO, New Delhi, 1985.

Nahar, S. L., Gobar Gas Plants in Developing Rural Economy, *Khadi Gramodyog*, **22(6)**, 271-275 (1976)

NAI, Methane Digesters for Fuel Gas and Fertiliser, *Newsletter No. 3*, New Alchemy Inst., West Pescadero, California, Spring, 1973.

Narayanagowda, K., Biogas - A Decentralised Rural Energy Source : Its Futures Perspectives for Policy Alternatives, *Proc. Seminar on Rural Energy Future: Perspectives for Policy Alternatives*, Bangalore, Nov. 24-26, 1983.

NCST, *Draft Status Report on Utilisation and Recyling of Wastes*, Nat. Comm. on Sci. and Tech., Technology Bhavan, New Delhi, Aug., 1975.

NCST, *All India Coordinated Project on Biogas Technology and Utilisation*, Project Outline, New Delhi, May, 1976.

Neelkantan, S., Perennial Source of Power, *Hindustan Times*, Nov. 10, 1955.

Neelkantan, S., Assessment of the Availability of Cow Dung for Potential Gobar Gas Production in India, *Agricultural Situation in India*, **30(9)**, 633-637 (1975).

Neelkantan, S. and M.S. Sondhi, Potential Biogas Production from Sewage and Animal Excreta in India, *paper presented at the 63rd Session of the India Sci. Congress (Section of Agri. Sci.*, Waltair, 1976.

Neelkantan, S., Project Proposal on Research and Action Planning for Biomass and Biogas Technology, *Proc. of Research Planning Workshop on Energy for Rural Dev.*, Ahmedabad, Dec. 27-29, 1979.

Neelkantan, S., Status of Biogas Programme in India, *Proc. of the Seminar-cum-Workshop*, Karnal, Oct. 3-10, 1980, S. Neelkantan (ed.), NDRI, Karnal, 1981, p. 4-8.

Nicolon, A., Producing Methane Gas from Animal Wastes in France, *Eco-Development News*, **34**, 38-44 (1985).

Nisanka, S. K., M. K. Misra and N. C. Sahu, Economy of Fuel Supply in an Indian Village, *Bioresource Tech*, **39(3)**, 249-261 (1992).

NOBDE, *Biogas Development in China*, A Report, National Office for Biogas Development and Extension, Beijing, China, Jul., 1980.

NRC, *Development and Application of Biogas Technology in Rural Areas of Egypt*, Final Project, Nat. Res. Centre, Cairo, 1980.

Nyns, E. J. and H. P. Navean, Status of European Work on Methane Fermentation from Biomass, in 'D. L. Wise (ed.), *Fuel Gas Production from Biomass*, CRC Press, Florida, 1981', p. 35-145.

Nyns, E. J., *Biogas Plants in the European Community and Switzerland*, Univ. of Louvain, Belgium, 1985.

Obel, E., Biogas for Rural Energy, *Kengo News (Kenya)*, **1(21)**, 10-11 (1985).

Odeyemi, O., Resource Assessment for Biogas Production in Nigeria, *Proc. of the ISES Congress*, Brighton, UK, Aug. 23-28, 1981, Vol. II, Pergamon, Oxford, 1345 (1982).

ONGC, *Energy Profile*, A published brochure, Economic Environment Scanning Group, ONGC, New Delhi, 1987.

Oppenlandes, J. C., *et al.*, ASAE Paper No. 75-4545, Economic Analysis of Dairy Manure Biogas System, ASAE Paper No. 75-4545, *American Soc. of Agri. Eng.*, St Joseph, MI, 1975.

Osada, Y., *et al.*, Studies on Fuel Gas Production from Organic Wastes with Bioconversion in Japan, *Proc of the Inter. Congress on Human Environment*, Kyoto, Nov., 1975, Yoichi Fukushima (ed.), Pergamon, NY, 1977.

Ostrouski, C. H., Feasibility Study of Biogas Production in Individual Farms in South Western Ontario, *Sharing the Sun*, **7**, 129-145 (1976).

Ossiandwski, P. and D. Pluschke, Biogas Technology : Advanced Training and Dissemination Programme, *Proc. 2nd EC Conf., Berlin*, Sept. 20-23, 1982, Applied Science, London, 757-760 (1982).

Oza, A., Biogas : A Friend of the Environment and Women, *Exchanges*, Dec. No 3, Actionaid, 3 Rest House Road, Bangalore, 1993.

Pachauri, R. K. (ed.), *Energy Policy for India*, Macmillan, Mumbai, 1980.

Pachauri, R. K. and L. Srivastava (eds.), *Energy Policy Issues*, Wiley Eastern Ltd. New Delhi, 1986.

Pai, K. K., Role of Commercial Banks in Promoting Gobar Gas Plants, *Khadi Gramodyog*, **20(70)**, 353-357 (1974).

Palz, W. and P. Charties, *Energy from Biomass in Europe*, Applied Sci. Pub., 1980.

Panditrao, Y. A., Energy Crisis and Gobar Gas, *Khadi Gramodyog*, **23**, 187-189 (1977).

Panichjaroon, S. and R. Deemark, *Technical Report on Biogas Plant in Thailand*, FAO/SIDA Workshop on Organic Fertilisers in Asia, 1976.

Panigrahi, N. R., Review of Biogas Programme in Orissa, *paper presented at the Nat. Workshop on Biogas Dev.*, Vigyan Bhavan, New Delhi, Feb. 26-27, 1982.

Pankhurst, E. S., Biogas, *Energy Digest*, **9(3)**, June, 28-33 (1980).

Pankhurst, E. S., The Prospects for Biogas : A European Point of View, *Biomass (UK)*, **3(1)**, 1-42 (1983).

Pant, K. C., *Address to the Inter. Workshop on Community Biogas Plants*, BORDA, Bremen, 1984.

Paramasivam, R., Economic Utilisation and Disposal of Cattle Dung in Rural Communities, *Symp. on Community Water Supply and Waste Disposal*, Nagpur, Vol. **2**, 62-67 (1966).

Parikh, J. K. and K. S. Parikh, Mobilisation and Impacts of Biogas Technology, *Energy*, **2(4)**, 441-445 (1976).

Parikh, J. K. and K. S. Parikh, *Potential of Biogas Plants and How to Realise it*, UN Institute for Training and Research Seminar on Microbiological Energy Conversion, (erstwhile) Fed. Rep. of Germany, Oct., 1976, Pergamon Press, Oxford, 555-592 (1977).

Parikh, J. K., Renewable Energy Options : What Could Developing Countries Expect from Them, *Energy*, **4(5)**, 989-994 (1979).

Parikh, K. S., *Cost-Benefit Analysis of Biogas Plants in India*, Ph. D Thesis, MIT, Cambridge, 1963.

Parikh, M., *Community Biogas Plant* (Centralised Alternative and Renewable Sources of Energy for Domestic Purpose and Power Generation), Agri. tools Res. Centre, Bardoli, 1979.

Parikh, M., Installation of Biogas Units on Turn-key basis, *paper presented at the Nat. Workshop on Biogas Dev.*, Vigyan Bhavan, New Delhi, Feb. 26-27, 1982.

Park, Y. D., *Biogas Research and Development in Korea*, Office of Rural Dev., Inst. of Agri. Sci., Suweon, Republic of Korea.

Patankar, G. L. and K. Chandra, Methane : Non-Poisonous, *Jagrithi*, **12**, Mar 6, 1974.

Patankar, G. L., Role of Gobar Plants in Agro-Industries, *Khadi Gramodyog*, **20(7)**, 351 (1974).

Patel, A. R., Problem of Implementation of Gobar Gas Plant Scheme, *Financial Express*, Mumbai, April 18, 1975.

Patel, A. R., An Alternative Source of Fuel and Fertilisers, *Khadi Gramodyog*, **21 (9)**, June, 423-426 (1975).

Patel, A. R., Gobar Gas Obtained Through the Fermentation of Cattle Dung Anaerobically : Potential Source of Energy, *Agri. Agro-Indus. Jour*, **10(10)**, Oct. 3-7, 1977.

Patel, A. R., Gobar : A Potential Source of Energy, *Yojna*, **21(10)**, 25-28 (1977).

Patel, H. C., Biogas Plants and Rural Development, *Khadi Gramodyog*, **26(5)**, 229-233 (1980).

Patel, J. J., *Present Position of Biogas Plant Research*, Jamnalal Bajaj Central Res. Inst. for Village Indus., Mumbai, 1959.

Patel, J.J., Small Cow Dung Gas Plant, Jamnalal Bajaj Central Res. Inst. for Village Indus., Mumbai, 1959.

Patel, J. J., Improved Cow Dung Gas Plants and Their Economics, in '*Radio Isotopes, Fertilisers and Cow Dung Gas Plants*, ICAR, New Delhi, 1961', p. 653-663.

Patel, J. J., Gobar Gas Plant at Datlapur, *Khadi Gramodyog*, **8**, 674-675 (1962).

Patel, J. J., Gobar Gas Plants, *Khadi Gramodyog*, May, 1963.

Patel, J. J., Constraints of Gobar Gas Plants, *Khadi Gramodyog*, Aug., 1963.

Patel, J. J., Gobar Plants : Its Development, Present Status and Future, *ESCAP Workshop*, New Delhi, 1975.

Patel, J. J., An Approach to Waste Utilisation in Rural India, *UNEP-FAO Seminar on Residue Utilisation Management of Agri. Wastes*, FAO, Rome, 1977.

Patel, J. J., Chinese Biogas Scene, *Jour. of Biowaste Treatment*, **1(2)**, 49-52 (1981).

Patel, S. M., *A Study of Performance of 100 Biogas Plants in Gujarat*, Inst. of Coop. Management, Ahmedabad, 1974.

Patel, S. M., Performance of Biogas Plants in Gujarat, *Khadi Gramodyog*, **21(11)**, Aug., 493-502 (1975).

Pathak, B. N. and G. Rajagopal, Studies on the Aerobic Digestion of Night-Soil at Ernakulam, *Health*, **4(1)**, 68-73 (1962).

Pathak, P. V. and Colah, J. J., Methane from Biogas, *Science Today*, **11(6)**, 45-46 (1976).

Pathak, P. V., Utilisation of Gobar Gas Plant, *World Science News*, **14**, 175-178 (1977).

Pathania, M. S. and K. N. S. Sharma, An Analysis of Constraints for Non-Functional Biogas Plants in Haryana, *Khadi Gramodyog*, **41(4)**, 175-180 1995.

Patil, S. J. and M. P. Dhangade, Economics of Gobar Gas Plants in Ahmedabad District of Maharashtra, *Khadi Gramodyog*, 195-198 (1984).

Paul, N, *Energy from Solid Wastes*, Marcel Dekker, NY, 1976.

Paul, N. J., Natural Resources : A Substitution for Crude Oil and Fertiliser with Focus on Gobar Gas Plant Mechanism, *Chemical Take Off*, **4(12)**, 6-7 (1975).

Paul, S., Biogas Systems : The Chinese Miracle, *Financial Express*, June 6, 1981.

Paul, S., Biogas Power : Chinese Experience, *Financial Express*, Dec. 17, 1981.

Paul, T. M., Potentialities of Gobar Gas Under Bombay Conditions, *paper presented at the 8th Dairy Industry Conf.*, Trivandrum, 1970.

Paul, T. M., Disposal of Gobar Gas Slurry, *paper presented at a Seminar on Gobar Gas*, Varanasi, Mar., 1971.

Paul, T. M., Utilisation of Dung Produced in Bombay Suburbs, *Khadi Gramodyog*, **20(7)**, 348 (1974).

Paul, T. M., Gobar Gas Plant for Bombay Suburbs, *Khadi Gramodyog*, **21(2)**, Nov., 110-113 (1974).

Paul, T. M., An Appraisal of Gobar Gas Plants in India, *paper presented at the Seminar on Appropriate Tech.*, Pune, Feb. 25, 1975.

Paul, T. M., Food, Fodder, Fertiliser and Fuels from Animal Wastes, *paper presented at the Seminar on Utilisation of Farm Wastes for Rural Industries Growth*, Bangalore, Dec., 1975.

Paul, T. M., Studies on Biogas Source, *Current Science*, **45(20)**, 1976.

Paul, T. M., Role of Biogas in Rural Development, *paper presented at the Seminar on Education in 2001*, IIT, Chennai, Mar. 12-14, 1976.

Paul, T. M., Socio-Economic Benefits of Integrated Waste Management by Recycling, *Summer Inst. on Waste Management by Recycling*, Mumbai, 94-98 (1976).

Perumal, S., Review of Biogas Programme in Tamil Nadu, *paprer presented at the Nat. Workshop on Biogas Dev.*, Vigyan Bhavan, New Delhi, Feb. 26-27, 1982.

Pillai, C. R., *et al.*, Potentialities of Using Biogas and Solar Energy in Textile

Industry, *paper presented at the Symp. on 'Energy Saving : Engineering and Wet Processing Aspects*, Bombay Textile Res. Assoc., Mumbai, Oct. 19-20, 1981', p. 65-73.

Pimetal, D. and M. Pimetal, *Food, Energy and Society*, Edward Arnold Publishers, London, 1979.

Pimplikar, B. D., Cow Dung Gas Plant, *Proc. Symp. on Radioisotopes, Fetilisers and Cow Dung Gas Plant*, ICAR, New Delhi, 395-397 (1961).

Poland, F. G. and S. R. Harper, *Biogas Developments in North America*, Georgia Inst. of Tech., Atlanta (USA), 1985.

Pokharel, R. K. and R. P. Yadav, *Application of Biogas Technology in Nepal: Problems and Prospect*, MIT Series No II, ICIMOD, Kathmandu, Nepal, 1990.

Polak, M. H. and Rabin, G., Biogas Programme in Dominico Caribbean, *Biogas Forum (BORDA)*, **III (42)**, 4-8 (1990).

Pollock, B. L., Energy from Urban Wastes, *Proc. of Inter. Soc. Energy Conver. Eng. Conf.*, San Francisco, Aug. 26-30, 525-530 (1974).

Posmowski, P., Biogas Seen Meeting Indian Village Energy Needs, *UNESCO Features*, 695-696 (1976).

PPH, Experiences on Health Management of Dung in Rural Areas, Peoples Pub. House. Beijing, 1974.

PR, Marsh Gas Used in Rural China, *Peking Review*, 16, Jan, 1973, p. 22.

PR., Popularising the Use of Marsh Gas in Rural Areas, *Peking Review*, **18**, Jul. 25, 1975, p. 15.

PRAD, *Scope of Activities in Biogas Technology and Utilisation*, State Planning Institute, Lucknow, 1977.

PRAD, *Bench-Mark Survey Report on Community Biogas Plant Project*, State Planning Institute, Lucknow, 1978.

PRAD, *Community Biogas Plant*, State Planning Institute, Lucknow, 1978.

PRAD, *Janata Biogas System : An Evaluation*, State Planning Institute, Lucknow, March, 1978.

Prakasam, T.B.S., Application of Biogas Technology in India, *Proc. of the Seminar on Biomass Energy for City, Farm and Industry*, Oct. 25-26, 1979, J. G. Press, Emmaus, 202-211 (1980).

Prakasam, T.B.S., Some Aspects of Biogas Technology in India and Egypt, *Proc. of the Seminar on Biogas and Alcohol Fuels Production*, Chicago, Mar. 12, 1981, J. G. Press, Emmaus, p. 229-248.

Prakash, G., *et al.*. Comparison of Various Renewable Energy Options to Meet Energy Demand of an Urban Community, *Proc. Nat. Solar Energy Conv., Solar Energy Society of India*, Madurai, Sept. 13-15, 1986.

Prakash, O. M., Development of Alternative Source of Energy for Rural Sector, *Electrical India*, **17(3)**, 15 (1977).

Prasad, A. and G. Gupta, Gobar Gas Plants in Rajasthan : A Study, *Changing Villages*, **4(6)**, 31-48 (1982).

Prasad, C. R., K. K. Prasad and A. K. N. Reddy, Biogas Plants: Prospects, Problems and Tasks, *Eco. and Pol. Weekly*, **9(32-34)**, Special number, 1974.

Prasad, C.R., Utilisation of Organic Wastes in Biogas Plants, *Khadi Gramodyog*, Sept., 514-518 (1985).

Prasad, G.V.V.S.D.S., Community Biogas Plants : A Case Study, *Moving Technology*, **4(5)**, 23-25 (1989).

Prasad, H. K., Gobar Gas Plant : An Empirical Study, *Khadi Gramodyog*, Jul., 1967.

Prasad. H. K. and A. N. Sharma, Biogas Plants and Their Prospects, *Agricultural Situation in India*, **37(2)**, May, 93-95 (1982).

Protacio, A. C., Feasibility of Using Biogas as an Energy Source, *Proc. Seminar-Workshop on Alternative Sources of Energy for Dev.*, Cavite, Philippines, Aug., 1978.

Puri, G. G., Gobar Gas Plants as Cheap Energy Source, *Inaugural Souvenir of M. P. Regional Centre of Inst. of Public Health Engg.*, 1977.

Pyakural, K. A., Techno-Socio Economic Study on Biogas, *Biogas Newsletter*, Shakti Pub., P. O. 1309, Kathmandu, Nepal, 1978.

Pyle, D. L., *Technical Options in Anaerobic Digestion*, A Background paper, Dept. of Chem. Engg. & Chem. Tech., Imperical College, London, Nov, 1976.

Pyle, D. L. and C. Freeman, *Methane Production by Anaerobic Fermentation : An Annotated Bibliography*, Intermediate Tech. Dev. Group, London, 1977.

Pyle, D. L., Anaerobic Digester Designs in the Third World, *Proc. Inter. Symp. on Anaerobic Digestion*, Cardiff, Wales, Sept. 17, 1979, Applied Science Publishers, 345-376 (1980).

Pyle, L., Objectives, Testing and Evaluation of Biogas Systems, *Report of the Training Workshop on Monitoring and Evaluation of Biogas Systems*, Lusaka, Zambia, Aug. 29 - Sept. 1, 1983, Commonwealth Sci. Council, London.

Qu, G., China Turns to Biogas, *Mazingira*, **12(1)**, 61-64 (1979).

Qureshey, W. A., Biogas Plants and Rural Economy, *Financial Express*, Jan. 13, 1977.

Qureshi, M. A. and V. P. Kharbanda, China's Energy Policy-Progress on the Biogas Technology Front, *Jour. of Scientific and Indus. Res.*, **41**, 413-416 (1982).

Qureshi, N. H., Development of Biogas in Pakistan, *Workshop on Biogas Tech. and Utilisation*, New Delhi, Sept., 1975. Also published in *World Science News*, **12(19)**, 407 (1975).

Rairekar, R. R., Role of Agriculture Finance Corporation (AFC) in Biogas Programme, *paper presented at the Nat. Workshop on Biogas Dev.*, Vigyan Bhavan, New Delhi, Feb. 26-27, 1982.

Rajabapiah, P. and H. N. Chanakya, Alternative Approaches for the Indian Biogas Programme, *Proc. of the Seminar on Rural Energy Futures, Perspective for Policy Alternatives*, Bangalore, Nov. 24-26, 1983.

Rajagopal, L. S., Biogas Plant for Rural India, *Summer Institute on Waste Management by Recycling*, NDRI, Aeray Milk Colany, Mumbai, June, 112-116 (1976).

Rajendran, K., *et al.*, Techno-Economic Survey for Setting up of Gobar Gas Plants in a Typical Village in Coimbatore District, *Energy Management*, **6(4)**, Oct.-Dec., 293-300 (1982).

Rejeswaran, R., *et al.*, Importance of Biogas Plants on Energy Use Pattern of rural Households : A Case Study, *Urja*, **28(3)**, 41-42 (1991).

Ramarao, A., Biogas Programme in India, *Khadi Gramodyog*, June, 405 (1988).

Rana, K. N., Problems of Gobar Gas Technology and Utilisation in Nepal, *Proc. Workshop on Biogas Tech. and Utilisation*, New Delhi, Sept., 1975.

Rana, R. S. and S. R. Varma, A Survey of Biogas Plants in Punjab, *Proc. Nat. Seminar on Biogas Tech.*, Ludhiana, July 9-11, 59-63 (1984).

Ranganathan, V. and S. Shanta, Economic Valuation of Biogas Plants : Case Studies in Thanjavur and Periyar Districts, *Pacific and Asian Jour. of Energy*, **2(2)**, 93-106 (1992).

Rao, A. R. and R. K. Malik, Biogas as an Option for Dung Utilisation, *Urja*, **12**, 321-332 & 368 (1982).

Rao, A. R., Biogas Programme in India- I, *Khadi Gramodyog*, June, 399-410 (1987).

Rao, A. R., Biogas Programme in India-II, *Khadi Gramodyog*, Dec., 165-168 (1988).

Rao, E. G. K., Advantage of Biogas, *Indian Farming*, **9(4)**, 1959.

Rao, E. G. K., Community Biogas Plants, in '*Radio Isotopes, Fertilisers and Cowdung Gas Plant*, ICAR, New Delhi, 1961.'

Rao, J. R. N., Technical Report: Biogas Process, *paper presented at the Workshop on Biogas Tech. and Utilisation*, New Delhi, Sept. 16-22, 1975.

Rao, K. S., Role of Cow Dung Gas Plant in Agricultural Dev., *Proc. Symp. on Radioisotopes, Fertilisers and Cow Dung Gas Plants*, New Delhi, 1961, p. 373-380.

Rao, K. S. and D. Vaja., Community Biogas Plant : GEDA Experiment, *Moving Technology*, **4(5)**, 26-30 (1989).

Rao, R. V., *et al.*, Some of the Factors for Reduction of the Cost of Biogas Plant, *paper presented at the Seminar on Utilisation of Farm Wastes for Rural Industrial Growth*, Bangalore, Dec. 31, 1975.

Rao, V. A., Gobar Gas Plant : Source of People's Power, *People's Action*, **7(3)**, 42-43 (1973).

Ravindranath, N. H. and Hall, D. O., *Biomass, Energy and Environment*, Oxford University Press, Oxford, 1995.

Rawat, B. S., Economic Aspects of Cattle Yard Waste Recycling, *Summer Institute on Waste Management by Recycling*, Mumbai, June, 37-45 (1976)

Ray, S., Shifting Urban Emphasis : An Energy Alternative, *Khadi Gramodyog*, **27(3)**, 205-208 (1980).

Rayarikar, R. R., Role of AFC in Promoting Biogas Development with Industrial Credit Support, *Financing Agriculture*, **14(2-3)**, 33-39 (1982).

RBI, *Report of the Inter-Institutional Group of Financing Gobar Gas Plants by Bank*, Credit Planning Cell, RBI, Mumbai, 1976.

Reddy, A. K. N., I. Rajaramanian, D. K. Subramanian and P. Rajabapaiah, *A*

Community Biogas Plant System for Pura Village, Karnataka State Council for Sci. and Tech., Bangalore, 1979.

Reddy, A. K. N., Karnataka's Biogas Potential, *paper presented at the Technical Seminar on Energy Planning for Karnataka*- 1980, Nov. 10-11, 1979.

Reddy, D. C., Energy from Gobar Gas, *Kurushetra*, **26(15)**, 23-24 (1981).

Reeps, V., UNDP/ESCAP Regional Workshop on Integrated Rural Energy/ Environmental Planning for Sustainable Rural Development, *Biogas Forum*, **I** No **60**, 23-24 (1995).

Rele, S. J., International Cooperation in Biogas Technology, *Yojna*, **21(2)**, 27 and 34 (1977).

Revelle, R., Energy Use in Rural India, *Science*, **192**, 969-976 (1976).

Reynolds, G. F., Generation of Methane From Waste Materials, *Appropriate Technology*, **2(2)**, 11-13 (1975).

Richards, K., Gas from Waste Makes Growing Contribution to Fuel Supplies, *Energy Management (U. K)*, Issue **1**, Apr., 25 (1988).

Rijal, K., Resource Potential and Economic Evaluation of Biogas Utilisation in Nepal, *Energy*, **11(6)**, 545-550 (1986).

Rijal, K., An Approach for Evaluation of the Biogas Programme Development in Nepal, *Economic Jour. of Nepal*, **10(3)**, 45-54 (1986).

Roy, R., Community Biogas Plants : Implementation in Rural India, *Appropriate Technology*, **8(1)**, 17-20 (1981).

Roy, R., Family and Community Biogas Plants in Rural India, *Appropriate Technology*, Intermediate Tech. Pub., London, 1981.

Rozaria, R. D., Gobar Gas : Its Time for a Second Look, *Science Today*, 37-46 (1975).

Ru-Chen, C., Development of Biogas Utilisation in China, *Natural Resources Forum*, **5(3)**, 277-282 (1981).

Sagar, B., and N. S. Grewal, Gobar Gas : An Alternate Source of Energy, *Symp. on Rural Development*, Chandigarh, Sept., 1974.

Sagar, G. J., Low Cost Technology for Disposal of Human Excreta with Community Latrines, *Jour. of the Inst. of Engrs. (India), Mech. Engg.*, **62(2)**, 47-50 (1982).

Saint, K., Biogas Technology for Tribal, Small and Marginal Farmers and Labourers : Seva Mandir's (Udaipur) Experience, *Changing Villages*, **42(2)**, 20-26 (1982).

Sanghi, A. J. and D. Dey, A Cost-Benefit Analysis of Biogas Production in Rural India : Some Policy Issues, in 'W. Lockertz (ed.), *Agriculture and Energy*, Academic Press, New York, 1977', p. 607-624.

Sanghi, A. K. and D. Dey, *A Cost-Benefit Analysis of Biogas Production in Rural India : Some Policy Issues*, Center for the Biology of Natural Systems, Washington Univ., St. Louis, Missouri 63130, 1975.

Sankar, T. L., Fuelling India's Future : The Perspectives, *Chem. Indus News*, **18(10)**, 900-903 (1974).

Sarkar, A. N., Biogas Brings Cheers in Gulaothi, *Intensive Agriculture*, **19(80)**, 45-46 (1981).

Sasse, L., *Brief Report on International Conference on Biogas Technology and Implementation Strategies*, Pune, 1990, *Biogas Forum* **1(4)**, 13-19 (1990).

Sasse, L., Biogas : The Foolish Flame, *Biogas Forum*, **IV** No **43**, 13-15 (1990).

Sasse, L., Comparison of Utilising Biogas Plants by Farmers in India and Indonesia, *Biogas Forum*, **IV (55)**, 20-21 (1993).

Sasse, L., Low Maintenance Waste Water Treatment Systems (LOMWATS) : An Attempt to Treat More Waste Waters in Developing Countries, *Biogas Forum*, **II** No **61**, 4-7 (1995).

Sathyamoorthy, A., Financing of Biogas Plants, *Changing Villages*, **6(4)**, 284-290 (1984).

Schmid, L. A., Feedlot Wastes to Useful Energy : Fact or Fiction, *Jour. Env. Eng. Div.*, **101**, 787-793 (1975).

Sethna, H. N., Development of Energy Sources, *Khadi Gramodyog*, Apr., 1974.

Shah, A. C., Impact of Drought on Biogas Programme, *Economic Times*, Mumbai, June 13, 1988, p. 11.

Shah, H. C., Biogas Development in Gujarat State, *Changing Villages*, **6(4)**, 310-318 (1984).

Shande, P. Y., Development and Extension Work with Cowdung Gas Plant, in '*Radio Isotopes, Fertiliser and Cowdung Gas Plant*, ICAR, New Delhi, 1961,' p. 370-372.

Shankar, R., *Story of Biogas (in Hindi)*, Agricole Pub. Academy, New Delhi, 1984.

Sharma, D., Biogas : Slow-Track Answer to Energy Needs, *Hindustan Times*, Oct. 19, 1995, p. 16.

Sharma, J. P., P. Vasudevan, N. K. Bansal, An Economic Evolution of Community Size Biogas Plant, *Energy Management*, **10(3)**, 165-168 (1986).

Sharma, M. and K. P. Sukumaran, Biogas Technology : A Renewable Energy Source for the Future, *Chemical Age of India*, **31(12)**, 8 (1980).

Sharma, N. L., *Economic Viability of Biogas Plants in Una District of H. P.* M. Sc. (Agri. Econ.) thesis, Himachal Pradesh Krishi Vishwavidyalaya, Palampur.

Sharma, S. C., Bioconversion of Solar Energy into Fuel Gas, *paper presented at the 8th Annual Conf. of Assoc. of Microbiologists of India*, Madurai, Dec. 21-23, 1977.

Sharma, U. P., Biogas Systems, *Khadi Gramodyog*, **27(3)**, 195-198 (1980).

Shelat, R. N. and G. L. Karia, Utilisation of Biogas in South Gujarat : A Case Study, *Proc. Nat. Seminar on Tech. for Agricultural Dev.*, Institute of Engineers (India), Chandigarh, Apr., 1977.

Shelat, R. N. and G. L. Karia, Biogas Technology for Rural Development : A Case Study, *paper presented at the Inter. Conf. on Rural Dev. Tech. : An Integrated Approach*, AIT, Bangkok, June, 1977.

Shelke, P. P. and V. M. Kurkure, Community Biogas : Case Studies in Vidarbha, *Jour. of the Inst. of Engrs. (India)*, **72** (PTI **D1**), 21-22 (1991).

Shende, P. Y., Development and Extension Work with Cow Dung Gas Plant, *Symp. on Radioisotopes, Fertilisers and Cow Dung Gas Plant*, New Delhi, 370-372 (1961).

Shetty, N. P. N., *et al.*, Non-Polluting Energy and Their Uses in Rural Areas, *Khadi Gramodyog*, **23(1)**, 38-41 (1976).

Shingari, B. K. and J. Singh, Use Gobar to Reduce Breeding Cost, *Invention Intelligence*, **11**, 251-253 (1976).

Shukla, J. P., *Biogas Plant*, A Report, National Sugar Institute, Kanpur, 1965.

Sikka, P., Biogas Technology Utilisation in India : Efficient Use of Organic Waste Materials, *Science and Culture*, **44(1)**, Jan 24-26, 1978

Silwal, B. B., *Biogas Plants in Nepal : An Economic Analysis*, Res. Report Series No. 14, P. O. Box. 1312, Kathmandu, Nepal, Oct., 1991.

Silveria, R., Gas Recovered from Urban Refuse in Rio de Janeiro : A Case Study, *paper presented at the 12th World Energy Conf.*, New Delhi, Sept. 18-23, 1983.

Simpson, M. H., Demonstration of Anaerobic Digesters in Developing Countries: *Part II, Env. Sci.*, **23(2)**, March-April, 20-24 (1980)

Simpson, M. H., Demonstration of Anaerobic Digesters in Developing Countries: Part III - the Philippines, *Env. Sci.*, **23(2)**, March-April, 20-24 (1980).

Singh, H. P., *Proc. Annual Report of the Six Coordination Committee Meetings of the All India Coordinated Research Project on Renewable Energy Sources for Agriculture and Agro-based Industries*, CIAE, Bhopal, 1995.

Singh, H. P. and Maheshwari, R. C., Indian Advances in Biogas Technology : Review of Work Done under All India Coordinated Research Project (AICRP) on Renewable Energy Resources, *Biogas Forum*, **I** No **60**, 1995.

Singh J. B., Involvement of Voluntary Agencies in Biogas Programme, *paper presented at the Nat. Workshop on Biogas Dev.*, Vigyan Bhavan, New Delhi, Feb. 26-27, 1982.

Singh, J. B., Biogas Energy Problems : The Problems of Development, *Bio-Energy Re-News*, **1(6)**, 37-42 (1983).

Singh, J. B., Myles, R. and Dhussa, A., *Manual in Deenbandhu Biogas Plant*, Tata Mc Graw Hill Pub. Co. Pvt. Ltd., New Delhi, 1987.

Singh, K. K., Review of Inappropriate Economic Analysis of an Appropriate Alternative Energy System for Villages, *Proc. Nat. Solar Energy Conv.*, Delhi, Dec. 17-19, 1982.

Singh, K. N., Extension Aspects of Cow Dung Gas Plant, *Proc. Nat. Symp. on Biogas Tech. and Usage*, IARI, New Delhi, Nov. 29-30, 1977.

Singhal, O. P., Development of Biogas Technology in India, *Steam and Fuel Users Jour.*, **33(3&4)**, 51-57 (1983).

Singhal, O.P., Energy Alternatives for Irrigation in India, *Urja*, **21(2)** 125-128 (1987).

Singh, P. N., Biogas Solution to the Rural Energy Problems, *Urja*, **4(9)**, 218-220 (1979).

Singh, R., and R. C. Anand, Comparative Performance of Indian Small Scale and Conventional Anaerobic Digester, *Bioresource Tech.*, **47(3)**, 235-238 (1994).

Singh, R. B. and K. K. Gupta, Practical Aspects of Gobar Gas Production, *Proc.*,

Symp. om Perspectives in Indus. Microbiology, Mumbai, April, 1977.

Sinha, S., Involvement of Command Area Development in Biogas Programme, *Proc. Nat. Workshop on Biogas Dev.*, Vigyan Bhavan, New Delhi, Feb. 26-27, 1982.

Siren, G., *Biomass for Energy*, UK-ISES, London, July, 1979.

Siraj, W.U., Biogas Programme in Pakistan, *Biogas Forum*, **II(49)**, 10-12 (1992).

Sirohi, A. S., Economics of Gobar Gas Plant, *Proc. Symp. on Biogas Tech. and Usage, IARI*, New Delhi, Nov. 29-30, 1977.

Sirohi, A. S. and I. Singh, Biomass Energy : Economic Feasibility, *Eastern Economist*, **77**, 264-268 (1981)

Sivagnanam, R.V.R., Barriers to the Growth of Biogas Systems in India, *Energy Management*, **8(4)**, Oct.-Dec., 271-278 (1984).

Skrinder, R. T., Review of Inter. Biogas Programme, *Proc. Symp. on Fuels from Biomass*, Troy, NY, June, 823-858 (1978).

Skrinder, R. T., Observations of Fuel Gas Productivity in Asia, in 'D. L. Wise (ed.), *Fuel Gas Production from Biomass*, Vol I CRC Press, Florida, p. 155-171 (1981).'

Slane, T. C., *Economic Analysis of Methane Generation : Internal Costs and External Benefits,* A Report, Massachusetts Agricultural Experiment Station, 1975.

Slane, T. C., Economic Analysis of Methane Generation : Feasibility on Commercial Egg Farms, *Proc. of the 3rd Inter. Symp. on Livestock Wastes*, Michigan, 1975.

Slane, T. C., *et al.*, An Economic Analysis of Methane Generation Feasibility on Commerical Egg Farms, in '*Managing Livestock Wastes, Amer. Soc. of Agri. Engrs.*, St. Joseph, MI, 1975'.

Smil, V., *China's Energy : Achievements, Problems, and Prospects*, Praeger Publishers, NY, 1976, p. 106.

Smil, V., Energy Solution in China, *Environment*, **19(7)**, 27-31 (1977).

Smil, V., Chinese Biogas Programme Sputters, *Soft Energy Notes*, **5(3)**, 88-89 (1982).

Smith, W. H. and J. R. Frank, *Methane from Biomass : A System Approach*, Elsevier Applied Science, London, 1988.

Smith, W. H., *et al.*, Methane from Biomass and Waste : A Programme Review, *TIDE*, **2(1)**, 1-20 (1992)

Sobel, A. T. amd R. E. Muck, Energy in Animal Manures, *Energy in Agriculture*, **2**, 161-176 (1983).

Sodhiya, H. C. and K. K. Jain, Rural Energy Crisis and Biogas Technology, *Jour. of Rural Dev.*, **7(4)**, 427-432 (1988).

Sohal, B. S., M. Singh and N. S. Grewal, Survey of Family-sized Biogas Plants Installed in Punjab, *SESI Jour.*, **1(1)**, 67-72 (1991).

Sohal, T. S. and R. M. Fulzele, Transfer of Biogas Technology Function and Training Needs, *proc. Nat. Workshop on Biogas Tech.*, PAU, Ludhiana, July 3-4, 1980.

Solly, R. K. and R. Yarrow, Biogas in Fiji : Economic and Social Factors and

Future Prospects, *Proc. 2nd Inter. Workshop on Biogas Tech. and Utilisation*, Manila, Oct., 1975.

Solly, R. K., Biogas Potential in Island Countries of the Pacific, *Proc. South Pacific Commission Regional Technical Meeting on New and Renewable Sources of Energy*, Papeete, French Polynesia (France), Sept. 28-Oct. 2, 1982.

Soubolle, B. R., *Fuel Gas From Cowdung*, Shayogi Prakashan, Tripureshwar, Kathmandu, Nepal, 1976.

Sparling, A. B., Energy Recovery from Livestock Wastes, *Proc. of an Inter. Conf. on Biomass*, Biomass Energy Inst., Winnipeg, Canada, 1973.

Srinivasan, H. R., History of Development of Gobar Gas Plant, *Khadi Gramodyog*, **20(3)**, 176-181 (1973).

Srinivasan, H. R., Gobar Gas Plants : Promises and Problems, *Indian Farming*, Feb., **23 (11)**, 29-33 (1974).

Srinivasan, H. R., Revolutionary Promise of Gobar Gas, *Commerce*, Annual No., 85-87 (1977).

Srinivasan, H. R., Gobar Gas Scheme, *Proc. Nat. Symp. on Biogas Tech. and Usage*, IARI, New Delhi, Nov. 29-30, 1977.

Srinivasan, H. R., Management of Agricultural and Agro-Industrial Wastes, *UNEP/FAO Seminar on Residues Utilisation*, FAO, Rome, 1977.

Srinivasan, H. R., Biogas in India, *Proc. Expert Group Meeting on Biogas Development*, Bangkok, June, 1978.

Srinivasan. H. R., Biogas Development in India, *Mazingira*, **5(1)**, 54-59 (1981).

Srinivasan, H. R., Biogas Development by KVIC, *Changing Villages*. **6(4)**, 221-232 (1984).

Srinivasan, T. G. and U. Ghosh, Socio-Benefit Cost Appraisal of Community Gobar Gas Plant, *Energy Management*, **1(4)**, 243-248 (1977).

Sriram, N., M. Krishnan and R. T. Thampy, Organic Waste as Source of Energy, *Chemical Age of India*, **27**, 1052-1054 (1976).

Srivastava, P., Population Free Energy for Villages : Gobar Gas (in Hindi), *Gram Shilpa*, **3(30)**, 65-70 (1982).

Srivastava, S. N. and R. H. Richharia, Utility of Cow Dung Gas Plants, *Proc. Bihar Acad. of Agri. Sci.*, **6(7)**, 104-107 (1957).

Srivastava, U. K. and T. K. Moulik, Energy Consumption in Rural Households and Biogas Plants : Tasks Ahead, *Vikalpa*, **2(1)**, 1977.

Srivastava, U. K., Economic Aspects of Biogas Plants, *paper presented at the Research Planning Workshop on Energy for Rural Dev.*, Ahmedabad, 1979.

Steward, J. and B. Richmond, Biogas for Javanese Villages : A Simple Unit, *Appropriate Technology*, Sept., **7(2)**, 15-17 (1980).

Steward, J and B. Richmond., Biogas for Javanese Villages : The Products of Digestion, *Appropriate Technology*, **7(3)**, 1980.

Stuckey, D. C., *Biogas in Developing Countries : A Critical Appraisal*, Imperial College of Sci. and Tech., London, 1984.

Subramanian, M. N., Biogas Plant in Gujarat, *Changing Villages*, **4(5)**, 38-48 (1982).

Subramanian, S. K. and S. Ganesh, *Biogas Technology Performance and*

Prospects in India, A Report, NCST, Dept. of Science and Tech., New Delhi, 1974.

Subramanian, S. K, *Biogas Systems in Asia*, Mgt. Dev. Inst., Gurgaon, 1977.

Subramanian, S. K., Biogas Systems in Asia : A Survey in 'B. Andrew, *et al.*, *Biogas Technology in the Third World*, Multidisciplinary Development Research Centre, Ottawa, 1978,' p. 97-121.

Sudirjo, G. P. and S. Kismomihardjo, Report on Utilisation of Biogas in Indonesia, *paper presented at ESCAP Biogas Workshop*, New Delhi, 1975.

Sumicad, R. C., Biogas Technology and its Economic Impact, *paper presented at the Workshop on Biogas Tech. and Utilisation*, Manila, Oct., 1975.

Sudaresan, D., Canara Bank's Strategies and Experiences in Financing Biogas Plants, *Khadi Gramodyog*, **30(8)**, 335-337 (1984).

Susheela, H., On Production of Gobar Gas in Rural Areas, *Kurukshetra*, **40(12)**, 20-21 & 23 (1992).

Suval, P. N., An Overview of Biogas Programme in Nepal, *Proc. of the Conf. On Energy from Biomass and Wastes*, Univ. of Peradeniys, Sri Lanka, Nov. 26-30, 1979. Also published in *Regional Jour. of Heat and Mass Transfer* (India), **2(1)**, 9-12 (1980).

Swaminathan, R. K., Implementation of Biogas Technology in Village : A Case Study, *Proc. of the Seminar on Rural Energy Futures : Perspectives for Policy Alternatives*, Bangalore, Nov. 24-26, 1983.

Swanwick, J. D., *Survey of the Performance of Sewage Sludge Digesters in Great Britian*, Water Poll. control Board, 639-669 (1969).

Tam, D. M. and Thanh, N. C., Biogas Technology in Developing Countries, *Env. Sanitation Review*, AIT, Bangkok, 1982.

Tarbaghia, T. M., Design of Biogas Plant with Special Reference to Applications at Benghazi, Libya, *Renewable Energy (U.K)*, **3(2 & 3)**, 207-209 (1993).

Tasdemiroglu, E., Review of Biogas Technology in Turkey, *Biomass*, **0144-4565**, 137-149 (1989).

Tasdemiroglu, E., Economics of Biogas Space Heating Systems in Rural Turkey, *Bioresource Technology*, **36(2)**, 147-155 (1991).

Taylor, C., Critical Factors in Economic Evaluation of Small Decentralised Energy Projects, *paper presented at the Workshop on Biogas Systems*, New Delhi, Feb. 28-Mar 4, 1977.

Tejnarayan, S. D. and R. Moorthy, Case Study of a Cow Dung Gas Plant, *Khadi Gramodyog*, **10**, 815-817 (1964).

TERI, Biogas Programmes Directory, Mumbai, 1980.

TERI, Biogas Potential in India : An Overview, *Science for Villages*, No. **63-64**, 23-24 (1982).

TERI, *Biogas : A Source of Rural Employment*, New Delhi, 1995.

TERI, *Biogas Technology : An Information Package*, New Delhi, 1994.

TERI, *Managing Biogas Programme*, New Delhi, 1994.

TERI, *Biogas Technology : A Promise for the Future*, New Delhi, 1994.

TERI, *Community Biogas System in Methan (Gujarat) : A Case Study*, Habitat Place, Lodhi Road, New Delhi, 1993.

Terrado, E. N., Biogas Development : The Philippines, *Biomass Energy Project: Planning and Development*, Pergamon, New York, 40-163 (1980).

Thery, D., Biogas Programme in India and China : An Ecologically Appropriate Technology in Two Highly Contrasting Socio-Economic Settings, *Resource Mgt. and Optimisation*, **1(4)**, 289-314 (1981).

Thom, C and Banks, D. I., Biogas as a Rural Domestic Energy Source : A Pilot Study, *Jour. of Energy in Southern Africa*, **5(4)**, 121-125 (1994).

Thomas, A.M., Biogas Technology and KVIC, *Khadi Gramodyog*, **27**(3), 149-150 (1980).

Thomas, A. M., Biogas to Meet Energy Needs of Rural Areas, *Commerce*, Ann. No., 67-70 (1980).

Thornton, T., *Assessment of Anaerobic Digestion in US Agriculture : Final Report*, Economics, Statistics and Cooperatives Service, Washington, DC, 1978.

Tietjen, C., From Biodung to Biogas : Historical Review of European Experience, in 'Jewell W. J. (ed.), *Proc. of the Cornell Waste Management Conf.*, Ann Arbor Sci., Michigan, 1975'.

Tran, V.V., Biogas in Vietnam, ESCAP, *RERIC News*, **5(3)** (1968).

Trevelyan, W. E., The Methane Fermentation : A Discussion Paper, *Tropical Science*, **17(4)**, 193-209 (1975).

Tripathi, S., China's Great Leap in Biogas Technology, *Indian Express*, (Published from Delhi), March 31, 1980.

Tucker, J.B., Biogas Systems in India : Is the Technology Appropriate ? *Environment*, **24(8)**, 12-20 (1982)

Tuladhar, S. M. and T. N. Veziroglu, Evaluation of Biogas Plants in Nepal, *Proc. of the Seventh Miami Inter. Conf. on Alternative Energy Sources*, Miami Beach, Dec. 9, 1985, Coral Gables, Florida, 1986, p. 315.

Tyner, W. E. and J. Adams, Rural Electrification in India : Biogas Versus Large Scale Power, *Asian Survey*, **17**, 724-737 (1977).

Umesh, B.V., *Biogas : The MCRC Way*, Monograph Series No. 8. on Engg. of Photosynthetic System, Shri AMM Murugappa Chattiar Res. Centre, Tharamani, Chennai, 1981.

UNEP, *Biogas Fertiliser System*, Technical Report on a Training Seminars in China, United Nations, 1981.

UNIDO, *Appropriate Industrial Technology for Energy for Rural Requirements, United Nations*, New York, 1979.

UNIDO and NCST, *Report of the Workshop on Biogas Technology and Utilisation*, UNIDO and NCST, Dept. of Sci. and Tech., New Delhi, Jul.-Aug. 1975.

US-ASEAN, Philippine Country Paper on New and Renewable Sources of Energy, *paper presented at the US-ASEAN Seminar on Energy Technology: Biomass, Coal, Solar, Wind Energy Planning*, Bandung, Indonesia, June 7-18, p. 182.

Vaclav, S., Biogas Production in China, *Dev. Digest*, USAID, Washington, 1979.

Vaidya, G. D., *Gas Light for Your Villages from Cowdung*, Local Self Govt., Hyderabad, 1951.

Van Brakel, J., *The Ignis Fatus of Biogas*, Delft Univ. Press, The Netherlands, 1986.

Van-Nes, W. J., *Technical Biogas Potential per District in Nepal*, A Report, Research Unit., Gobar Gas Company, Butwal, Nepal, Sept., 1992.

Vas-Nes, W. J., amd J. Lam, Desired Future Structure of the Biogas Sector in Nepal, *Biogas Forum*, **1** No **56**, 1994.

Van Nes, W. J., Biogas Support Programme in Nepal, SNV-Nepal, P.O. Box 1966. Kathmandu, Nepal, *Biogas Forum*, **III (54)**, 4-7 (1983).

Varma, B. P., Energy from Rural Wastes, *Economic Times*, New Delhi, Sept. 22. 104, p. 6.

Varshney, A. C., *et al.*, Evaluation of Crop Residues in North Gujarat, *Biological Wastes*, **19(3)**, 227-231 (1987).

Veena, D. R., *Biogas Technology: A Study of Community Biogas Plant*, Ashish Pub. House, 8/81, Punjabi Bagh, New Delhi, 1986.

Veerakyathaiah. V. D., Review of Biogas Programme in Karnataka, *paper presented at the Nat. Workshop on Biogas Dev.*, Vigyan Bhavan, New Delhi, Feb. 26-27, 1982.

Venkatramana, P., National Project on Biogas Development, *Urja Bharti*, **3(3)**, 33-35 (1993).

Venkatramana, P., Success Through Cooperation : Community Biogas Plant at Methan, *Urja Bharti*, **3(3)**, 36-37 (1993).

Venkataramana, P., *Community Biogas System in Methan (Gujarat) : A Case Study*, A Report, Biomass Users Network, San Jose, Costa Rica, 1993.

Ventura, P. F. and J. Mircen, Status of Biomethanation Development in the Philippines, *Applied Microbial Biotechnology*, **4(10)**, 151-158 (1988)

Vergara, W. and D. Pimental, Study of the Energy Potential of Fuels from Land Biomass in Five Countries; *Proc. of the Institute of Gas Technology Symposium on Energy from Biomass amd Wastes*, Washington, Aug., 87-105 (1978).

Verghese, B. G., Integrated Biogas Algal System, *Commerce*, **133(3422)**, 1090 (1976).

Verma, J. N., Cow Dung Gas Plant for Fuel and Manure, *Kurukshetra*, **22(12)**, 7-8 (1974).

Verma, O. S., Gobar Gas Plants : Rate of Return, *Economic Times*, **7(12)**, 1976.

Verma, O. S., Gobar Gas-cum-compost Making : Cost Benefit Analysis, *Economic Times*, Dec., **16 (5 & 8)**, 1976.

Verma, O. S., Gobar Gas Plant : An Innovative Posture, *Khadi Gramodyog*, **25(2)**, Nov., 136-140 (1978).

Verma, O. S., Gobar Gas Plant : Installer's Scientific Parsimony, *Khadi Gramodyog*, **25(8)**, 365-368 (1979).

Verma, O. S., Profitability of Biogas Plant, *Proc. Seminar-cum-Workshop*, Karnal, Oct. 3-10, 1980, S. Neelkantan (ed.), NDRI, Karnal, 63-67 (1981).

Verma, O. S., Operational System of Gobar Gas in Rural India, *Proc. Seminar-*

cum-Workshop, Karnal, Oct. 3-10, 1981, S. Neelkantan (ed.), NDRI, Karnal, 39-42 (1981).

Verrier, D., French Programme on Methane Fermentation, *Biomass*, **2(1)**, 17-28 (1982).

Vidyarthy, G. S., Prospect and Achievements of Biogas Programme, *Agri. Today*, **1(9)**, 2-4 (1977).

Vidyarthi, G. S., A Status Report on Biogas Production Technology and Utilisation, *Proc. of the Workshop on 'Biogas Utilisation in Engines'*, IIP, Dehradun, Apr. 19-20, 1978.

Vidyarthi, V., Appropriate Technology for Meeting Rural Energy Needs : An Appraisal of Biogas Plants and Their Scope, *Decision*, **6(4)**, 443-449 (1979).

Vidyarthi, V., *Biogas Technology Development and Extension in India : An Overview*, Discussion paper, Appropriate Tech. Dev. Assoc., PB 311, Gandhi Bhavan, Lucknow, Dec., 1980.

Vidyarthy, G. S. amd S. S. Mahdi, National Project for Biogas Development, *paper presented at the Nat. Workshop on Biogas Dev.*, Vigyan Bhavan, New Delhi, Feb. 26-27, 1982.

Vijay, V. K., S. Kothari and J. K. Gaur, Environmental Upgradation of a Village by Mass Adoption of Biogas : A Case Study, in 'N. K. Bansal (ed.), *Proc. Nat. Solar Energy Conv.*, Dec, 28-30, 1992, Tata McGraw Hill Pub. Co. Pvt. Ltd., New Delhi, 1993, p. 403-406.

Villa, A., Biogas in Columbia, *Changing Villages*, **4(6)**, 47-52 (1982).

Vimuktanand, S., Fuel Gas from Cow Dung, *Symp. on Radioisotopes, Fetilisers and Cow Dung Gas Plants*, New Delhi, 402-403 (1961).

Vishnoi, S. L. and Bose, S. P., Effect of Educational Exposures on the Acceptance of Cow Dung Gas Plant by Farmers, *Indian Jour. of Agronomy*, **5**, 284-291 (1961).

Vishnoi, P. S., Energy Farming Through Bioconversion, *Energy Management*, **2(4)**, Oct.-Dec., 267-275 (1978).

Vyas, S. K., *et al.*, Biogas Plant Failture : Causes and Remedies, *Rural Technology Jour.*, **7(2)**, 13-16 (1990).

Vyas, S. K., Economic Analysis of Large Size Biogas Plant, *Rural Tech. Jour.*, **8(4)**, 1-7 (1992).

Wadhera, B. L. and S. S. Saluja, Biogas as an Alternative Energy Source in India, *Proc. of the 4th Miami Inter. Conf.*, Dec., 1981, Ann Arbor, 387-398 (1982).

Walde, K., Macro-Economic Effect of Biogas Plants, *Biogas Forum*, **IV** No **59**, 1994.

Ward, R. F., Digesters for the Third World, *Proc. 2nd Inter. Symp. an Anaerobic Digestion*, Tvavemunde, Sept. 6-11, 315-341 (1981).

Wase, D. A. J., Biogas : Fact or Fantasy ?, *Biomass*, **4(2)**, 127-142 (1984).

Wen, W., Biogas Utilisation in China, *Proc. of the 1980 Bio-Energy World Congress and Exposition*, Atlanta, Apr., 434-435 (1980).

Werner, E., *Biogas : A Decentralised Economically Feasible Alternate Energy*

Source, A Report, Centre for Energy and Res., Univ. of Puerto Rico, Puerto Rico, USA, 1980.

Wook, D. H., Methane Gas Utilisation as a Rural Fuel in Republic of Korea, *Proc. of the Workshop on Biogas Technology and Utilisation*, New Delhi, Aug., 1975.

Wright, A. J., Methane Digesters : Waste into Energy, *Sunworld*, **6**(**2**), 34-37 (1982).

Xun, M., *et al.*, The Latest Evaluation of Hygienic Effect of Biogas Plants in China, *Biogas Forum*, **II** No **49**, 14-17 (1993).

Yadav, L. S. and P. R. Hesse, *The Development and Use of Biogas Technology in Rural Areas of Asia* : A Progress Report, Project Field Document No. 10, FAO, Vienna, 1981.

Yant, C., Anaerobic Treatment and Utilisation of Animal Manures and Agricultural Wastes in Taiwan, *Regional Jour. of Energy, Heat and Mass Transfer*, **5**(**3**), 215-230 (1983).

Yizhang, Z., The Development of Biogas Technology and International Co-operation in China, *Proc. 2nd Symp. for New and Renewable Energy*, Tripoli, Libya, 1992.

Zhensheng, Y., *et al.*, The Effect of Biogas Management on Improvement of Environment and Preventing Diseases, *Biogas Forum*, Special Issue - China II, I, 32-38 (1992).

Zimzombe, I. M., Biogas Generation in Zimbabwe : Policies, Developments, Problems and Tasks, *Proc. of the Conf. on Making Use of Renewable Energy*, Harare, Zimbabwe, March 21, 1984, J. C. Johnson (ed.), Professional Conference, Zimbabwe, Harare, 7-9 (1984)

Acknowledgement

This bibliography was compiled after scanning numerous journals, reports, newspapers and books individually available with the author and several libraries. In particular the author would like to convey gratitude to the authorities and library staff of Tata Energy Research Institute, New Delhi, National Productivity Council, New Delhi, UN Asian and Pacific Centre for Transfer of Technology, New Delhi, and Administrative Staff College of India, Hyderabad for varied assistance and cooperation.

under stated conditions of pressure and temperature (equal to 252 calories, 778 footpounds, 1055 joules and 0.293 watt-hours).

BULK DENSITY : refers to the mass (weight) of a material divided by the actual volume it displaces as a whole substance, expressed in lb/ft^3, kg/m^3, etc.

CALORIE : Amount of heat required to raise the temperature of one gram of water by one degree centigrade.

CARBOHYDRATE : These are certain neutral compunds made up of carbon, hydrogen and oxygen which include cellulose, starches and sugars. They have general fromula as C_x $(H_2O)_y$.

CARBON-TO-NITROGEN RATIO (C:N) : It is the ratio of carbon to nitrogen by weight in organic materials. Anaerobic bacteria produce maximum biogas when fed with organic materials having a C:N ratio compatible with their metabolic requirements. If the ratio lies between 20 to 30, it is regarded as favourable.

CATALYSIS : Effect caused by a small qunatity of a substance (catalyst) on a chemical reaction in which catalyst remains unchanged.

CELLULASE : An enzyme capable of decomposing cellulose into simpler carbohydrates.

CELLULOLYTIC : A property helpful for hydrolysis of cellulose by rendering it water-soluble.

CELLULOSE : A biological polymer which is made of sugar molecules that provides basic building materials of plant fibre. Cellulose is basically the main polysaccharide in living plants that forms skeletal structure of plant cells and can be hydrolysed to glucose. It has a genral chemical formula $(C_6\ H_{12}\ O_6)_n$.

CELLULOSIC FEEDSTOCK : Materials such as wood, crop stalks etc. containing sugar units linked by bonds which are not easily rupturable.

CELSIUS (CENTIGRADE) : It is a temperature scale commonly used in all scientific work. At sea level water freezes at 0°C. and boils at 100°C.

CENTRIFUGE : A rotating device for separating liquids of different specific garvities or for separating suspended colloidal particles according to particle-size fractions by centrifugal force.

CETANE NUMBER : It is a measure of a fuel's ease of self-ignition; the higher the number the better the fuel suited for combustion.

COD (CHEMICAL OXYGEN DEMAND) : It is measure of water pollution which determines oxygen consuming capacity of inorganic and organic matter present in water. This test gives a measure of the oxygen absorbed by a sample from boiling acidic potassium dichromate solution. Mercuric sulphate is normally added to form an unoxidisable compound with any chloride present in the sample which would otherwise interfere. Excess dichromate is titrated with ferrous sulphate or ferrous ammonium to the potassium sulphate solution, the amount of oxidisable organic matter being proportional to the potassium dichromate consumed.

COLUMN : A vertical cylindrical vessel containing a series of perforated plates or packed with materials through which vapurs may pass, used to increase the degree of separation of liquid mixtures by distillation.

COMPOUND : A chemical term denoting a specific combination of two or more distinct elements.

CONCENTRATION : It is given by ratio of mass or volume of solute present in a solution to the amount of solvent.

CONTINUOUS FERMENTATION : It is a steady state fermentation process which occurs without interruption, due to continuous feeding of plant every day accompanied by corresponding withdrawal of digested effluent at the same rate. In this process feed rates are adjusted to suit the desired retention time.

CONTINUOUS PLANTS : Biogas plants in which feedstock is continusously added or withdrawn with volume of slurry kept at a desired level to ensure a continuous gas supply. However, in such plants floating material creates a scum which hinders the gas flow. Continuous plants are specified according to the period of digestion and the digester capacity depending upon temperature and withdrawal rate.

C. P. : Candle power which denotes light intensity.

DIGESTER : The sealed tank or container in which the biological requirements of anaerobic digestion are controlled to achieve digestion and optimise biogas yield. The volume of digester is a function of plant capacity and retention time.

DIGESTION : A process by which complex organic molecules are broken down into simpler ones. In anaerobic fermentation anaerobic bacteria cause decomposition.

DILUTION RATE : It is the reciprocal of the mean retention time of the flowing medium in the digester.

EFFLUENT : It is digested substrate which exits the digester. During anaerobic fermentation, organic content of the substrate is reduced and stabilised making effluent as an almost odourless, homogenised liquid mass which does not attract rodents or flies. The biogas plant effluent is a resource as it contains almost all of the nitrogen that was in the raw material. The effluent is an excellent fertiliser as its nitrogen is more readily absorbable by plants than nitrogen in raw manure. Its application thus improves biological, physical and chemical properties of soil.

ENZYMES : The group of catalytic proteins that are produced by living microorganisms. Enzymes facilitate decomposition of complex organic molecules into simpler ones without themselves being altered or destroyed.

EVAPORATION : It is conversion of a liquid to vapour state by adding latent heat of vapourisation. It generally refers to vapourisation into the atmosphere.

FACULTATIVE (ANAEROBE) : A microorganism that grows equally well under aerobic and anaerobic conditions.

FAHRENHEIT SCALE : A temperature scale in which boiling point of water is 212° and its freezing point 32°. To convert °F to °C, substract 32, multiply by 5, and divide the product by 9 as given by the following relationship.

$$°C = [°F-32]\frac{5}{9}$$

FEEDSTOCK : The basic raw material such as animal manure which is fed to a plant to produce biogas through anaerobic fermentation. Feed concentration is expressed as percentage of dry matter or total solids in the feed.

FERMENTATION : Usually viewed as synoym with digestion. Biological enzymatic transformation by which organic material is broken down into simpler constituents by micro-organisms which generates gas.

FLAME LIFT OFF : refers to a flame which does not touch flame port of the stove. It can lift off so much that the flame goes out.

FLAME SPEED FACTOR : Measure of the speed at which a flame travels along a column of the gas.

FLASH POINT : Temperature at which a combustible mixture ignites when a flame is introduced.

FLOCCULATION : It refers to the aggregation of fine suspended particles to form floating clusters or dumps.

GOBAR GAS : Alternative name for biogas which is formed from cow or buffalo dung (*Gobar* is the Hindi and Nepalese word used to denore dung).

HEAT EXCHANGER : A device that transfers heat from one fluid (liquid or gas) to another or to the environment.

HEAT VALUE : Amount of heat that can be obtained from a fuel and generally expressed in kcal/kg or BTU/lb. Calorific value is the synonym of heat value.

HYDROLYSIS : It refers to the decomposition or alteration of a substance by chemically adding a water molesule to the unit at the point of bonding.

INFLUENT : It is the basic raw material such as animal manure or crop residues which is used to produce biogas through anaerobic fermentation. The term is interchangably used for feed or feedstock.

INOCULUM : A small amount of bacteria produced from a pure culture which is used to start a new culture. As it contains active micro-organisms, its addition to substrate helps in starting and activating fermentation.

JOULE (J) : The metric unit of energy with one BTU equal to 1055 joules. As the joule is a very small unit, it is also expressed in kilojoules (KJ) (a thousand joules) or in megajoules (MJ) or million joules.

LIGNIN : A polymeric and noncarbohydrate constituent of wood that functions as binder and support for cellulose fibres. It is highly resistant to biodegradation.

LIPIDS : Fatty materials.

LIPOLYTIC : A property helpful for hydrolysis of liquid materials by making them water-soluble.

LOADING RATE : refers to the rate at which amount of volatile solids is fed to a digester. The loading rate is usually expressed as weight of volatile solids per unit of digester volume per unit time. The volatile solids are the organic portion of the waste solids which constitute about 80 per cent of the raw waste (total solids). It is briefly expressed as kg. VS/m^3. day.

LIQUEFACTION : refers to the change in phase of a substance to liquid state. In the context of fermentation it refers to the conversion of water-insoluble carbohydrate to water-soluble carbohydrate.

MANURE : It refers to animal faeces and urine, waste feed, bedding and antislip material available from a barn and yard.

MARSH GAS : An alternative name for biogas when it is formed in marshes.

MESOPHILIC : Bacteria which thrive in a temperature range close to 95°F (35°C).

METHANE : The simplest hydrocarbon consisting of one carbon atom and four hydrogen atoms (CH_4); a flammable and odourless gas. It is the main constituent of both biogas and natural gas.

METHANOGENIC : A property helpful in producing methane.

MOLECULE : The chemical term for the smallest particle of matter which is chemically the same as the whole mass.

MUNICIPAL SOLID WASTE (MSW) : refers to combined household and industrial wastes generated within a municipal area, consisting of any materials including food wastes that are discarded or rejected as spent, worthless, useless or in excess and that are not wet enough to be free-flowing. MSW is also sometimes referred to as municipal refuse.

NET ENERGY BALANCE : It represents the amount of energy available from fuel when it is burnt less the amount it takes to produce the fuel.

NIGHT-SOIL : It refers to human faeces which are highly biodegradable and ideal for anaerobic fermentation. They are however difficult to handle due to stinking smell.

OCTANE NUMBER : It specifies a rating which indicates tendency to knock when a fuel is used in a standard internal combustion engine under standard conditions.

PATHOGENS : Organisms which can cause disease in both animals and human being e.g. typhoid. During anaerobic fermentation, most of the pathogens of night-soil are destroyed.

pH : A measure of acidity or alkalinity of a solution that describes free hydrogen ion concentration of a system; a solution of pH less than 7 is acidic; pH of 7 is neutral; and pH over 7 is alkaline.

PHYTOTOXIC : A characteristic which is poisonous to plants.

POLYMER : A class of molecules comprising long chains or cross-linked simple molecules.

POUNDS PER SQUARE INCH ABSOLUTE (psia) : It is the measurement of pressure in complete vacuum or zero pressure.

POUNDS PER SQUARE INCH GAUGE (psig) : It is the measurement of pressure above atmospheric pressure.

PROTEIN : refers to a class of high-molecular weight polymer compounds comprising a variety of amino acids joined by peptide linkages.

PROTEOLYTIC : A property helpful for hydrolysis of proteins by making them water-soluble.

RENWABLE ENERGY RESOURCES : Energy resources that can be renewed naturally for repeated applications such as solar, hydro and wind power.

RETENTION TIME : It represents average time slurry remains in digester. The volume of slurry in the digester remains constant and incoming slurry displaces an equal amount of processed slurry from the digester each time it is loaded. As volume is constant, fraction of digester's liquid volume replaced each day determines the retention time. For example, if slurry equal to one-tenth of the digester's liquid volume is added daily, digested slurry has an average retention time of 10 days. A very brief retention time does not allow the bacteria enough time to digest the feed and a long retention time does not furnish enough fresh slurry to promote bacterial growth and a high gas production rate. Retention time is generally estimated by dividing total weight of material in system by weight removed per unit time such as hour, day, week, etc.

SCRUBBING : refers to the process of cleaning a gas by removing chemical contaminants and particulates.

SCUM : It is waste material formed on top of liquid slurry, and can pose a severe problem in digesters receiving much fatty or floatable material. Methane formation can be inhibited if the scum is allowed to accumulate and to set hard. Scum builds localised zone of acid accumulation which inhibits anaerobic fermentation.

SLUDGE : Preccipitated solid matter that gravitates at digester bottom.

SLUDGE GAS : An alternative name for biogas especially the one formed in sewerage.

SLURRY : The liquid feed that flows into the plant which is primarily a mixture of manure and water.

SPECIFIC GRAVITY : It is the ratio of the mass of a solid or liquid to the mass of an equal volume of distilled water at 4°C.

STOICHIOMETRIC RATIO : It is the ratio of chemical substances necessary for a reaction to occur completely.

SUBSTRATE : refers to material supplied for microbial action in a digester. it is used by the microorganisms for their growth and metabolism ultimately leading to gas formation.

SUPERNATANT : refers to the liquid material which collects at the surface when a substrate is allowed to settle.

THERMOPHILIC : Bacteria that are most active in temperature range of 120° to 140° (49° – 60°C).

TKN : Total Kjeldahl Nitrogen, the amount of nitrogen obtained by Kjeldahl method of digesting organic material with sulfuric acid.

TOTAL SOLIDS : Weight of solid matter remaining after a sample is dried to constant weight at (100 ±1)°C.

TOXINS : Poisonous material which inhibit anaerobic fermentation by killing microorganism.

VOLATILE ACIDS : Intermediate materials produced in a digester by acid-forming bacteria and used by methane-forming bacteria. They are essentially low molecular weight fatty acids.

VOLATILE SOLIDS : Portion of solids volatilized at (550 ± 50°C); the difference between total solids content and ash remaining after ignition at (550 ± 50°C). These are primarily the organic constituents of feed.

Acknowledgement

Various definitions and notes included herein have been framed or culled out from numerous sources such as books, reports and papers published in various journals. Gratitude is conveyed to all the original authors who framed these definitions first. In particular gratitude is conveyed to the following sources from which information was largely gathered.

1. Barnett, A., L. Pyle and S. K. Subramaniam, *Biogas Technology in the Third World*, Inter. Dev. Res. Centre. Box 8500, Ottawa K16 3 HP, Canada.
2. BORDA, *Biogas : Manual for the Realisation of Biogas Programmes*, GATE, P. O. Box 5180, 6236 Eschborn 1, Germany, 1980.
3. Chawla, O. P., *Advances in Biogas Technology*, Pub. and Inf. Div., ICAR, New Delhi, 1986.
4. Khandelwal, K. C., *Some Information about Biogas Technology and Development in India*, A Report, Ministry of Agriculture, Krishi Bhavan, New Delhi, Sept., 1978.
5. Tarr, T. D., *Fuels from Farms : A Guide to Small Scale Ethanol Production*, SERI, Golden, Colorado, 1980, p. GI-G7.
6. Volunteers in Technical Assistance (VITA), *Energy Directory*, 1815 North Lynn Street, Suite 200 Arlington, Virginia 22209-2079, USA.

Appendix III

IMPORTANT CONVERSION FACTORS

1. Weight

1 oz = 28.35 g
1 lb = 0.453592 kg = 0.009 cwt.
1 kg = 2.20462 lb = 0.02 quintal
1 jin (chinese) = 0.550 kg
1 cwt = 112 lb = 50.802 kg
1 metric ton = 0.98421 long ton = 1.10231 short tons = 2204.6 lb
1 English or long ton = 1.01605 metric tons = 1.12 short tons
1 short ton = 0.892857 long ton = 0.907185 metric ton = 2000 lb

2. Length

1 inch = 0.0254 m
1 ft = 0.333 yard = 12 inch = 0.305 m
1 yard = 3 ft = 36 inches = 0.914 m
1 m = 1.094 yards = 3.281 ft = 39.37 inches = 0.001 km
1 ft^2 = 0.093 m^2
1 square yard = 0.836 square metre
1 square metre = 1.194 square yard
1 mu (Chinese)* = 660 m^2 = 0.066 hectare

3. Volume

1 $inch^3$ = 16.387 cm^3
1 ft^3 = 28.317 litres = 7.4805 American gallons
= 6.2288 Imperial gallons = 0.17811 American barrel
= 0.028317 m^3
1 m^3 = 1000 litres = 264.17 American gallons
= 219.97 Imperial gallons = 6.2898 American barrels
= 35.315 ft^3
1 litre = 1000 cm^3 = 0.001 m^3 = 61.024 $inch^3$
= 0.26417 american gallon = 0.21997 Imperial gallon
= 0.035314 ft^3
1 kilolitre = 1000 litres = 6. 2898 American barrels
1 hectolitre = 100 litres
1 American gallon = 231 $inch^3$ = 0.133681 ft^3 = 0.0037854 m^3
= 0.0238095 American barrel
= 0.83268 Imeprial gallon = 3.785 litres
1 Imperial gallon = 277.42 $inch^3$ = 0.160544 ft^3 = 0.0045461 m^3

= 1.20094 American gallons = 0.028594 American barrel
= 4.55 litres

1 American barrel = 9702 $inch^3$ = 5.6146 ft^3 = 0.15899 m^3
= 158.99 litres = 42 American gallons
= 34.9726 Imperial gallons

4. Miscellany

1 kcal = Amount of heat required to heat up 1 litre of water by 1°C
= 4187 Ws = 4187J (1Ws = 1J = 1 Nm) = 3.968 BTU

1 kwh = Performance of 1kw for a period of 1 hour
= 860 kcal = 3.6 x 10^6 Ws = 3411 B. T. U. = 1340 h. p. hour

1 kw = 1000 watts = 1.340 h.p. = 737 foot pounds per second

1 kgf/cm^2 = 0.0283 m^3 = 28.3 litres

1 m water column = 1 m WS = 1 Mp/m^2 = 10 kp/cm^2 = 1MN/m^2

1 atm = 1 kp/cm^2

1 lb/in^2 (psi) = 6.89 kN/m^2 = 0.0703 kgf/cm^2 = 27.7 in water gauge
= 0.06805 atmospheres

1 kg/cm^2 = 14.22 lbs/m^2 = 0.9678 atm = 10,000 mm water gauge

1 mile = 1760 yards = 1.60934 km

1 international nautical mile = 6076.12 ft = 1.852 km

1 sq.km = 100 hectares = 0.386102 sq. miles

1 sq. mile = 640 acres = 258.999 hectares = 2.58999 km^2

1 acre (British) = 4047 m^2 = 4840 sq. yards = 0.40486 hectares

1 hectare = 10^4 m^2 = 2.47105 acres = 0.01 km^2

1 electron volt (eV) = 1.6021892 x 10^{-19}J

1 cal = 4.18 J

1 lcal/min = 69.783 W = 9.351 × 10^{-2} h.p. = 3.9865 BTU

1 joule = 2.389 × 10^{-4} kcal = 2.389 × 10^{-7} kwh = 9.486 × 10^{-4} BTU

1 MJ = 238.8 kcal = 947.9 BTU

1 MJ/m^3 = 26.75 BTU/ft^3

1 BTU = 1.055 KJ = 0.252 kcal

1 BTU/ft^3 = 37.2 KJ/m^3 = 8.894 kcal/m^3

1 BTU/hour = 0.293 W

1 BTU/minute = 17.58 W = 2.356 × 10^{-2} h.p. = 0.252 kcal/min.

1 BTU/second = 1.06 kW

1 watt = 1.36 × 10^{-3} h.p. = 1.433 × 10^{-2} kcal/min
= 5.688 × 10^{-2} BTU/min.

1 h.p. = 745.7 W = 10.68 lcal/min = 42.44 BTU/min

1 therm = 10^5 BTU = 25200 kcal = 105.5 MJ

1 tonne of crude oil = 7.5 barrels of crude oil

20,000 barrels of crude oil per day = 1 million tonne of oil per annum

1000 m^3 of natural gas = 1 tonne of oil equivalent = 35300 ft^3 of natural gas

1 tonne coal equivalent = 29.3 GJ (UN standard)

1 tonne oil equivalent = 42.6 GJ (UN standard)

1 ft^3/1b = 0.0624 m^3/kg

1 m^3/kg = 6.02 ft^3/1b

Lumens = Candle power × 2

25 watt bulb	≅ 230 lumens
40 watt bulb	≅ 430 lumens
60 watt bulb	≅ 730 lumens
1 man hour	= 250 kcal
1 woman hour	= 200 kcal
1 child hour	= 120 kcal
1 bullock hour	= 2300 kcal
1 kg fuelwood	= 3800 kcal
1 litre kerosene	= 8980 kcal
1 barrel	= 200 litres = 44 gallons
1 billion	= 1000 million = 10^9 = 100 crores
1 trillion	= 1000 billion = 10^{12}
°F	= °C × 1.8 + 32°
K	= t°C + 273

Note : (1) KJ is the short form of kilojoules which is equal to thousand joules.

(2) MJ is the short form of megajoules which is equal to million joules.

(3) GJ is the short form of gigajoules which is equal to 10^3 joules.

(4) In Sichuan province of China mu is equal to 1.3 times that in other provinces, thus in this province 1 mu = 858 m^2 or 0.086 hectares.

Appendix IV

DIRECTORY OF AGENCIES ASSOCIATED WITH THE DEVELOPMENT OF BIOGAS TECHNOLOGY

ARGENTINA

Centre de Estudios Fotosintticos
Y Bioguismicos, Suipacha 531,
2000 Rosario, Argentina

AUSTRALIA

Department of Chemical
Engineering, University of Sydney,
New South Wales - 2006, Australia

Food Research Laboratory
CSIRO, Delhi Road, P. O. Box 52,
North Ryde, NSW 2113,
Australia

Division of Chemical Technology
Commonwealth Scientific and
Industrial Research Organisation,
Division of Chemical Technology,
P.O. Box 310, South Melbourne,
Victoria 3205, Australia

School of Agriculture and Forestry,
University of Melbourne,
Parkville, Victoria 3052,
Australia

AUSTRIA

Inter. Inst. for Applied Systems
Analysis, A 2361 Laxenburg,
Austria.
Joint FAO/IAEA Division of Isotope
and Radiation Wagramerstrasse
5, P. O. Box 100, 414000 Vienna,
Austria.

BANGLADESH

Department of Chemical
Engineering, Bangladesh University
of Engineering and Technology,
Dacca, Bangladesh.

Institute of Fuel Research and
Development, Dacca, Bangladesh.

BELGIUM

Directorate - General XVII, Energy,
European Commission,
Rue de la Loi 200,
B - 1049 Brussels, Belgium.

Faculty of Agricultural Sciences,
State University of Gent,
Coupure, 9000 Gent, Belgium.

International Association for Rural
Development (IARD), Rue du
Commerce 20, Bte 3, 1040
Brussels, Belgium.

Unit of Bio-Engineering, University
of Louvain, 1 Place Croix due Sud,
B-1348, Louvainla-Neuve,
Belgium.

BRAZIL

Biomass Users Network,
Rua Francisco Dias Velho 814,
Brooklin Novo - Sao Paulo SIP
Brazil - CEP 04581.

COMGAS : Companhia de Gas de Sao Paulo, CEP 01304
Sao Paulo SP, Brazil

Gerente Nacional de Biogas
Embrater, Empresa Brasileira de Assistaencia Tecnica e Extensa O Rural Brazilia, Brazil.

Laboratorio de Energia Biomassa,
Universidade Federal Da Paraiba,
Centro de ciencias Agrarias,
Areia-Pb 58397, Brazil.

Superintendente de Planejamento e Desenuopuimento Companhia de Gas de Sao Paulo Rua Augusta 1600
CEP 01304, Sao Paulo S. P., Brazil

EMBRATER:Empresa Brasileiira de Assistencia Tecnia e Extensao Rural Vinculada ao Ministerio da Agricultura SEP/NORTE - W3 - Q515 - Lote 03 - Brasilia - DF, Brazil.

Institute of Electrotechnics and Energy Av. Prof. Luciano Gualberto, 1289, Cidade Universitaria
05508 - 900 - Sao Paulo - S. P., Brazil.

Institute de Pesquisas : Technologicas (IPT) do
Estado de Sao Paulo S/A IPT
Cidade Universitaria, Sao Paulo SP,
Caixa Postal 7141 CEP 01000,
Endereco, Brazil.

Nucleus of Non-Conventional Energy Sources, Federal University of Ceara, Centro de Ciencias,
Campus do pici,
Caixa Postal 3010, 6000
Fortaleza - CE, Brazil.

Universidado Estadual De Compines,
Faculdade De Engenharia de Campinas, Cidade Universitaria,
Campinas - SP, Brazil.

BURUNDI

Project Biogas Programme on Special Energy, Bujumbura,
Burundi.

CANADA

ADI Limited, 1115 Regent Street,
P. O. Box 44, Frederiction,
NB E3B 4Y2, Canada.

Alberta Energy and Natural Resources 9th Floor, 9915 108 Street, Edmonton, Alberta,
Canada.

Centre de Recherche
Industrielle du Quebec,
333 rue Franquet,
C. P. 9038, Ste-Foy,
Quebec GIV 4C7, Canada.

Department of Bio-Resource Engineering, University of British Columbia, 2075 Wesbrook Mall,
Vancouver, B. C. V6T IW5,
Canada.

Division of Biological Services,
NRC of Canada, Ottawa, Ontario
K14 OR6, Canada.

International Environmental
Consultants, 52233 Dundas Street

W., Islington, Ontario M9B 146, Canada.

Urget Delisle and Associates, 426 Chemin des Patriotes, St. Charles Sur Richelieu, Quebec, BP N 60, JOH 2 GO, Canada.

Biomass Energy Institute, Inc., 1329 Niakwa Road, Winnipeg, Manitoba R2J 9Z9, Canada.

Canadian International Development Agency, 200 Promenade du Portage Hull, Quebec KIA 064, Canada.

Brace Research Institute, Faculty of Engineering, P. O. Box 900, Macdonald College, Mc Gill University, Ste. Anne de Bellevue, Quebec H9 X 3V9, Canada.

Department of Agricultural Engineering, University of Manitoba, Winnipeg, Manitoba R3T 2N2, Canada.

Department of Chemical Engineering, University of Waterloo, Ontario, N2L 261, Canada.

International Development Research Center, P. O. Box 8500, Ottawa KIG 3H9, Canada.

Manitoba Research Council, Industrial Technology Centre, 1329 Niakwa Road, Winnipeg, Manitoba R2J 3T4, Canada.

Waste Water Technology Centre, Environment Canada, Burlington, Canada.

CHILE

Campex Semillas Baer, Fundo Elcanto, Casilla 943, Gorbea, Chile.

Corporation Industrial Para EI Desarrollo Regional Del (CIDERE), Biobio, Concepcion, Chile.

DENMARK

Environmental Engineering Laboratory, Dept. of Civil Engineering, University of Aelborg, DK - 9000, Aelborg, Denmark.

DOMINICANA

Universidad Autonoma De Santo Domingo, Cindad Universitaria, Santo Domingo, Republica Dominicana.

EGYPT

Soil and Water Research Institute, Agricultural Research Centre, Giza, Egypt.

Water Pollution Laboratory, National Research Centre, Al Tahrir Steet, Dokki, Giza, Cairo, Egypt.

ENGLAND

Alternative Technology Group, Open University, Faculty of Technology, Milton Keynes MK7 6AA, UK.

A. O. SMITH (Harvestore) Ltd., Eye, Suffolk IP23 7HS, UK.

Biomechanics Ltd.,
Caxton House, Wellesley Road
Ashford, Kent TN 24 8ET, UK.

British Petroleum Company Ltd.,
BP Research Centre, Chertsey
Road, Sunbury-on-Thames,
Middlesex TW16 7LN, UK.

Centre for Env. Tech.,
Imperial College,
48 Princes Gardens,
London SW7 2PE, UK.

Chemapec Ltd., Block IR, Milton
Trading Estate, Abingdon, Oxon
OX14 4RR, UK.

Commonwork Enterprises Ltd,
Bore Place, Chiddingstone,
Edenbridge, Kent, UK.

Department of Biological Sciences,
Pertsmouth Polytechnic, Portsmouth,
Hampshire, UK.

Department of Industry, Warren
Spring Laboratory, P. O. Box 20,
Gunnels Wood Road, Stevenage
Hertfordshire SGI 2BX, UK.

Energy Conversion and Pollution
Control, S.C.S. Biotechnology Ltd.,
Elmsfield, Chipping Norton,
Oxford OX7 5X7, UK.

Farm Gas Ltd, Heath Workshop,
Lydham, Bishop's Castle, Salop
SY9 59 5HB,UK.

Hamworthy Engineering Limited,
Pumps and Compressors Division,
Fleets Corner, Poole, Dorset BH17
7LA, UK.

IMES Ltd. (ADBAC Ltd.),
the Great Yard, Marridge Hill,
Rambury, Marborough, Wills,
UK.

Industrial Engine Division, J. Mann
and Son Ltd., Soxham, Bury St.
Edmunds, Suffolk 1P 28 6QZ, UK.

International Institute of Energy
Development (IIED),
10 Percy Street,
London WIP ODR, UK.

London Research Station, British
Gas Corporation, Michael Road,
London SW6 2AD, UK.

Microbiology Department, Queen
Elizabeth College, University of
London, Campden Hill Road,
London W8 7AH, UK.

Reading Industrial Engines Ltd,
Unit 5, Stadium Way, Reading,
Berks RG3 6BX, UK.

Shell International Petroleum Co.,
Non-Traditional Business Division,
Shell Centre, London SEI, UK.

Wright Rain Limited, Ringwood,
Hampshire BH24 IPA, UK.

Anaerobic Digestion and Biomass
Advancement Company Ltd.,
(ADBAC), the Great Yard,
Marridge Hill Ramsbury,
Marlborough Wilts,
Mortimer Hill, Mortimer, Nr.
Reading, Berkshire RG7 3PG, UK.

Applied Energy Systems Ltd, 1
Whippendell Road, Wetford, Herts
WDI 7LZ, UK.

British Anaerobic and Biomass Association Ltd. (BABA), White House. Little Bedwyn, Marlborough, Wittshire SN8 3JP, UK.

Cavendish Laboratory, Univ. of Cambridge, Madingely Road, Cambridge CB3 OHE, UK.

Cardiff University Industry Centre, P. O. Box 78, University College, Cardiff CFI IXL, UK.

Commonwork Construction Ltd., Bore Place, Bough Beech, Edenbridge, Kent TN 8 7AR, UK.

Denis Evers and Associates Ltd, Hastings House, 43 Foregate Street, Worcester WRI IEE, UK.

Department of Chemical Engineering, University of Manchester, Institute of Science and Technology, P. O. Box 88, Manchester M60 1QD, UK.

Department of Science, Polytechnic of Wales, Mid-Glamorgan, CG37 IDL, UK.

Energy Economics Centre, Univ. of Surrey, Guildford Surrey GU2 5XH, UK.

Energy Programme, Science Policy Research Unit, Mantell Building, Univ. of Sussex, Falmer, Bringhton BNI 9RF, UK

Energy Research Group, Aston UNIV. Aston Triangle, Birmingham B47 ET, UK.

Energy Research Centre, School of Env. Studies, Univ. of Ulster, Coleraine County Londonderry BT52 1SA, Northern Ireland, UK.

Energy Technology Support Unit, Department of Energy, AERE Harwell, Oxfordshire OXII ORA, UK.

Farrow Irrigation Ltd, Horseshoe Road, Spalding Lines PE11 3JA, UK

Helix Multi-Professional Services, Mortimer Hill, Mortimer, Reading, Berkshire RG & 3PG, UK.

Industrial Development Department, Tropical Products Institute, Culham, UK.

Intermediate Technology Development Group, 9, King Street, London WC 2E 8HN, UK.

Kilbees Slurry Digesters Limited, Winkfield, Windsor Forest Berkshire SL4 2EU, UK.

Microbial Biochemistry Department, The Rowett Research Institute, Green Burn Road, Bucksburn, Aberdeen, AB2 9SB, UK.

PAQUES (UK) Ltd., Shire Hill Industrial Estate,

Thaxted Road, Saffron Walden
Essex, CB11 3AQ, UK.

SCS Biotechnology Ltd, Elmsfield, Chipping Norton, Oxford OX 5X7, UK.

Warsen Spring Laboratory, Dept. of Trade and Industry, P. O. Box 20, Gunnels Wood Road, Stevenage, Herts SGI 2BX, UK.

GERMANY

ARTEFACT,
Bahnhofsstrabe 6 2392 Glucksburg, Germany

Bremen Overseas Research & Development Assoc. (BORDA), Breitenweg 55, D-28195, Bremen, Germany.

GTZ/GATE-ISAT
P. O. Box 5180, D - 65760
Eschborn, Germany.

Institute of Farm Economics, Federal, Research Centre of Agriculture, Breaunschweig, Germany (erstwhile FRG).

Energy Research and New Technology Development, Research and Dev. Division, Volkswagenwerk
AG, 318 Wolfsburg,
Germany (erstwhile FRG).

Institute fur Verfahrenstechnik, RWTH Aachen, Templegraben 55, D 5100 Aachesn,
Germany (erstwhile FRG).

Interdisciplinary Group of Energy Technology, Technical University Berlin, Postanschrift, T. V. Berlin, IPAT, Sekr TH-2, Lentzeallee 86, D - 1000 Berlin 33,
Germany (erstwhile FRG).

FIJI

Directorate of Energy Planning and Development,
Ministry of Energy,
P. O. Box 2256,
Government Buildings, Suva, Fiji.

School of Natural Resources,
University of South Pacific,
P. O. Box 1168,
Suva, Fiji.

GHANA

Technology Consulting Centre, University of Science and Technology, Kumasi, Ghana.

GUATEMALA

Central American Research Institute for Industry, Applied Research Division, P. O. Box 1552, Guatemala City, Gautemala.

INDIA

Aga Khan Foundation, Sorojini House, 2nd Floor,
6, Bhagwan Das Road,
New Delhi 110001.

Aga Khan Rural Support Programme (AKRSP), Choic Premises, Swastik Cross Road, Navarangpura, Ahmedabad 380009.

Agricultural Engineering Wing, Govt. Implement Factory, Satyanagar, Bhubaneswar 751007.

Agricultural Finance Corporation Ltd., Dhanraj Mahal, 1st Floor, Chhatapati Shivaji Maharaj Marg, Mumbai 400039.

Agricultural Tools Research Centre, Suruchi Campus, P.O. Box 4, Bardoli 394601.

Agriculture Prices Commission, Krishi Bhawan, New Delhi.

Appropriate Technology Development Association, Post Box 311, Gandhi Bhavan, Lucknow 226001.

Association of Voluntary Agencies for Rural Development, Kailash Colony, New Delhi 110048.

Bharatiya Agro Industries Foundation, Uruli-Kanchan, Pune 412202.

Biochemistry Division, National Sugar Institute, Kanpur.

Biomass Energy Development Centre, Court Road, New Colony, Etawah, UP.

BTRS, Gujarat Agricultural University, Anand 388110.

C. C. Shroff Research Institute, S.V. Road, Goregaon (West), Mumbai 400062.

Central Building Research Institute, Roorkee.

Central Mechanical Engineering Research Institute, Mahatma Gandhi Avenue, Durgapur 713209.

Central Salt and Marine Chemicals Research Institute, Waghwadi Road, Bhavnagar 364002.

Centre for Rural Development and Appropriate Technology, IIT, Hauz Khas, New Delhi 110016.

Centre for Energy Studies, IIT, Hauz Khas, New Delhi 110006

Centre for Water Resources Development and Management, Kunnamangalam, Kozhikode 673571, Kerala

College of Technology and Agricultural Engineering, Sukhadia University, Udaipur 313001.

Cotton Technological Research Laboratory, Matunga, Mumbai 400019.

Dairy Bacteriology Division, National Dairy Research Institute, Karnal, Haryana.

Department of Agricultural Engineering, G. B. Pant University of Agriculture and Technology, College of Technology, Pantnagar 263145, UP.

Department of Agri. Engg. & Plant Physiology, H. P. Agricultural Univ., Palampur 17062, HP.

Department of Chemical Engineering, Andhra University, Visakhapatnam.

Department of Chemical Engineering, IIT, Powai, Mumbai 400076.

Department of Civil Engineering,
College of Engineering,
Sri Venkateswara University,
Tirupati, AP.

Department of Civil Engineering,
Punjab Agricultural University,
Ludhiana.

Department of Economics and
Sociology, Punjab Agricultural
University, Ludhiana.

Department of Management Studies,
University of Madras Chepank,
Chennai 600005.

Department of Mechanical
Engineering, Indian Institute of
Technology, Powai,
Mumbai 400078.

Department of Mechanical
Engineering,
Institute of Technology,
Banaras Hindu University,
Varanasi 221005.

Department of Mechanical
Engineering,
Karnataka Regional Engineering
College, Srinivasnagar 57417,
Karnataka.

Department of Microbiology,
College of Agriculture,
Dharwar, Karnataka.

Department of Microbiology,
College of Basic Sciences and
Humanities, Haryana Agricultural
University, Hissar 125004.

Department of Microbiology,
Maharashtra Association for
Cultivation of Science, Pune
411004.

Division of Agricultural Economics,
Indian Agricultural Research
Institute, New Delhi 110012.

Energy Division, Jyoti Limited,
Tandalja, Vadodara 391410.

Environmental Survey Laboratory,
Kalpakkam Post, Chingelpet.

Food Technology and Biochemical
Engineering Dept., Jadavapur
University, Calcutta 700032.

Gobar Gas and Fertiliser Production
Unit, Naya Tola, Muzaffarpur
842001.

Gobar Gas Research Station,
Ajitmal, Etawah 20611, UP.

Gobar Gas Youth Centre,
Ajitmal, Etwah, UP.

Indian Institute of Management,
Vastrapur, Ahmedabad 380015.

Bureau of Indian Standards,
Manak Bhavan,
9 Bahadur Shah Zafar Marg,
New Delhi 110001.

Institute of Economic Growth,
University of Delhi,
University Enclave,
Delhi 110007.

Jyoti Solar Energy Institute,
Birla Vishakarma Mahavidyalaya,
Vallabh Vidyanagar, Vadodara.

Karnataka State Council for
Science and Technology,
Government of Karnataka,
Bangalore 560012.

Maharashtra Arogya Mandal, 435,
Hadapsar, Pune 411028.

Action for Food Production
(AFPRO) 25/1A, Institutional Area,
D Block, Pankha Road,
Janakpuri, New Delhi 110058.

Agricultural Refinance and
Development Corporation,
Shrinektan, Dr. Annie Besant Road,
PB No. 6552,
Worli, Mumbai 400018.

Agriculture, Forests and
Cooperative Dept., Government of
Gujarat, *Sachivalaya*,
Gandhinagar 382010, Gujarat.

Application of Science and
Technology to Rural Areas,
Indian Institute of Science,
Bangalore 560012.

Appropriate Technology Unit,
Allahabad Polytechnic,
Allahabad 211002.

Asian and Pacific Centre for
Transfer of Technology,
Technology Bhavan Complex,
New Mahrauli Road,
P. O. Box 4575, New Delhi
110016.

Asian Energy Institute,
India Habitat Centre,
Lodhi Road, New Delhi 110003.

Bhagubhai Mafatlal Polytechnic,
Vile Parle, Isla Juhu Road,
Mumbai 56.

Biocenter, Shahibag House,
Shahibag, Ahmedabad 38004.

Biogas Cell,
Special Schemes Organisation,
Govt. of Rajasthan, Secetariat,
Jaipur.

Birla Vishvakarma *Mahavidyalaya*,
Engineering College, Kaira, Gujarat.

Carburettors India Limited,
29, Mount Road, Chennai

Cement Research Institute of India,
South Extension - I, New Delhi.

Central Institute of Agricultural
Engineering,
Shri Guru Tegh Bahadur Complex,
T. T. Nagar, Bhopal 462003.

Central Power Research Institute,
Bangalore 560012, Karnataka.

Centre for Development of Rural
Technology, Allahabad Polytechnic,
Allahabad 211002.

Centre for the Study of Science,
Technology and Development,
CSIR, Complex - NPL,
Hilliside Road, New Delhi 12.
College of Agricultural
Engineering, Punjab
Agricultural University, Ludhiana.

Cooper Engineering Limited,
Satara Road, Maharashtra.

Credit Planning Cell, Central Office, Reserve Bank of India, Mumbai 400001.

Department of Agricultural Engineering, College of Agriculture, Dharwad 580005.

Department of Biochemical Engineering, IIT, Hauz Khas, New Delhi 110016.

Department of Chemical Engineering, A. C. College of Technology, PAUT, Chennai 600025.

Department of Chemistry, University of Gorakhpur, Gorakhpur 273001.

Department of Civil Engineering, Nagarjunsagar Engineering College, Saifabad, Hyderabad 500488.

Department of Civil Engineering, University of Roorke, Roorkee.

Department of Farm Machinery/ Agro Energy, Tamil Nadu Agricultural University, Coimbatore 641003.

Department of Mechanical Engineering, Indian Institute of Science, Bangalore.

Department of Microbiology, Sri Parasakthi College for Women, Courtlam, Tirunelveli 627802, Tamil nadu.

Department of Processing and Agri. Structures, College of Agri. Engg., Punjab Agri. Univ., Ludhiana 141004.

Department of Science and Technology, Technology Bhavan, New Mehrauli Road, New Delhi 110029.

Division of Soil Science and Agricultural Research Institute, New Delhi 110012.

Environment Science and Engineering Group, IIT, Powai, Mumbai 400076.

Food and Agriculture Organisation of the United Nations, 55, Lodhi Estate, New Delhi 110003.

Gandhi *Gram* Trust, Gandhi *Gram,* Madurai 624302.

Gobar Gas Research and Development Centre, Kora *Gramodyog Kendra*, Borivli (W), Mumbai 400092.

Gobar Gas Scheme, Khadi and Village Industries Commission, *Gramodya*, 3 Irla Road, Vile Parle (West), Mumbai 400056.

Indian Council of Agricultural Research (ICAR), Krishi Bhavan, New Delhi.

Malanadu Development Society, Kanjivapally 686507, Kerala.

M/s Sunflame Industries Pvt. Ltd., Shed No. 2, Plot No. 58, 12/6 Mathura Road, P.O. Amar Nagar, Faridabad 121003

M/s Gas Chemical Industries Pvt. Ltd., Budshah Chowk, Amnakadal, Srinagar, J&K.

M/s Sweet Home Appliances Pvt. Ltd., 3-E/16, B.P.N.I.T., Faridabad 121001.

M/s Associated Engineering Works, Chivatam Road, Tanuku 534211, AP.

M/s Tulsi Domestic Appliances, 30A, Old Industrial Area, Alwar 301001. Rajasthan.

M/s Agriculture Associates, Station Road, Alwar 301001.

M/s Amar Singh and Sukhdeo Singh, Manufactures of Agricultural Equipment, Divyapur, Etawah, U.P.

M/s Gobar Gas Plant Consulting Centre, 46, Narsingh Rao Tori, Jhansi 284002.

M/s Santosh Engineers, Manufacturers of Gobar Gas Equipment & Appliances, 11, Swami Vivekanand Nagar, Kaligarh, Jagatganj, Varanasi.

National Environmental Engineering Research Institute, Nehru Marg, Nagpur 440020.

National Institute of Waste Recycling Technology, A-18, Juhu Apartments, Juhu Road, Mumbai 400049.

Photosynthesis and Energy Division, AMM Murugappa Chettiar Research Centre, Tharamani, Chennai 600042.

Management Development Institute, Sukhrali, Gurgaon 122001, Haryana.

M/s Rupak Enterprises, 1/146 Vishwas Nagar, Shahdara, Delhi 110032.

M/s Mech-ci-co 1-7, GIDC, Indus. Township Vatwa, Ahmedabad 382445.

M/s Batra Investments Pvt. Ltd., 14/1 Mathura Road, P.O. Amar Nagar, Faridabad 121003.

M/s Baroda Appliances, 886/A GIDC, Makarpura, Baroda 390010.

M/s Mitaso Appliances Ltd., Plot No. 63, Sector 6, Faridabad 121006

M/s Agro Pumpsets and Implements Limited, 10-2-317/3 Vijaynagar Colony, Hyderabad 500457.

M/s Gadepally Bio-Industries, Kambham Choulty Street, Rajahmundry 533101, AP.

M/s Patel Gascrafters Pvt. Ltd., 20, Shree Sai Bazzar, Mahatma Gandhi Road, Santcruz (West), Mumbai 400054.

National Dairy Research Institute, Karnal.

National Institute of Rural Development, Rajendra Nagar, Hyderabad 500030.

Paradise Garden and Agricultural Seed Farm, Village Kachroli, Babarpur, Haryana.

Planning Research and Action Division, State Planning Institute, Kalankar Bhavan, Lucknow.

Propulsion Division, National Aeronautical Laboratory, P.B. No. 1779, Bangalore 560017

Indian Institute of Petroleum, Dehradun 248005.

Institute of Cooperation Management, 32, Parimal Society, Ellisbridge, Ahmedabad 380006.

Internal Combustion Engines Laboratory, Department of Mechanical Engineering, Indian Institute of Technology, Chennai 600036.

Kapur Solar Farms, Bijwasan Najfgarh Road, P. O. Kapas Hera, New Delhi 110037.

Khadigram Prayog Sangh, Harijan Ashram, Ahmedabad.

Maharashtra Gandhi *Smarak Nidhi*, Pune 400029.

Rajiv Gandhi Foundation, Jawahar Bhawan, Dr. Rajendra Prasad Road, New Delhi 110001.

Postgraduate Department of Economics, Sardar Patel University, Vallabh Vidanagar 38820, Gujarat.

R & D Centre, Indian Oil Corporation, Sector - 13, Faridabad.

R & D Division, Danfoss (India) Limited, New Delhi 110001.

Regional Biogas Experimental Centre, Gandhigram, Madhurai, Tamil Nadu.

Regional Biogas Experimental Centre, Mansabal, Srinagar, J & K.

Regional Centre for Energy, Heat and Mass Transfer for Asia and the Pacific, Department of Mechanical Engineering, IIT., Chennai 600036.

Regional Research Laboratory, Jorhat 780052.

Rural Development and Cooperation Department, Govt. of Karnataka, Multistorey Building, Dr. Ambedkar Road, Bangalore 560001.

Rural Electrification Corporation, New Delhi.

Sarang Ferroconcrete, P. O. Pen, Raigad. Maharashtra.

School of Energy & Env. Studies, Devi Ahilya Univ., Indore 452001.

Shivadan Renewable Energy Research Institute, S. No. 192/3, Industrial Estate, Sangli 416416.

Sircilla Cooperative Electric
Supply Society Ltd,
Sircilla, Karimnagar, AP.

Tata Energy Res. Inst., Darbari
Seth Block, India Habitat Centre,
Lodhi Road, New Delhi 110003.

Thapar Polytechnic
Patiala 147001.

Tractor Training and Testing
Station, Tractor Nagar,
Bundi 466445, MP.

Vispute Engineering and Research
Works, Opposite Jangli Maharaja
Temple, Shivajinagar, Pune 411005.

Western Regional Station of
National Dairy Research Institute,
Aarey Milk Colony, Mumbai
110065.

Ministry of Non-Conv.
Energy Sources (MNES)
14, C. G. O. Complex,
Lodhi Estate, New Delhi 110001.

R & D Centre,
Kirloskar Oil Engines Ltd.,
Khadki, Pune 411003.

Regional Biogas Experimental
Centre, Allahabad.

Regional Biogas Experimental
Centre, Khadasli, Bhavnagar,
Gujarat.

Regional Biogas Experimental
Centre, Nasik, Maharashtra.

Regional Research Laboratory,
CSIR, Trivandrum 695019,
Kerala.

Rural Housing Wing,
Birla *Vishvakarma Mahavidyalaya*,
P. O. Vallabh Vidyanagar 388120,
Kaira, Gujarat.

Scientific Research Centre,
ESCORTS Ltd., 25 Mathura Road,
Faridabad 121002.

Shri Bhagubhai Mafatlal
Polytechnic, Irla Road,
Vile Parle (West),
Mumbai 400056.

Society of Science for the People,
Newabpet, Nellore, AP 521002.

Structural Engineering Research
Centre, Ghaziabad

Tata Research Development and
Design Centre, 1, Mangaldas Road,
Pune 411001.

Tractor Training Centre,
Hissar, Haryana.

UNDARP,
945 Bottee Street,
Pune 411001.

Vigyan Club,
Singhamau, Lucknow.

Western India Industries,
(Regional Distributors of Kirloskar
Oil Engine Ltd, Pune)
17, Navyug Market,
Ghaziabad 201001.

IREDA,
India Habitat Centre
Lodhi Road,
New Delhi 110003.

Various Nodal Agencies Identified for Implementing NPBD

Non-Conventional Energy
Dev. Corp. of A. P.,
5-8-207-2, Pisgah Complex,
Nampally, Hyderabad 500001, AP.

Directorate of Agriculture,
Itanagar 791111,
Arunachal Pradesh.

Dept. of Rural Dev.,
7th Floor, G. S. Road,
Bhangagarh,
Guwahati 781001,
Assam.

Bihar Renewable Energy
Development Agency,
18 Aniket Housing Society,
Kidwaipuri,
Patna 800001
Bihar.

Directorate of Agriculture,
Govt. of Goa,
Panji 403001, Goa.

Gujarat Agro-Industries Corp.,
Khet Udyog Bhavan,
Opp. High Court, Navrangpura,
Ahmedabad 380014, Gujarat.

Directorate of Agriculture,
Govt. of Haryana,
Chandigarh 160001, Haryana.

Directorate of Agriculture,
Govt. of Himachal Pradesh,
Nalagarh House, Shimla 171002,
HP.

Department of Agriculture,
Govt. of Jammu & Kashmir,
Srinagar 190001, J & K.

Dept. of Rural Dev. & *Panchayati*
Raj, Govt. of Karnataka,
Multistoreyed Secretarial Bldg.,
Bangalore 560001, Karnataka.

Department of Agriculture,
Govt. of Kerala, *Vikas* Bhavan,
Thiruvanthapuram 695001.

Dept. of Rural Dev.,
Mantralaya, Mumbai 400032,
Maharashtra.

MP Agro-Industries Corp.,
GTB Complex, T. T. Nagar,
Bhopal 462003, M. P.

Dept. of Sci. Tech. & Env.
New Secretariat Complex,
Govt. of Manipur,
Imphal 795001, Manipur,

Meghalaya Non-Conv. Energy &
Rural Development Agency,
Lower Lachaumierre,
Shillong 793001, Meghalaya.

Directorate of Animal Husbandry
and Veterinary Services,
Govt. of Mizoram,
Aizawl 796001, Mizoram.

Dept. of Rural Development,
Govt. of Nagaland,
Kohima 797001, Nagaland.

Orissa Renewable Energy Dev.
Agency, 559 Mancheshwar
Industrial Estate,
Bhubaneswar 751010, Orissa.

Directorate of Agriculture,
Govt. of Punjab,
Chandigarh 160001, Punjab.

Special Schemes & Integrated
Rural Dev. Dept., *Sachivalaya,*
Govt. of Rajasthan,
Jaipur 302005, Rajasthan.

New & Renewable Sources of
Energy Dept., Govt. of Sikkim,
Gangtok 737101. Sikkim.

Dept. of Rural Dev.
Govt. of Tamil Nadu,
Kuralagam 108, Chennai
600001, Tamil Nadu.

Department of Agriculture,
Govt. of Tripura,
Agartala 799001, Tripura.

Department of Rur. Dev.,
Govt. of Uttar Pradesh,
Jawahar Bhawan,
Lucknow 226001, UP.

Dept. of Sci., Tech &
Non-Conv. Energy Sources,
New Secretariat Building,
Calcutta 700001, WB.

Development Commissioner,
Port Blair 744001,
A & N Islands.

Dept. of Rural Dev.,
Chandigarh Administration,
Chandigarh 160001.

Animal Husbandry-cum-
Veterinary Officer,
Silvasa 396230,
Dadra & Nagar Haveli.

Delhi Energy Dev. Agency,
11 Lancers Road,
Timarpur, Delhi 110007.

Directorate of Agriculture,
Pondicherry 605001.

Addresses of KVIC Offices Associated with Biogas Programme

Khadi and Village Industries
Commission, Ahmedabad.

Khadi and Village Industries
Commission, Bangalore.

Khadi and Village Industries
Commission, Bhubaneshwar.

Khadi and Village Industries
Commission, 23, Chittaranjan
Avenue, 6th Floor, Calcutta 12.

Khadi and Village Industries
Commission, Hyderabad.

Khadi and Village Industries
Commission, Lajpat Rai Bhawan,
Kaisarbagh, Lucknow.

Khadi and Village Industries
Commission, New Delhi.

Khadi and Village Industries
Commission, Patna.

Khadi and Village Industries
Commission, Thiruvanthapuram.

Khadi and Village Industries
Commission, P. B. No 34, Ambala.

Khadi and Village Industries
Commission, Bhopal.

Khadi and Village Industries Commission, Vile Parle, Irla Road, Mumbai 56.

Khadi and Village Industries Commission, GN Bardoli Road, Ambari, Guwahati 1.

Khadi and Village Industries Commission, Jaipur.

Khadi and Village Industries Commission, Avuai Shammujham Road, Goplapuram, Chennai 86.

Khadi and Village Industries Commission; Goa, Daman and Diu.

Cedmar Apartment, Swami Vivekanand Road, Panaji 403001, Goa.

Khadi and Village Industries Commission, Srinagar.

Addresses of State KVIBs Associated with Biogas Programme

Andhra Pradesh Khadi and Village Industries Board, 2-4-111-4 Farahatkada, Kachiguda, Hyderabad.

Bihar State Khadi and Village Industries Board, Gandhi Maidan (East), Patna 4.

Gujarat Rajya Khadi Gramodyog Board, Green Villa, Juna Wadaj Ahmedabad 13.

Himachal Pradesh Khadi and Village Industries Board, Laj Bhavan, Court Road, Simla 1.

Karnataka Khadi and Village Industries Board, 5, Seshadri Road, Bangalore 9.

Madhya Pradesh Khadi and Village Industries Board, Govindpura Estate, Bhopal.

Manipur Khadi and Village Industries Board, Paona Road, Imphal, Manipur.

Nagaland State Advisory Board for Khadi and Village Industries, Kohima, Nagaland.

Punjab Khadi and Village Industries Board, 41 - 2A, Chandigarh 1.

Tamil Nadu Khadi & Village Industries Board, Kuralagam Building, Chennai.

UP Khadi & Village Industries Board, 7/105C, Tilak Nagar, Kanpur.

Assam Khadi and Village Industries Board, Chandmari, Guwahati 3.

Union Territory Khadi and Village Industries Board, Secretariat, Chandigarh.
Haryana Khadi and Village Industries Board, Bungalow No 53, Sector 2, Chandigarh.

Jammu and Kashmir Khadi and Village Industries Board, 5, Sesadri Road, Srinagar.

Kerala Khadi and Village Industries Board, Thiruvanthapuram.

Maharashtra Khadi and Village Industries, Board, Bhatia *Balrakshak Vidyalaya,* 19-21 Manohar Das Street, Mumbai 1.

Meghalaya Khadi and Village Industries Board, Shillong, Meghalaya.

Orissa Khadi and Village Industries Board, Unit - III, Bhubaneswar.

Rajasthan Khadi and Village Industries Board, Station Road, Jaipur.

Tripura Khadi and Village Industries Board, Agartala.

WB Khadi & Village Industries Board, 12 B. D. D. Bag, Calcutta.

Regional Biogas Development and Training Centres

Location (States/UTs Covered)

1. AP Agricultural University, Hyderabad. (Andhra Pradesh).
2. Assam Agricultural University, Jorhat. (Assam & NE States).
3. Rajendra Agricultural University, Pusa Samastipur. (Bihar).
4. Gujarat Agricultural University, Anand. (Gujarat, D & N).
5. HP *Krishi Vishwa Vidyalaya,* Palampur. (HP, J & K, & Chandigarh).
6. University of Agricultural Sciences, Bangalore (Karnataka).
7. *Kasturbagram Krishi Kshetra,* Indore. (Madhya Pradesh).
8. Centre of Science for Villages, Wardha. Maharashtra (Marathwada & Vidarbha).
9. Institute of Renewable Energy, Nasik. (Maharashtra).
10. Orissa Univ. of Agri. & Tech., Bhubaneshwar. (Orissa).
11. College of Tech. & Agri Engg., Udaipur. (Rajasthan, & Delhi).
12. TN Agricultural University, Coimbatore. (TN, & Pondicherry).
13. Regional Inst. of Rural Dev., Ghazipur, Gurukul Kangri (sub-centre of RIRD), Hardwar. (Uttar Pradesh).
14. Indian Institute of Technology, Kharagpur. (WB, & A & N Island).
15. Dept. of Agriculture, Thiruvanthapuram. (Kerala).
16. Haryana Agricultural Univ., Hissar. (Haryana).
17. Punjab Agricultural Univ. Ludhiana. (Punjab).

INDONESIA

Centre for Research on Energy, Institut Teknologi Bandung, Jalan Ganesha O, Bandung 40132, Indonesia.

Pusat Teknologi Pembangunan Institut Teknologi Bandung JI Ganesha 10 Building Sem a Surat Menyrat Ke Kotak Pos 276, Bandung, Indonesia.

World Vision of Asia, P. O. Box 101, Malang, East Java, Indonesia.

IRELAND

Department of Microbiology,
University College, Galway,
Ireland.

Institute for Industrial
Research and Standards,
Biomass Conversion,
Technical Information Service,
Ballymun Road,
Dublin 9, Ireland.

ITALY

Dept. of Energy, Univ. of Florence,
Via S. Marta, 3-150139 Firenze,
Italy.

ENI Group Research Association,
00015, Monterotondo,
Rome, Italy.

Politecnico di Torino,
Dipartimento de Energetica
Corso Duca degli Abruzzi 24,
10129 Torino, Italy.

Renewable Energy Sources Dept.,
Italian Agency for New
Technology, CRE Casaccia via
Anguillavese, 30100060
Rome, Italy.

JAPAN

Department of Animal Hygiene,
Nippon Veterinary and
Zootechnical College, Japan.

Global Industrial and Social
Progress Research Institute,
7th Floor, Mori Building No 33,
3-8-21 Toranomon,
Minato-ku, Tokyo 105, Japan.

Laboratory of Animal Waste
Management, National Institute
of Animal Industry,
Trukuba Norindanchi, PO Box 5,
Ibaraki 305, Japan.

Fermentation Research Institute,
Agency of Industrial Science &
Technology, Yatabe-Machi,
Ibaraji, Japan-300-21.

Institute of Energy Economics,
Shuwa Kamiyacho Building,
10th Floor, 3-13, Toranomon,
4 - Chome Minato - ku,
Tokyo, Japan.

MALAYSIA

Asian and Pacific Development
Centre, Pesiaran Duta,
P. O. Box 12224
Kula Lampur, Malaysia.

Department of Environment
Sciences, Universiti Pertanian
Malaysia,
P. O. Box 203, Sunegi Besi,
Selangor, Malaysia.
Standards and Industrial
Research Institute of Malaysia,
P. O. Box 5, Shah Alam,
Selangor, Malaysia.

Division of Microbiology,
Fakulti Sains,
University of Kebangsaan Malaysia,
Kwala Lumpur 22-14, Malaysia.

NEPAL

Baluja Yantra Shala,
Sanitary Engineering Limited,
P. O. Box 2070, Kathmandu, Nepal.

Biogas Company,
Kupondol, Lalitpur, Nepal.

Gobar Gas and Krishi Yantra Vikash Private Limited,
Research and Development Unit.
Butwal, Rupendelhi,
Lumbini Zone, Nepal.

Biogas Consultancy Unit,
Development and Consulting
Services, Butwal, Nepal.

Biogas Support Programme,
SNV - Nepal, P. O. Box 1966,
Kathmandu, Nepal.

Gobar Gas Company
Kathmandu, Nepal.

THE NETHERLANDS

Department of Biotechnology,
Division for Nutrition and Food
Research, TNO P. O. Box 360,
3700 A. J. Zeist, The Netherlands.

Gist-Brocades NV
Wateringseweg 1, NL- Delft,
The Netherlands.

Department of Water Pollution
Control, Agricultural University,
Biotechnion, De Dreijen 12, 6703
BC Wageningen, The Netherlands.

Research Institute of Public
Health Engineering,
Box 214, Delft, The Netherlands.

NEWZEALAND

Invermay Agricultural Research
Centre, Private Bag,
Mosgiel, New Zealand.

NIGERIA

Department of Soil Science,
University of Ife, Ife-Ife,
Nigeria.

PAKISTAN

Directorate of Research,
Faculty of Engineering,
University of Peshawar,
Peshawar, Pakistan.

Pakistan Council of Appropriate
Technology,
I-B, Street 47, F-7/1
Islamabad, Pakistan.

Merin Limited,
Dada Chambers,
M. A. Jinnah Road,
Karachi 2, Pakistan.

PEOPLES REPUBLIC OF CHINA

Academia Sinica,
Chengdu Institute of Biology,
Bioenergy Laboratory,
Husai Campus, Chengdu,
Sichuan, P. R. China.

Asia-Pacific Regional Biogas
Research & Trg. Centre (BRTC),
4 Duan 13 Renming South Street,
Chengdu 610041, Sichuan,
P. R. China.

Biogas Experimental Station,
Shanghai Municipality Institute
of Indus. Microbiology,
Shanghai, P. R. China.

Biogas Office of Shantou City,
Guandong Province,
P. R. China.

Biogas Research Division,
Agri. Academy of the
Hubei Province, P. R. China.

Chengdu Biogas Institute,
No 13 Block 4,
Peoples South Street,
Chengdu 610041,
Sichuan, P. R. China.

Chengdu Xingguang Dev. Corp. for Appropriate Sci. & Tech.,
No 17, Yulin East Street,
People South Road,
Chengdu 610041, Sichuan,
P. R. China.

China Biogas Society,
4 Duan Renning South Street,
Chengdu 610041, Sichuan,
P. R. China.

Env. Engg. Dept.,
Qinghura Univ., Beijing,
P. R. China.

Guanazhen Institute of
Energy Conversion,
Chinese Acad. of Sciences,
Guanzhen, P. R. China.

Laboratory of Biomass Energy,
Chekiang Agri. Univ.,
Hangehow, Chekiang,
P. R. China.

Rural Energy Office of
Hangzhou County,
Zhejiang Province,
P. R. China.

Rural Energy Office of
Nimgxiang County,
Human Province,
P. R. China.

Rural Energy Office of
Qu Xian County,
Zhejiang Province.
P. R. China.

Rural Energy Office of
Yigang District, Human Province,
P. R. China.

PHILIPPINES

Batek Centre, Salitran, Desmarinas,
Cavite, Philippines.

Bureau of Animal Industry,
Ministry of Agriculture,
Manila, Philippines.

FKM Management Corporation,
7th Floor Liberty Building,
Pasay Road, Makati, Metro Manila,
Philippines.

National Institute of Biotechnology
& Applied Microbiology,
Laguna, Philippines.

Research Division,
Bureau of Animal Industry,
Albang, Rita Legarda Bldg.,
Sta Mesa, Manila,
Philippines.

Biogas Unit, Non-Conv.
Energy Res. Div.,
Bureau of Energy Dev.,
Ministry of Energy,
Manila, Philippines.

Maya Farms Division,
Liberty Flour Mills, Inc.
Metro Manila,
P. O. Box 2235, Manila,
Philippines.

National Institute of Science and
Technology, P. O. Box 774,
Pedro Gil Street, Corner Taft
Avenue, Manila,
Philippines.

REPUBLIC OF SINGAPORE

Inst. of South East Asian Studies,
Heng Mui Keng Terrace,
Pasir Panjang,
Singapore 0511,
Republic of Singapore.

SCOTLAND

North of Scotland
College of Agriculture,
Aberdeen, Scotland.

Rowett Research Institute,
Greenburn Road, Bucksburn,
Aberdeen Ab2 9 SB, Scotland.

SOUTH AFRICA

National Institute for Water
Research, South African Council
for Scientific and Industrial
Research, P. O. Box 395,
Pretoria,
Republic of South Africa.

Dept. of Microbiology & Plant
Pathology, Univ., of Natal,
P.O. Box 375,
Pietermarizburg 3200,
Republic of South Africa.

SOUTH KOREA

Rural Energy Resources Research
Division, Institute of Agricultural
Sciences, Office of Rural
Development, Suweon, South
Korea.

Korea Energy Economics Institute,
665 - 1, Naeson - Dong
Euiwang - Si, Kyungki - Do,
Seoul, South Korea.

SRI LANKA

Ceylon Institute of Scientific
and Industrial Research,
363, Bauddhaloka Mawatha,
Colombo 7, Sri Lanka.

Engineering Industries Group,
Industrial Development Board of
Ceylon, 615, Galle Road,
Katubedda, Sri Lanka.

Department of Civil Engineering,
University of Peradeniya,
Peradeniya, Sri Lanka.

Industrial Development Board of
Ceylon, 615, Galle Road,
Katubedda, Moratuwa, Sri Lanka.

SWEDEN

Centre for Research on Natural
Resources and Environment,
Stockholm Univ.,
S - 10691, Stockholm
Sweden.

Division of Applied Microbiology,
Royal Institute of Technology,
Stockholm, Sweden.

Stockholm Environment Institute,
P. O. Box 2142, 103 14 Stockholm,
Sweden.

Swedish Institute of Agricultural
Engineering, Box 7033,
S - 75007 Uppsala,
Sweden.

Metangruppen Bioquest,
Skargardsgatan 4,
S - 414 58 Goteborg, Sweden.

SWITZERLAND

Inter. Academy of the Environment,
4 Chemin de Conches,
CH - 1231 Conches
Geneva, Switzerland.

International Reference Centre for
Wastes, WHO/IRC, Ueberlandstr,
133 Ch - 8600 Dubendorf,
Switzerland.

Project Biogas, Clausiusstr 47,
ETH - Zentrum Ch - 8092, Zurich,
Switzerland.

National Energy Research
Foundation (NERF),
Ch - 8001, Zurich, Switzerland.

SEDESA,
Societe d 'Etude de l'
Environmental Rue du Midi 33,
1800 Vevey, Suisse,
Switzerland.

Swiss Federal Research Station
for Farm Management and
Agricultural Engineering,
CH - 8355 Tanikon,
Switzerland.

TAIWAN

Council for Agricultural
Planning and Development,
37, Nan Hai Road,
Taipei, Taiwan.

Industrial Technology Research
Institute, Engineering and Mining
Research Service Organisation,
1. Sec. I., Fu-Hsin S. Road,
Taipei, Taiwan.

Taiwan Livestock Research
Institute, 112, Mu-Chang Road,
Hsinhau, Tainan, Taiwan.

Department of Chemical
Engineering,
National Taiwan University,
Taipei 107, Taiwan.

Taiwan Institute of Environmental
Sanitation, Taipei 207, Taiwan.

Union Chemical Laboratories,
Industrial Technology Research
Institute. 1021 Kwang-Fu Road,
Hsinchu, Taiwan.

TANZANIA

Centre for Agricultural
Mechanisation and Rural
Technology.
P. O. Box 764, Arusha, Tanzania.

THAILAND

Agricultural Engineering Division,
Department of Agriculture,
Bangkhen, Bangkok 9, Thailand.

Division of Environmental
Engineering, Asian Institute of
Technology, P. O. Box 2754,
Bangkok, Thailand.

Environmental Engineering
Laboratory, Thailand Institute of
Scientific and Technological
Research, Bangkhen, Bangkok 9,
Thailand.

Kasetsart University
Bangkok - 9, Thailand.

Division of Energy Technology,
Asian Institute of Technology,
GPO Box 2754, Bangkok 10501,
Thailand.

Energy Research Institute,
12th Floor, Institute Bldg. III,
Chulalongkorn Univ.,
Phyathai Road, Bangkok 10330,
Thailand.

National Energy Administration,
Chulalongkorn Univ.,
Bangkok, Thailand.

National Energy Administration,
Pibultham Villa, Kasatsuk Bridge,
Bangkok 5, Thailand.

FAO Regional Office for
Asia and the Pacific, Maliwan
Mansion, Phra Atit Road,
Bangkok, Thailand.

TURKEY

Dept. of Energy Systems,
P O Box 21, Gebze 411470,
Kocaeli, Turkey.

Solar Energy Institute,
Ege Uni., Bornova,
Izmir, Turkey.

USA

The Alliance to Save Energy,
1725 K. Street NW, No 509
Washington DC 20006 - 1401, USA

AID Document and Information
Handling Facility,
7222 47th St. Suite 100,
Chevy Chase, MD 20815, USA.

Applied Physics Laboratory,
John Hopkins University,
Baltimore, Maryland 20810, USA.

Battelle,
Pacific Northwest Laboratories,
PO Box 999, Richland,
Washington 99352, USA.

Bioenergy Council,
825 A-1625 Eye Street,
Washington, Dc 20006, USA.

Biological Energy Research,
Department of Energy,
Washington, DC 20545, USA.

Biomass Programme Office,
Solar Energy Research Institute,
1617, Cole Boulevard,
Golden, Colorado 80401, USA.

Conservation and Renewable
Energy, Enquiry and Referral
Service, Franklin Research Centre,
P. O. Box 8900, Silver Spring,
MD 20907, USA

Department of Agricultural and
Chemical Engineering,
Colorado State University,
Fort Colin, Colorado 80523, USA.

Department of Agricultural
Engineering,
University of California,
Davis, California, USA.

Department of Agricultural
Engineering, University of
Minnesota, St. Paul, MN 55108,
USA.

Department of Biology
College of the Holy Cross,
Worcester, MA 01610, USA.

Department of Chemical
Engineering, University of
Arkansas, Fayetteville,
Arkansas 72701, USA.

Department of Civil Engineering,
Stanford University, Stanford,
CA 95305, USA.

Department of Environmental
Systems Engineering,
Clemson University,
Clemson, SC, USA.

Department of Microbiology,
University of Florida,
Gainsville, Florida 326001, USA.

Department of Soil and
Environmental Sciences, University
of California, Riverside,
California 92521, USA.

Division of Biological Energy
Research, Office of Basic Energy
Sciences, Department of Energy,
Washington,
DC 20545, USA.

Division of Environmental and
Nutritional Sciences,
University of California,
Los Angeles, California 90024,
USA.

East - West Centre,
Resource Systems Institute,
Honolulu, USA.

Energia Global, Inc.
1400 16th Street NW No 710
Washington DC 20036, USA.

Energy Centre, SRI International,
333 Ravanswood Avenue,
Menlo Park, CA 94025, USA.

Energy & Env. Analysis, Inc.
1655, North Fort Myer Drive,
Arlington, Virginia 22209,
USA.

Energy and Environment Policy
Center, John F. Kennedy School of
Government, Harvard Univ.,
79 Boyston Street Cambridge,
MA 02138, USA.

Energy Technology Department,
Stanford Research Institute,
Menlo Park, California, USA.

Getty Synthetic Fuels, Inc.,
2750 Signal Parkway,
Signall Hill, California 90806, USA.

Hamilton Standard,
Division of United Technologies
Corporation, Windsor Locks,
Connecticut 06096, USA.

Imperial Valley Biogas project,
5449 Kershaw Road, P. O. Box
1227, Brawlwy, CA 92227, USA.

Institute for Energy Studies,
Building 500, Stanford Univ.,
CA 94305, USA.

Institute of Gas Technology,
Chicago, Illinois, USA.

International Fund for Renewable
Energy and Energy Efficiency,
750 First Street NE No 930,
Washington DC 20002, USA.

Jet Propulsion Laboratory,
4800 Oak Grove Drive,
Pasadena CA 91103, USA.

Lawrence Berkeley Laboratory,
University of California,
Berkeley, CA 94720, USA.

Martin Marietta Energy Systems,
Inc., Box 2002, Oak Ridge,
Tennessee 37831 - 6501, USA.

Mason Dixon Farms,
1750 Mason Dipon Road,
Gettysburg, PA 17325, USA.

Midwest Research Institute,
425 Volker Boulevard,
Kansas City. MO 64110, USA.

National Centre for the Analysis of Energy Systems,
Brookhaven National Laboratory,
Uptron, New York 11973, USA.

National Fertiliser Development Centre, Tennessee Valley Authority,
Muscle Shoals, Alabama 35660, USA.

National Renewable Energy Laboratory,
1617 Cole Boulevard
Golden, CO 80401 - 3393, USA.

Oak Ridge National Laboratory,
Oak Ridge, Tennessee 37830, USA.

Pacific Gas and Electric Co,
77 Beale Street, San Francisco,
CA 94106, USA.

Sanitary Engineering Research Laboratory, College of Engineering,
University of California,

Richmond Field Station Bldg. 112,
47th and Hofffman Blvd.,
Berkeley, Richmond,
California 94804, USA.

Solar Energy Research Institute,
1536, Cole Boulevard,
Golden, Colorado 80401, USA.

Stockholm Environment Institute
9514 Garwood Street
Silver Spring, MD 20901, USA.

United States Agency for International Development,
(USAID), 320 21 Street NW,
Washington D. C. 20523, USA.

Woods Hole Oceanographic Institute, Woods Hole
Massachusetts 02543, USA.

Western Research Institute,
The Univ. of Wyoming Res. Corp
P. O. Box 3395, Univ. Station,
Laramie, WY 82071 - 3395, USA.

Alternative Energy Division
Office of Energy,
Puerto Rico, USA.

Associates, 3702 East Lake Street
Suite 202, Minneapolis,
Minnesota 55406, USA.

Battelle Columbus Laboratories,
505 King Avenue, Columbus,
Ohio 43201, USA.

Biogas of Colorado, Inc.
5611 Kendall Court Arvada,
Colorado 80002, USA.

Biomass Fuels Programme,
Tennessee Valley Authority,
Muscle Shoals, Alabama, USA.

Centre for the Biology of Natural Systems, Washington University, Box 1126, St. Louis,
Missouri 63130, USA.

Centre for Energy and Environmental Studies,
Schenley Park, Carnegie Mellon Univ., Pittsburgh, PA 15213, USA.

Center for Energy.Management,
Campus Box 428,

Univ. of Colorado, Boulder,
CO 80309 - 0428, USA.

Centre for Env. Information,
50 West Main Street,
Rochester NY 14614 - 1218,
USA.

Centre for International
Energy Development,
Argonne National Laboratory,
Argonne, USA.

Centre for Strategic and
International Studies,
1800 K. Street NW No 400
Washington DC 20006, USA.

Climate Institute,
324 4th Street NE, Washington
DC 20002 - 5821, USA.

Department of Agricultural
Engineering, Alabama Agricultural
Experiment Station, Auburn
University, Alabama 36849, USA.

Department of Agricultural
Engineering, Riley - Robb Hall,
New York State College of
Agriculture and Life Sciences,
Cornell University, Ithaca,
New York 14853, USA.

Department of Agricultural
Engineering, University of Hawaii,
Honolulu, Hawaii 96822, USA.

Department of Chemical
Engineering, Cornell University,
Ithaca, New York 14853, USA.

Department of Chemical
Engineering, University of
Missouri, Rolla, Missouri 65401,
USA.

Department of Civil Engineering,
University of Missouri,
Rolla, Missouri 65401, USA.

Department of Microbiology/
Chemistry,
Montana State University,
Bozeman, Montana 59717, USA.

Department of Public Works,
Bureau of Sanitation, Room 1410,
City Hall East, 200 North Main
Street, Los Angeles,
California 90012, USA.

Department of Social Sciences,
Oregon State University,
Cornallis, Oregon 97331, USA.

Division of Biomedical and
Environmental Research, Energy
Research and Development
Administration,
Washington DC 20545, USA.

Dynatech R & D Co.
99, Erie Street,
Cambridge, MA 02139, USA.

Ecotope Group
2332 East Madison,
Seattle, WA 98112, USA.

Energy Conversion Devices Inc.,
1675 West Maple Road,
Troy, MT 48084, USA.

Energy Engg. Institute,
San Diego Univ.,
California 92182 - 0191,
USA.

Energy Institute,
Fairfax Station,
Virginia, USA.

Energy and Resource Planning Group,
Lawrence Livermore National Laboratory, P. O. Box 5510, Livermore,
CA 94551 - 0620, USA.

Energy Systems Division,
National Aeronautics and Space Administration (NASA),
600 Independence Avenue SW,
Washington DC 20546,
USA.

Energy Systems Lab.,
Texas A & M Univ.
College Station, Texas,
TX 778 43 - 3123, USA.

Energy, Environment and Technology, USAID,
320 Twenty First Street, N. W.,
Washington, DC 20523, USA.

Gas Research Institute,
8600 West Brynmawr Avenue,
Chicago, Illinois 60631, USA.

G. V. Olsen Associates,
170 Broadway, NY 10038
New York, USA.

IE Associates, 3702, East Lake Street Suite 202, Minneapolis,
Minnesota 55406, USA.

International Institute for Energy Conservation,
750 First Street NE, No 940
Washington DC 20002, USA.

International Biomass Institute,
Suite 209, 1000 Vermont Avenue,
NW, Washington DC 20005, USA.

Institutional Industries, Capitol Centre Building, GN - 61, P. O. Box 9699, Plympia,
Washington 98504, USA.

Larmie Energy Tech. Centre,
P. O. Box 3395, Univ. Station,
Laramie, NY 82071, USA.

Lewis Research Centre,
21000 Brook Park Road,
Cleveland, OH 44135, USA.

Los Alamos National Laboratory,
Los Alamos, New Mexico 87545,
USA.

Meat Animal Research Centre,
Agricultural Research Services,
Department of Agriculture,
Clay Centre, Nebraska 68933, USA.

National Centre for Energy Management and Power,
University of Pennsylvania,
Philadelphia, PA 19104, USA.

National Space Technology Laboratory, Bay St. Louis,
Missouri 39529, USA.

Omega Albha Recycling Systems,
Rt. 1, Box 51, Orma,
West Virginia 25268, USA.

Perennial Energy, Inc Box 15-Dora,
Missouri 65637, USA.

Pittsburgh Energy Tech. Centre,
4800 Forbes Avenue,
Pittsburgh, PA 15213, USA.

Programme for Energy Research, Education and Public Service,
Oxley Hall, 1712 Neil Avenue,
Columbus, OH 43210, USA.

School of Forestry, Environmental Science, Duke University,
Durham, North Carolina 27706
USA.

Tata Energy and Resources Institute,
1600 Wilson Boulevard,
Suite 500, Arlington,
VA 22209, USA.

US Asia Environment Partnership,
1133 20th Street NW 300,
Washington DC 20036, USA.

US Environmental Protection Agency, 401 M Street, Washington DC 20460, USA.

US Department of Energy,
1000 Independence Avenue SW,
Washington, DC 20585, USA.

Volunteers in Technical Assistance,
3706 Rhode Island Avenue
Mt. Ranier, MD 20822, USA.

World Resources Institute,
1709 New York Avenue, N. W.
Washington, DC 20006, USA.

World Wildlife Fund
1250 24th Street, NW,
Washington, DC 20037, USA.

WEST INDIES

Jamaica Scientific Research Council
Hope Garden P.O. Box 350,
Kingston 6, Jamaica,
West Indies.

University of the West Indies,
Kingston, Jamaica,
West Indies.

UN Agencies Involved With Biomethanation

Asian Development Bank,
P. O. Box 789, 0980 Manila,
Philippines.

Economic Commission for Africa,
P. B. No. 3001,
Addis Ababa, Ethiopia,

Economic Commission for Europe,
Palais des Nations
1211, Geneva - 10, Switzerland.

Economic Commission for
Latin American and the Caribbean,
Casilla 179D - Santiago, Chile.

Economic and Social Commission for Asia and the Pacific,
UN Building, Rajadamnern Avenue,
Bangkok 10200, Thailand.

Food and Agriculture Organisation,
Viale delle Terme di Caracalla-
00100 Rome, Italy,

International Fund for Agricultural Development,
Via del Serafico,
107 - 00142 Rome, Italy.

Sustainable Energy and Environment Division (SEED),
UNDP, 1 UN Plaza,
New York 10017, USA.

United Nations,
New York 10017, USA.

UNIDO,
Vienna Inter. Centre,
P. O. Box 300,
A - 1400 Vienna, Austria.

UN Research Institute
for Social Development,
Palais des Nations,
CH - 1211, Geneva 10,
Switzerland.

United Nations,
Palais des Nations,
1211 Geneva 10,
Switzerland.

United Nations,
Vienna Inter. Centre,
P. O. Box 500,
A - 1400, Vienna, Austria.

UN Environmental Programme,
P. O. Box 30552,
Nairobi, Kenya.
UNESCO, 7, Place De Fontenoy,
75007 Paris, France.

UN Fund for Sci. & Tech.
for Development (UNFSTD),
1, UN Plaza, New York,
NY 10017, USA.

UNICEF, UNICEF House,
3 UN Plaza, New York,
NY 10017, USA.

UN University, Toho Seimei Bldg.,
15-1 Shibuya 2- Chome
Shibuya-ku, Tokyo 150, Japan.

WHO, CH-1211, Geneva 27,
Switzerland.

The World Bank, 1818 H Street,
N. W., Washington,
DC 20433, USA.

Acknowledgement

This directory of source of information on biogas systems was developed after scanning numerous journals and reports available with libraries of TERI, NPC, APCTT and ASCI (Admini- strative Staff College of India). Gratitude is conveyed to the concerned authorities for their courtesies and help.

—Author

Subject Index